SOCIAL COMPLEXITY IN PREHISTORIC EURASIA
MONUMENTS, METALS, AND MOBILITY

Social Complexity in Prehistoric Eurasia challenges current interpretations of the emergence, development, and decline of social complexity in the steppe region of China and the former Soviet Union. Through a thematic investigation of archaeological patterns ranging from monument construction and use to the production and consumption of metals and the nature of mobility among societies, the essays in this volume provide the most up-to-date thinking on social and cultural change in prehistoric Eurasia. Collectively, they challenge broader theoretical trends in Anglo-American archaeology, which have traditionally favored comparative studies of sedentary agricultural societies over mobile pastoralist or agro-pastoralist communities. By highlighting the potential and limitations of comparative studies of social complexity, this volume sets the agenda for future studies of this region of the world. It emphasizes how the unique nature of early steppe societies can contribute to more comprehensive interpretations of social trajectories in world prehistory.

BRYAN K. HANKS is associate professor in the Department of Anthropology at the University of Pittsburgh and research associate at the Carnegie Museum of Natural History in Pittsburgh. He has been involved in collaborative archaeological research in the Russian Federation since 1998 and has received funding from the National Science Foundation and the Wenner-Gren Foundation for Anthropological Research.

KATHERYN M. LINDUFF is UCIS Professor of Art History and Anthropology at the University of Pittsburgh. She is the co-editor (with Karen S. Rubinson) of *Are All Warriors Male? Gender Roles on the Ancient Eurasian Steppes* and (with Sun Yan) *Gender and Chinese Archaeology*.

SOCIAL COMPLEXITY
IN
PREHISTORIC EURASIA

MONUMENTS, METALS, AND MOBILITY

EDITED BY

Bryan K. Hanks

University of Pittsburgh

Katheryn M. Linduff

University of Pittsburgh

CAMBRIDGE
UNIVERSITY PRESS

DAMAGED

CAMBRIDGE UNIVERSITY PRESS
Cambridge, New York, Melbourne, Madrid, Cape Town, Singapore,
São Paulo, Delhi, Dubai, Tokyo

Cambridge University Press
32 Avenue of the Americas, New York, NY 10013–2473, USA

www.cambridge.org
Information on this title: www.cambridge.org/9780521517126

First published 2009

Printed in the United States of America

A catalog record for this publication is available from the British Library.

Library of Congress Cataloging in Publication data

Social complexity in prehistoric Eurasia : monuments, metals, and mobility /
edited by Bryan K. Hanks, Katheryn M. Linduff.
p.   cm.
Includes bibliographical references and index.
ISBN 978-0-521-51712-6 (hbk.)
1. Social archaeology – Eurasia. 2. Steppe archaeology – Eurasia. 3. Prehistoric
peoples – Eurasia. 4. Monuments – Eurasia – History. 5. Metallurgy – Eurasia –
History. 6. Social mobility – Eurasia – History. 7. Excavations (Archaeology) – Eurasia.
8. Eurasia – Antiquities. 9. China – Antiquities. 10. Soviet Union – Antiquities.
I. Hanks, Bryan K., 1967– II. Linduff, Katheryn M. III. Title.
DS328.S63 2009
930–dc22   2008040587

ISBN 978-0-521-51712-6 Hardback

CONTENTS

PART TWO MINING, METALLURGY, AND TRADE

PART THREE FRONTIERS AND BORDER DYNAMICS

CONTRIBUTORS

Francis Allard is associate professor in archaeology at Indiana University of Pennsylvania. He has worked, lived, and studied in China at various times since the early 1980s and has conducted archaeological research in south China and Vietnam. Most recently he has directed a project in central Mongolia, focusing on the emergence and development of nomadic pastoralism in that region.

Chunag Amartuvshin is a senior research archaeologist at the Mongolian Institute of Archaeology, Academy of Sciences, and co-director of the Institute's National Cultural Resource Management Sector. Dr. Amartuvshin is also co-director of the Joint Mongolian-American Baga Gazaryn Chuluu Expedition, and his research interests include the emergence of social complexity among nomadic groups, the study of mortuary process, and the preservation of steppe nomadic heritage.

David W. Anthony is professor of anthropology and anthropology curator at the Yager Museum of Art and Culture at Hartwick College. Anthony was director with Dorcas Brown of the Samara Valley Project in Russia and has published *The Horse, the Wheel, and Language: How Bronze Age Riders from the Eurasian Steppe Shaped the Modern World*, which combines 20 years of research in Ukraine, Russia, and Kazakhstan in a study of the Proto-Indo-European language.

Thomas Barfield is professor of anthropology at Boston University and president of the American Institute for Afghanistan Studies. Barfield has conducted extensive fieldwork among nomads in northern Afghanistan and is the author of *The Central Asian Arabs of Afghanistan* and co-author of *Afghanistan: An Atlas of Indigenous Domestic Architecture.* He has written more broadly on the history and culture of nomadic pastoral societies in *The Perilous Frontier: Nomadic Empires and China* and *The Nomadic Alternative.* Barfield's current research focuses on the political economy of Afghanistan, the rule of law, and state building.

Emma C. Bunker is a research consultant to the Denver Art Museum's Asian Art Department and specializes in the arts of ancient China and Southeast Asia. Bunker is a well-known authority on personal adornment in China, the art of the horse-riding groups of the Eurasian steppe, and Khmer art of Southeast Asia. Her numerous publications have presented groundbreaking research on these subjects.

Evgenii N. Chernykh is a member of the Institute of Archaeology at the Russian Academy of Sciences in Moscow. He has carried out research for decades on ancient metallurgy and has published five volumes on the excavations at Kargaly, the best-known and only recently studied mining complex in Siberia.

Andrei V. Epimakhov is a Ph. D. research Fellow at the Institute of History and Archaeology, Russian Academy of Sciences, as well as assistant professor at Southern Ural State University. He has recently co-authored (with Ludmila Koryakova) *The Urals and Western Siberia in the Bronze and Iron Ages.*

William W. Fitzhugh, an anthropologist specializing in circumpolar archaeology, ethnology, and environmental studies, is director of the Arctic Studies Center and curator in the Department of Anthropology, National Museum of Natural History, Smithsonian Institution. He has spent more than 30 years studying and publishing on Arctic peoples and cultures in northern Canada, Alaska, Siberia, and Scandinavia. He has produced international exhibitions, *NOVA* specials, and several films.

Michael D. Frachetti is assistant professor of anthropology at Washington University in St. Louis. Frachetti's archaeological fieldwork has been conducted in Kazakhstan, and he has published numerous articles on Bronze Age pastoral societies. He has also conducted ethnographic studies of Kazakh pastoralists and carried out research on prehistoric rock art in the Italian Alps, Roman and Islamic landscapes in North Africa, and Neolithic hunter-gatherers in Finland.

Rubin Han is professor and past director of the Institute for Historical Metallurgy and Materials at the University of Science and Technology Beijing. Han has published widely on the production of alloyed metals and on the history of metallurgy in early China. She has taken special interest recently in the intersection between the imperial Chinese and their northern neighbors.

Bryan K. Hanks is associate professor in the Department of Anthropology at the University of Pittsburgh and research associate at the Carnegie Museum of Natural History in Pittsburgh. He has been involved in collaborative archaeological research in the Russian Federation since 1998 and has received funding from the National Science Foundation and the Wenner-Gren Foundation for Anthropological Research.

William Honeychurch is assistant professor of anthropology at Yale University. He has conducted field research in Mongolia for more than a decade and has written many articles on pastoralism in Mongolia in the Bronze and Iron Ages. His work in Egiin Gol was the first pedestrian survey conducted in Mongolia. He has also conducted a field survey in the southern, arid regions of Mongolia.

Jean-Luc Houle is a Ph.D. candidate in the Department of Anthropology, University of Pittsburgh. His fieldwork over the past five years has taken place in Mongolia, where he is conducting a pedestrian surface survey aimed at the reconstruction of pastoral lifeways during the Bronze and Iron Ages.

Philip L. Kohl is professor of anthropology and Kathryn W. Davis Professor of Slavic Studies at Wellesley College. He is the author of *The Making of Bronze Age Eurasia* and scores of articles on Bronze and Iron Age Eurasia and is the co-editor of *Nationalism, Politics, and the Practice of Archaeology.*

Ludmila Koryakova is professor at the Ural State University and Institute of History and Archaeology, Ural Branch of the Russian Academy of Sciences. She has received fellowships from the European Community (INTAS), the Russian Academy of Sciences, Centre National de la

Recherche Scientifique (CNRS), and the French Ministry of Foreign Affairs. She has published more than 80 articles in American, European, and Russian journals. She has recently co-authored (with Andrei Epimakhov) *The Urals and Western Siberia in the Bronze and Iron Ages*.

Xiaocen Li completed his Ph.D. in 2004 at the Institute of Historical Metallurgy and Materials, University of Science and Technology Beijing, in the Department of History of Science and Technology under the direction of Rubin Han. His research focuses on the metallurgical technology used by the bronze-using cultures of southwest China in the late first millennium BCE. He has recently joined the faculty at the University of Science and Technology Beijing.

Katheryn M. Linduff is UCIS Professor of Art History and Anthropology at the University of Pittsburgh. She is the co-editor (with Karen S. Rubinson) of *Are All Warriors Male? Gender Roles on the Ancient Eurasian Steppes* and (with Sun Yan) *Gender and Chinese Archaeology*. She has conducted field research in Inner Mongolia for many years.

Jianjun Mei is professor and director of the Institute for Historical Metallurgy and Materials at the University of Science and Technology Beijing. His research focuses on the transmission of metallurgy and the intersection of cultures in the Neolithic period and Bronze Age in northwestern China and eastern Kazakhstan.

David L. Peterson is assistant professor of anthropology and a scientist at the Center for Archaeology, Materials and Applied Spectroscopy at Idaho State University. He is co-editor (with Laura M. S. Popova and Adam T. Smith) of *Beyond the Steppe and the Sown* and the author of several articles on Bronze Age metallurgy in the Eurasian steppes and Caucasus. He is currently involved in geo-archaeological survey and source analysis of copper ores and copper and bronze artifacts in Armenia and laboratory investigations of the biological effects of copper production on Bronze Age populations in the southeastern Urals.

Laura M. S. Popova is an Honors Faculty Fellow, Barrett Honors College, Arizona State University. Her current research and publications focus on the politics of pastoral land use, past and present, highlighting the ways in which the socio-political, ecological, and cultural orders of pastoral societies shape and restructure global and local environments.

Colin Renfrew is Disney Professor Emeritus of Archaeology in the University of Cambridge and Senior Fellow of the McDonald Institute for Archaeological Research. He has worked in the field of European prehistory and is the author of *Archaeology and Language: The Puzzle of Indo-European Origins* and other works.

Gideon Shelach is professor and chair of East Asian studies and director of the Louis Frieberg Center for East Asian Studies at The Hebrew University, Jerusalem. He has engaged in fieldwork in northeastern China since 1994. His latest book is *Prehistoric Societies on the Northern Frontiers of China: Archaeological Perspectives on Identity Formation and Economic Change during the First Millennium* BCE.

Joshua Wright received his Ph.D. in Anthropology from Harvard University (2006), where he studied the relationship between community, subsistence, and landscape while writing on the transformation of the monumental landscape in northern Mongolia during the adoption of nomadic pastoralism. He continues to pursue his interests in the archaeological prehistory and history of Inner Asia as well as general anthropological, spatial, and natural-science-based studies of megaliths, monumentality, nomadism, and mobility through the rubric of landscape archaeology and human ecology.

FOREWORD

From Myth to Method
Advances in the Archaeology of the Eurasian Steppe

Colin Renfrew

I N RECENT years, the archaeology of the Eurasian steppe has seen some
remarkable advances. Up to a couple of decades ago, it seemed that
little progress was being made, despite important archaeological discoveries in a number of relevant countries. The same rather simple models,
based on an undifferentiated view of mobile steppe pastoralism and the
notion of a short yet significant episode in which the domestication of
the horse was achieved, had held sway since the early twentieth century.
The valid contrasts emphasized in *The Steppe and the Sown* by Peake and
Fleure (1928) led in the early work of Gordon Childe (1926) to a simplistic view of mounted nomad pastoralists, a view that has survived into
recent times, although it was later reassessed by Childe himself (1950).

Today the picture is completely transformed, as the present volume
emphasizes. In particular, recent discoveries have now allowed a clear
differentiation to be established between the developments of the Bronze
and Iron Ages in the steppes, in social and economic terms as much as
in metallurgy. The development toward a pastoralist economy in the
earlier Bronze Age, as well exemplified by the Sintashta culture of western Siberia with its chariot burials (Koryakova and Epimakhov 2007:
66–80; Parzinger 2006: 251–259, 338–342), was not accompanied by any
conspicuous evidence of horse riding for military purposes, although
horses are documented for drawing chariots as early as 2000 BCE and
were presumably ridden earlier than this for the purposes of herding
(see Renfrew 1998). It was not until the Iron Age, in the first millennium BCE, that Eurasian nomad pastoralism developed as a militarily
significant enterprise with a complex, hierarchical, and ramified social

structure utilizing effective military power based upon the deployment of mounted warriors (Koryakova and Epimakhov 2007: 209–220; Parzinger 2006, 679–692). This was the period of the first great kurgans, such as at Arzhan in the Tuva area (Parzinger 2006: 606–619; Koryakova and Epimakhov 2007: 327), which may be regarded as royal burials of a nomadic elite, anticipating by several centuries the Scythians as they appear in the writings of Herodotus.

Steppe archaeology is now one of the most dynamic fields in the whole ambit of prehistoric studies, as is reflected in the publications of some earlier conferences (e.g., Mair 1988; Levine et al. 1999; Boyle et al. 2002; Levine et al. 2003) and documented in the recent magisterial survey by Parzinger (2006). The reasons for this upsurge in interest and in productive research are several, and they are well exemplified here.

In the first place, the vast terrain of central Eurasia has opened up to scholarship. International meetings are being held within the area, at sites such as Arkaim or Gonur Tepe, as much as in Beijing or Pittsburgh. This new openness has facilitated publication in the West by major scholars who did not earlier enjoy a wide readership there (e.g., Chernykh 1992; Mei 2000) and the participation in the field of a whole new generation of younger workers, many of whom are represented in this volume.

Second, it is at last possible to compare and contrast the various cultures, across a terrain that reaches almost to the Pacific Ocean in the east to lands bordering the Mediterranean Sea in the west, with the benefit of a secure chronological framework. Radiocarbon dates, increasingly accompanied by tree-ring dates in some cases, are beginning, for the first time, to produce a coherent chronology (see Hanks et al. 2007). Already there have been some shocks. The relatively early date of the Sintashta culture, associated with the first use of the chariot, is now well documented. And at the conference whose papers are presented here, the early dates for the Maikop burial, presented by Chernykh, and discussed also by Kohl, offer not so much a refinement as a disruption of most earlier assumptions.

Through these new projects, new areas of research are opening up. Prominent among these is the development of trade. New research on the sources and early use of tin has offered this commodity as one salient vector for the rapid development of bronze metallurgy in the later Bronze Age. There, as in other areas of steppe archaeology, the work of colleagues from the German Archaeological Institute, often in collaboration with scholars from the steppe lands or neighboring countries, has been particularly important. Moreover, the ecology of the exploitation of the steppe lands is now the subject for sustained research. The basis for the early use of the area, before the development of the full system of mobile pastoralism seen during the Iron Age, is under investigation. And the much-debated question of the domestication of the horse is seen in a new light, especially when careful distinction is made between horses for food, to facilitate herding, for pulling chariots, and to support armed warriors. Molecular genetic research, applied to plant and animal species, is proving as relevant here as when applied to living human populations.

These approaches and the application of new models for change and of new explanatory frameworks have led to an exciting quickening in the pace of research, as the chapters in this volume document. A number of broad questions can now be posed rather more clearly. It is evident that the mounted warriors of the great chiefdoms of the Iron Age, some of them designated by classical writers as Cimmerians and Scythians, relied upon a social order and an economic system that were remarkably successful. They seem to have emerged in the first millennium BCE but were based on earlier antecedents. How can we better define the social and economic systems that sustained these prosperous mobile communities?

The communities of the Bronze Age of the second millennium BCE that preceded these clearly were themselves innovators, and it was during this time that the first great trading networks seem to have been established. We see the settlement archaeology of some of these communities in sites like Sintashta and Arkaim, in the so-called country of towns. But can we define more precisely the economies and societies at this time, including those of the Andronovo culture? The horse is documented as used for pulling chariots already at the beginning of the second millennium. But can we establish more clearly when horse riding became significant for military purposes? Yet the initial domestication of the steppes must have begun before this time. The evidence for plant and animal domesticates is not yet very abundant before 2000 BCE, yet by then some of the important transitions must have been occurring.

Issues need to be defined more clearly before we can hope to understand by what means, for instance, the horse-drawn chariot reached China. Early steppe metallurgy too needs further study, if we are to establish definitively whether the surprisingly late use in China of copper and of bronze was a technology learned from the West. Perhaps we are close to seeing answers to some of these questions.

The benefits to our understanding of world prehistory will then be immense. From a broad perspective, the degree and nature of the influence and which way the arrows of transmission point still have to be established conclusively on the basis of secure data. That goal is now within reach. There are also vast issues in linguistic prehistory. What was the role of the steppe communities in the dissemination of the languages of the Indo-European family? That vexed question has not yet been satisfactorily answered (see Anthony 2007), and some recent initiatives offer results that are disconcertingly inconclusive (Lamberg-Karlovsky 2002). In particular, the problem of how the Indo-Iranian languages (or their precursor) reached South Asia remains to be resolved. In a similar vein, we need to understand better the archaeological record to document the Mongol invasions and to explain the present-day distribution of languages in the area.

Such linguistic issues, however, simply serve to emphasize the critical role of the Eurasian steppe lands in world prehistory. At times, these vast tracts of land have served to separate two very active and sometimes independent heartlands of cultural activity: western Asia (with the eastern Mediterranean) to the west and China to the east. At other times, particularly with the more effective use of the horse and of the camel, they have formed an important zone connecting these two great centers (or congeries of centers) of domestication and later of civilization. The proper understanding of these changing interactions is now one of the major tasks that prehistoric archaeology has to address, and the essays here take some important steps in that direction. A few decades ago, the question of long-distance interactions across the Pacific Ocean was a puzzling and a much-disputed one. Today it seems largely resolved, and the complex question of trans-Eurasian interactions now seems more pressing. Significant interpretive problems remain, however, and continue to be controversial, as noted by Philip Kohl (2007: 133) in his recent thoughtful study of Bronze Age Eurasia:

Simplifying [two] starkly opposed interpretive models, one can say that the first group [of scholars] sees the basic direction of movements or cultural impulses even before the beginnings of the Bronze Age as proceeding east to west, whereas the latter group reverses the arrows and essentially interprets developments on the Pontic steppes and further east as ultimately dependent on innovations that were associated with the sedentary agricultural societies first of southeastern Europe, including the Cucuteni-Tripolye culture, and the mixed agricultural/transhumant societies of the Caucasus.

That expresses the dilemma, perhaps in its simplest form. It is further complicated by the additional role that the steppe lands may have played through their interactions with the Indian sub-continent, mediated by the arid yet potentially fertile lands that lie between, such as Turkmenistan and Serindia (including Xinjiang Province). These interactions varied dramatically with the changing nature of the societies in those different regions and with their assessment of the benefits of trade, travel, and conquest in the context of developing transport mechanisms and of the fluctuating range of commodities traded, not least metals and silk.

This timely volume addresses some of these important topics. It will make a significant contribution to the understanding of the prehistory and the cultures of the steppe lands and their neighbors.

References

Anthony, D.W. 2007. *The Horse, the Wheel, and Language: How Bronze-Age Riders from the Eurasian Steppes Shaped the Modern World.* Princeton: Princeton University Press.

Boyle, K., C. Renfrew, and M. Levine (eds.). 2002. *Ancient Interactions: East and West in Eurasia.* Cambridge: McDonald Institute.

Chernykh, E.N. 1992. *Ancient Metallurgy in the USSR: The Early Metal Age.* Cambridge: Cambridge University Press.

Childe, V. G. 1926. *The Aryans: A Study of Indo-European Origins.* London: Kegan Paul, Trench, Trubner.

1950. *Prehistoric Migrations in Europe.* London: Kegan Paul, Trench, Trubner.

Hanks, B.K., A.V. Epimakhov, and A.C. Renfrew. 2007. Towards a Refined Chronology for the Bronze Age of the Southern Urals, Russia, *Antiquity* 81: 353–367.

Kohl, P. L. 2007. *The Making of Bronze Age Eurasia.* Cambridge: Cambridge University Press.

Koryakova, L., and A. Epimakhov. 2007. *The Urals and Western Siberia in the Bronze and Iron Ages.* Cambridge: Cambridge University Press.

Lamberg-Karlovsky, K. 2002. Archaeology and Language: The Indo-Iranians. *Current Anthropology* 43(1): 63–88.

Levine, M., Y. Rassamakin, A. Kislenko, and N. Tatarintseva (eds.). 1999. *Late Prehistoric Exploitation of the Eurasian Steppe.* Cambridge: McDonald Institute.

Levine, M., C. Renfrew, and K. Boyle (eds.). 2003. *Prehistoric Steppe Adaptation and the Horse.* Cambridge: McDonald Institute.

Mair, V. H. (ed.). 1998. *The Bronze and Early Iron Age Peoples of Eastern Central Asia.* Washington, DC: Institute for the Study of Man.

Mei, J. 2000. *Copper and Bronze Metallurgy in Late Prehistoric Xinjiang.* BAR International Series 865. Oxford: Archaeopress.

Parzinger, H. 2006. *Die Frühen Völker Eurasiens*. Munich: Verlag C. H. Beck.

Peake, H., and H. J. Fleure. 1928. *The Steppe and the Sown*. New Haven: Yale University Press.

Renfrew, C. 1998. All the King's Horses: Assessing Cognitive Maps in Later European Prehistory. In S. Mithen (ed.), *Creativity in Human Evolution and Prehistory*. London: Routledge, pp. 260–284.

CHAPTER 1

Introduction
Reconsidering Steppe Social Complexity within World Prehistory

BRYAN K. HANKS AND KATHERYN M. LINDUFF

THIS VOLUME brings together a collection of essays that focuses specifically on themes connected with the analysis of social complexity in the third to first millennium BCE in the Eurasian steppe. This dialogue stems from a symposium held at the University of Pittsburgh in February 2006 that sought to evaluate current trends and to determine new directions for the study of Eurasian steppe archaeology. What became apparent during this meeting was that the steppe region has moved firmly into the spotlight of world prehistory and contemporary archaeological theory. No longer viewed as closed geopolitical spheres, the territories of the former Soviet Union and neighboring regions, and the traditions of research that have addressed these areas, have become promising new arenas of international collaboration. Important questions surrounding the emergence and diffusion of agricultural and pastoral adaptations, early metallurgical technologies and their use, and the role of mobile pastoralist societies in China, Central Asia, and Europe have become significant topics within scholarly discourse in recent years. Such issues are clearly reflected in the publication of three new, seminal books in 2007 on the Bronze and Iron Ages of the steppe region (Anthony 2007; Kohl 2007; Koryakova and Epimakhov 2007).

The chapters offered within this volume not only examine these important issues in steppe archaeology but also seek to contribute more specifically to a broader comparative theoretical analysis of early social complexity in world prehistory. Although it is undeniable that regional culture histories provide the basic foundation for descriptive and analytical archaeological patterns, such regional treatments also should be viewed from a broader theoretical perspective in order to build and refine models of understanding for the various trajectories of human development that have existed. As the chapters in this volume clearly indicate, steppe archaeology can contribute significantly to this agenda.

As Colin Renfrew discusses in the Foreword, several recent conferences and publications have added substantially to the growing corpus of literature on steppe archaeology. For example, international conferences held at the University of Cambridge (Boyle et al. 2002; Levine et al. 1999), University of Chicago (Peterson et al. 2006; Popova et al. 2008), and the Arkaim Heritage Center in the Russian Federation (Jones-Bley and Zdanovich 2002) represent important benchmarks in the rapidly developing field of Eurasian steppe archaeology.

Broader comparative treatments on early social complexity in other parts of the world have rarely turned to the Eurasian steppe region as a source for examining the emergence of hierarchy and heterarchy; scalar problems connected with socio-economic integration and organization; patterns of political centralization; and the role that subsistence and productive economies have in stimulating the emergence, development, and decline of socio-economic change. In contrast to this, two recent monographs published in English in the Russian Federation have sought to develop more encompassing comparative analyses of early states and the materialization of power within early civilizations (Grinin et al. 2004; Grinin et al. 2008). For example, *Hierarchy and Power in the History of Civilizations: Ancient and Medieval Cultures* (Grinin et al. 2008) brought together scholars addressing these themes for ancient states of the Old World, medieval Eurasian states, and the Maya region for New World states. Such dialogues being produced in English by Russian publishers are a welcome accomplishment in international scholarship and are clearly reinforcing the broader relevance of Eurasian steppe archaeology.

The rationale for our volume adds to this new paradigm by focusing more specifically on prehistoric developments connected with complex, non-state societies. These middle-range societies have been routinely categorized as tribes and chiefdoms, with various levels of complexity. Although ample debate has surrounded the use of neo-evolutionary terminology for the study of early societies, such terminology continues to be used when interpreting prehistoric steppe developments. Surprisingly, the earlier studies that produced such terminology rarely looked at mobile pastoralist or agro-pastoralist societies and instead routinely focused on sedentary, agricultural developments. We intend here to re-evaluate these trends in scholarship in order to determine whether such models have a place within studies of early social complexity in the steppe and to see if such applications contribute to, or perhaps challenge, the application of such modeling to the study of early social trajectories in world prehistory. Contributing authors to this volume, therefore, provide important discussions of historically contingent developments in the steppe and neighboring territories that are linked not only to unique social, economic, and environmental adaptations but also to broader theoretical themes that examine the nature of such developments. By addressing these issues from this perspective, a scholarly agenda is put forward that places steppe archaeology at the core of future studies that evaluate and interpret trajectories of change in the human past.

Volume Organization

KEY THEMES that long have contributed to steppe scholarship, stretching from the Soviet period up to the years since its collapse, include the modeling of social development and change, the role of metals in early societies, physical and cultural boundaries that characterize social landscapes, and the materialization of social power through monument construction and use. These themes are also broadly interwoven within world prehistory, and each can be seen as an important consideration within regional archaeologies. Therefore, themes that have both regional and broader comparative significance were selected as foci for the contributing authors of this volume to address in their independent chapters. As such, the volume consists of 15 full-length chapters, organized into four thematic parts with introductory essays: framing complexity; mining, metallurgy, and trade; frontiers and border dynamics; and social power, monumentality, and mobility. While the individual chapters are discussed in more detail within the introductory essays, we outline here the rationale for these specific themes.

FRAMING COMPLEXITY

In the past, mobile pastoral communities on the steppe were thought to follow a generally homogeneous pattern of social organization. Rather than making this assumption, the authors in this first part each bring a fresh approach, combined with newly gathered evidence from their own recent fieldwork, to bear on this issue. The contributors examine various definitions and theoretical approaches to the concept of *complexity* and its relationship to observable changes in economy, technology, and social organization within their research areas.

Distinct patterns of change in the archaeological record connected with the innovation and diffusion of new technologies, the emergence of warfare and military activities, and settlement patterning and mortuary practices are discussed. In recent years, archaeological evidence has been used to examine such distinct transitions through regional survey and geographic information systems (GIS) technologies, intensive settlement and cemetery excavations, and paleoenvironmental reconstruction. Much of this research has centered on identifying changes connected with the early, middle, and late phases of the Bronze Age, dating approximately from the third to the second millennium BCE. The chapters within this part focus on the conceptual problems associated with these developments and discuss broader theoretical strategies for evaluating the external and internal stimuli that contributed to these significant changes in the archaeological record.

Mining, Metallurgy, and Trade

Scholars have known for some time that the steppe region provided abundant resources to metal producers. However, the role these populations played in the emergence and maintenance of new industries and the "values" that became associated with metals and metal objects deserve much more attention. The set of chapters in this second part focuses on the importance of metal technology and its connection with the rise of new political and economic developments in the steppe zone and neighboring territories.

The mining, production, and trade of base metals for the production of bronze and other alloys as well as final products have been evaluated through various analytical models, including core-periphery relationships, multiple-core developments, and the emergence of metallurgical provinces of interaction and exchange. Although such models have illustrated the widespread and complex nature of early metallurgy in the steppe, the chapters within this part stress the importance of testing current understandings of the nature and extent of technological diffusion, the emergence of new social organization connected with mining and production communities, and inter-regional and intra-regional strategies connected with metals trade and exchange. Important issues addressed include the structure and organization of mining communities, elite strategies for political power foundations and their connection with trade and exchange patterns, the scale of interaction and diffusion of technology between steppe and non-steppe-based polities, and the cycling dynamics of regional prominence associated with the rise and collapse of metal production centers in the Eurasian steppe region.

FRONTIERS AND BORDER DYNAMICS

The only written records on steppe societies to be handed down are those produced by neighboring groups to the east, west, and south of the steppe region. As a result, steppe societies have been viewed as trade partners as well as irritants to the sedentary way of life. Much has been written since these ancient accounts about the interaction between the "steppe and sown" appeared, but until recently there has been little archaeological evidence to test the claims of the ancient authors.

The group of chapters in the third part examines the dynamics and results of that interaction. While some scholars have argued for the early emergence of state-level steppe societies, others have framed these developments as "supra-tribal" or "complex chiefdoms" (Kradin 2002, 2004). In recent years, the scholarship surrounding this issue has generated several key models for examining the emergence of new steppe socio-political orders and the fluidity connected with changing patterns of cultural identity and ethnicity. Such works have re-evaluated traditional static concepts of core-periphery relationships to illuminate the dynamics of "border" and "frontier" interaction, which clearly existed. Chapters in this part, therefore, investigate archaeological evidence in order to more effectively model the emergence of new patterns of steppe social complexity and identity.

SOCIAL POWER, MONUMENTALITY, AND MOBILITY

The fourth part of the volume focuses on the modeling of early complex societies in the eastern steppe zone and includes important new research and data from Mongolia, a region that has witnessed a surge of international collaborative projects in recent years. Chapters in this part address various strategies of power and centralization used by steppe polities, novel methods for investigating regional diachronic shifts in settlement organization and complexity, and the interpretation of new patterns of monument construction and funerary ritual practices. Many of the key political, social, and economic developments that occurred among steppe societies were situated at the non-state level of organization. Such developments have commonly been framed in terms of tribal and chiefdom levels of societal complexity, as discussed in the third part.

While the investigation of such middle-range societies has been effectively investigated through the comparative analysis of sedentary agricultural-based societies around the world, much less focus has been placed on evaluating the political and economic strategies and trajectories represented by pastoralist- or agro-pastoralist-based societies. With these problems in mind, this final part of the volume critically examines monolithic models and linear trajectories that have been used conventionally to interpret the organization of steppe pastoralist and agro-pastoralist societies and offers new perspectives on evaluating their pathways toward social change.

Concluding Remarks

THIS VOLUME has brought together many renowned scholars in the field of Eurasian steppe archaeology, and their contributions provide an important perspective on the vibrancy and optimism that exists within this field today and some of the challenges that lay before it. They also provide an important view on the state of the field and suggest not only where future work must be done but also what methods and theories may be particularly productive within such investigations.

The application of new scientific methods such as ancient DNA studies, bone isotope analyses, and the application of new absolute-dating chronologies are having a tremendous impact on current research programs in the region. In addition, greater cooperation among scholars from different disciplines (history, archaeology, etc.) may provide fuller, less-biased views of the human past. Coupled with this, a new generation of scholars is becoming active in larger collaborative archaeological programs. As a result of these important developments, steppe archaeology has made significant strides in the past two decades and shows every indication of being one of the most promising new territories for international research.

Nevertheless, as Colin Renfrew's discussion in the Foreword of this volume sets out, several distinct problems have been framed within the field of Eurasian steppe studies. These include the challenges surrounding the linguistic prehistory of the region, the emergence and spread of spoke-wheeled chariot technology, and the role of the horse in new patterns of social mobility and warfare. This volume stresses, rather, other questions that remain to be more fully understood and that can be productively coupled with the broader conceptual problems that continue to challenge the study of world prehistory. These include more-nuanced understandings of the relationship between technology and social practice, more effective modeling of processes of human migration versus diffusion in technologies and ideas, and various pathways to social and economic complexity that appear to be unique within pastoralist and agro-pastoralist orientations. We hope that this volume has gone some way in identifying these problems and in suggesting new approaches to solving them.

In closing, we would like to thank all the contributors for their lively discussions during the Pittsburgh symposium and for their probing analyses in the papers presented here. We would also like to thank Beatrice Rehl at Cambridge University Press for her sincere support in the publication of this monograph and two anonymous reviewers for very thoughtful suggestions on how it could be improved. As editors of this volume, we are honored to have had the opportunity to contribute to the exciting new agenda that is emerging in the scholarship of the Eurasian steppe region and we look forward to this volume's contribution to broader comparative studies of social complexity in world prehistory.

References

Anthony, D. W. 2007. *The Horse, the Wheel, and Language: How Bronze-Age Riders from the Eurasian Steppes Shaped the Modern World.* Princeton: Princeton University Press.

Boyle, K., C. Renfrew, and M. Levine (eds.). 2002. *Ancient Interactions: East and West in Eurasia.* Cambridge: McDonald Institute.

Grinin, L., D. Beliaev, and A. Korotayev (eds.). 2008. *Hierarchy and Power in the History of Civilizations: Ancient and Medieval Cultures.* Moscow: Uchitel.

Grinin, L., R. Carneiro, D. Bondarenko, N. Kradin, and A. Korotayev (eds.). 2004. *The Early State, Its Alternatives and Analogues.* Volgograd: Uchitel.

Jones-Bley, K., and D. G. Zdanovich (eds.). 2002. *Complex Societies of Central Eurasia from the 3rd to 1st Millennium BC.* Vols. 1 and 2. Washington, DC: Institute for the Study of Man.

Kohl, P. 2007. *The Making of Bronze Age Eurasia.* Cambridge: Cambridge University Press.

Koryakova, L. N., and A. V. Epimakhov. 2007. *The Urals and Western Siberia in the Bronze and Iron Ages.* Cambridge: Cambridge University Press.

Kradin, N. 2002. Nomadism, Evolution and World-Systems: Pastoral Societies in Theories of Historical Development. *Journal of World-Systems Research*, 8 (3): 368–388.

2004. Nomadic Empires in Evolutionary Perspective. In L. Grinin, R. Carneiro, D. Bondarenko, N. Kradin, and A. Korotayev (eds.), *The Early State, Its Alternatives and Analogues*. Volgograd: Uchitel.

Levine, M., Y. Rassamakin, A. Kislenko, and N. Tatarintseva. 1999. *Late Prehistoric Exploitation of the Eurasian Steppe*. Cambridge: McDonald Institute.

Peterson, D., L. Popova, and A. Smith (eds.). 2006. *Beyond the Steppe and the Sown: Proceedings of the 2002 University of Chicago Conference on Eurasian Archaeology*. Colloquia Pontica 13. Leiden: Brill.

Popova, L. M., C. W. Hartley, and A. Smith (eds.). 2008. *Social Orders and Social Landscapes*. Newcastle upon Tyne: Cambridge Scholars Publishing.

PART ONE

FRAMING COMPLEXITY

CHAPTER 2

Introduction

Ludmila Koryakova

I N THE past few years, several international conferences in the United
States (e.g., Chicago 2004; Pittsburgh 2006) have focused on the problems of social complexity in the vast region of Eurasia. I remember that
during my first trip to the United States in 1994, I met David Anthony,
Karen Rubinson, Adam Smith, Phil Kohl, and Karlene Jones-Bley. At
that time, these individuals were representative of a very small group
of American archaeologists whose academic interests were directed to
a better understanding of Eurasian steppe prehistory. In retrospect, it
has taken some time to overcome the consequences of the long academic
separation that existed between Anglo-American and Russian archaeology during the Soviet period. In recent years, the beginning of a much
better understanding between scholars of these regions and broader perceptions of Eurasian archaeological materials are being realized.

Both past and recent research has shown that Eurasian prehistory represents a number of socio-cultural phenomena not only of regional but
also of wider historical significance. The assessment of these phenomena,
particularly the character and level of social complexity of Eurasian cultures in light of modern theoretical models, forms a rather new agenda
in Eurasian studies. This volume, and the set of essays that forms part I,
contribute importantly to this new orientation.

All of the chapters in the first part focus on the Bronze Age.
Geographically, three of the four papers fully or partly concern the
Ural mountain region (Epimakhov, Anthony, Frachetti), one paper the
Caucasus (Kohl), and another also includes the Semirech'ye area of
southeastern Kazakhstan (Frachetti). These regions all played an active

role in some of the most important prehistoric socio-cultural processes that occurred in Eurasia. This is well represented by the richness and diversity of the archaeological sites discussed by these authors and the important problems they focus on connected with both the emergence and development of social complexity.

The concept of social complexity, as a paradigm, undoubtedly has good heuristic potential, but as a general theory it has been limited in terms of its application to the Eurasian steppe region, especially because the traditional criteria of complexity has long been "attached" primarily to the study of settled agricultural societies.

Traditionally, the development of social systems in archaeology was predominantly regarded in terms of different variants of evolutional theory that focused on levels of civilization that, in turn, associated states with the highest level of socio-political evolution (Trigger 1998). States are often seen as the result of the competition of various social and political elements over longer-term historical change. In recent decades, however, a great deal of research has shown that, in social evolution, hierarchical societal forms had many alternatives, which often were based on large demographic parameters and developed economies that were well adapted to specific ecological environments. Often, these developments did not have clear hierarchical structures. In some cases, the rise of social complexity can be accompanied by an increase in the role of vertical social relations (strongly hierarchical), but it can be also accompanied by an increasing development of horizontal connections (heterarchy). Therefore, it has become evident that stateless societies were not necessarily less complex than state societies (Bondarenko and Korotayev 2002; Bondarenko et al. 2006: 15–18).

This shift in thought has allowed scholars to conclude that in social development at least two broad strategies are tenable: hierarchical "vertical" (chiefdoms and states) and non-hierarchical "horizontal" (communities and polities) (Popov 1993, 1995; Kradin and Lynsha 1995; Korotayev 1997; Kradin 2001). As a result, in recent years theoretical discussions of social complexity have moved away from a focus on linear evolution, political centralization, decision making, and hierarchy toward alternatives such as heterarchy (Crumley 1995) and dual social strategies (Blanton et al. 1996; Feinman 2001).

As Kristiansen and Larsson (2005: 5) have noted, "the processual and postprocessual archaeologies of the last generation have one thing in common, an autonomous perspective. The local or regional unit is their favorite frame of theoretical and interpretative reference, and academic references consequently rarely transcend national or regional borders." These authors indicated the necessity to overcome this "regionalism," particularly by studying the process of interaction in all its forms and complexity. A similar view is sometimes visible through a number of publications emphasizing prehistoric interactions, transmissions, exchange, trade, large-scale social transformations, and networks (Harding 2000; Kohl 2007; Koryakova and Epimakhov 2007; Kristiansen and Larsson 2005; Kuz'mina 2007). On the other hand, this tendency is also accompanied by a growing interest in regional lines of development, and their comparative analysis, which has resulted in a willingness to consider societies not as isolated polities but rather as elements of larger cultural networks.

There is no doubt that the nature of complexity can be expressed in material culture differently. In this way, the concept of complexity cannot be narrowed down only to the political organization, stratification, and greater hierarchy of societies. The developed forms of rituals, mythology, art, some particular material attributes, and other cultural components can supplement and, in some cases, form the only source of information available to estimate early forms of prehistoric social complexity.

There are many examples in which poor archaeological evidence exists, and only through historical records are we able to ascertain the existence of larger complex social formations. As an example, historically known states in the Eurasian steppe and forest-steppe, such as the first Turkic Khanate (sixth century CE) and Siberian Khanate (fourteenth to fifteenth century CE), are not well represented by archaeological evidence. If we did not know about their existence from literary sources, it would not be possible to identify these societies as states, because their social power was not distinctively exhibited in material culture and because nomadic states and super-complex chiefdoms are characterized by certain observable traits.

All this serves as a background for the central theme discussed within the chapters of this first part of the volume. All of the authors stress the importance of defining and understanding various forms of prehistoric social complexity in the Eurasian steppe. For example, according to Frachetti, current archaeological models of complexity to date do not adequately fit the Bronze Age conditions evident across the Eurasian steppe zone, and many commonly cited corollaries of "complex" chiefdom- or state-level organization, such as agricultural surplus, centralized socio-political authority, and institutional mechanisms for control of

specialized production, are currently lacking in the Bronze Age archaeology of the steppe. At the same time, the archaeological data do not
allow one to consider Bronze Age societies as primitive or egalitarian, as
was stated not long ago in Russian archaeology, because of a whole range
of indicators of socio-political complexity detected by recent discoveries in the Eurasian steppe zone. In light of this, Frachetti proposes the
concept of "non-uniform complexity," which describes how institutions
are codified at local scales and how widely they condition the interactive
scale of heterogeneous communities. On the basis of his model, he compares the societies of three areas with different landscapes: the Trans-
Urals region (Sintashta culture), Semirech'ye (eastern Kazakhstan), and
the Margiana Oasis, each of which exhibited a distinctive trajectory of
social, economic, and political organization at the beginning of the second millennium BCE.

Anthony presents an elegant model for the origin and decline of the
Sintashta culture, an archaeological pattern that is usually regarded as
an example of the emergence of early social complexity in the Eurasian
steppe. Currently, twenty-two settlements of this culture with closed
circular, oval, or rectangular fortifications have been discovered within
a rather limited territory of about 60,000 square kilometers (Zdanovich
and Batanina 2002; 2007). The Sintashta archaeological complex is characterized by very specific attributes: systemic character of settlement
localizations; highly organized settlements with elaborated fortifications
and sectional architectural planning; burial sites with a high concentration of the remains of sophisticated ritual practice comprising several
variations in the association between human bodies and animals; and
the presence of metal objects, weaponry, wheeled transport of rather
advanced construction for that time, and an eclectic set of ceramics
(Koryakova and Epimakhov 2007).

Anthony builds his model on such powerful factors of emerging social
complexity as warfare, transport, and connections with developed centers in Central Asia. He argues that the intensification of warfare was
brought on by climatic changes and declining natural resources. His
chapter emphasizes important theoretical problems, such as society and
climate, prehistoric warfare and its indicators, and some specific characteristics relating to the Sintashta culture itself.

It is currently known that during the Holocene there were several serious climatic changes that greatly impacted steppe societies.
Paleoclimatologist V. V. Klimenko, a scholar who has systematically
worked on the correlation of historical events with distinct climatic

changes, has noted that, "the history of the ancient world is extremely determined by climate" (1998: 21). Klimenko has argued that the scale of regional temperature fluctuations differs from global fluctuations by both magnitude and character. Regional fluctuations can either precede or follow broader global changes. Therefore, in dealing with micro-regional cultural situations, it is very important to investigate both broader regional and local climatic conditions (Tairov 2003). In general, the Sintashta culture coincides in time with the early sub-boreal aridity that took place in the Eurasian steppe around 2700/2500–1700/1600 BCE, but its effect in the Trans-Urals is not well understood and is debated (Koryakova and Epimakhov 2007). Although the absolute chronology of the Eurasian Bronze Age is being ascertained by important new dating programs (Epimakhov et al. 2005), and scholars are more confident in the general chronology of the Sintashta culture, we still desperately need a precise internal chronology in order to understand the true character of Sintashta socio-cultural dynamics. It is also not clear how and why the descendants of mobile pastoralists, who did not know settlement (the Yamnaya-culture tribes) or knew only simple open villages (the Abashevo culture), started to create large and sophisticated settlements under poor ecological conditions. This development is obviously more complex than a simple ecological crisis, and the actual role of the fortified settlements within the southeastern Urals region is far from clear.

From all the chapters presented within Part I, it is clear that the Bronze Age of Eurasia is marked by great technological, economic, and cultural innovations that dramatically intensified social life when compared to the preceding Eneolithic period. Societal conflict and wars became important attributes of Bronze Age life in the steppe zone. However, their relationship to larger patterns of trade, cultural expansion, and political networks are not presently well understood. I think that Eurasian warfare, as a social phenomenon, still remains understudied in contrast to the historical emphasis that has been placed on formal studies of weaponry.

It is commonly accepted that such a factor as the rise of metallurgy is closely linked with social complexity in both local and interregional levels. The desire to possess necessary resources for metallurgy and control over distribution of metal production inevitably leads to the formation of large informational networks, on the one hand, and competition between societies, on the other. On the basis of the known major metallurgical traditions, such as the western zone connected with a Circumpontic origin and the eastern connected with the Seima-Turbino phenomenon, it is possible to hypothesize that in the beginning of the second millennium BCE two technological systems competed within Eurasia. Such a development provides an important area for a new range of cultural studies focusing on "globalization," which means that radical changes in material culture were also linked to changing social identities that occured in macro-regional phases of interaction – perhaps, in a sense, reminiscent of processes connected with contemporary globalization. Critical issues include the dichotomous relationship between brief time scales of rapid change in which new kinds of material culture, ideas, and technological knowledge spread over vast areas contrasting with long periods in which culture and society appear to be localized, persistent, repetitive, and marked by a routine character (Vandkilde 2004). The examination of the processes connected with short-term change and long-term continuity will help scholars to develop better models for understanding dynamic interaction between local strategies and global processes as well as the concept of "world-systems," if one uses the less rigid criteria of "informational networks" of interaction in such systems, as proposed by Chase-Dunn and Hall (1997). The latest research testifies to the invention and spread of the chariot complex within the Eurasian steppe in the early second millennium BCE. The studies of such phenomena perhaps can serve as important examples of such processes (see Anthony's and Epimakhov's chapters). In this case, acceptance of a chronological priority for the Eurasian spoke-wheeled chariot over the Near Eastern one will require that scholars revise conventional interpretations of prehistoric interactions between urban areas and what has been viewed as the steppe periphery.

Although such interactions in the Iron Age appear more clearly, this cannot be said of the Bronze Age. This theme is touched on in Kohl's chapter, which is devoted to an assessment of the Maikop cultural phenomenon. Despite its enigmatic character, the Maikop is undoubtedly responsible for "maturation" in Circumpontic techno-cultural networks, wherein the Maikop development played the role of a "center" during the second half of the fourth millennium BCE. Some elements of its heritage likely were transmitted to the north and northeast and into the steppe zone. In this respect, the Maikop and Sintashta cultures are comparable in terms of their singularity of development, which became an important foundation for later social, cultural, and technological patterns in the steppe region.

In conclusion, I would like to stress that the chapters in this first part highlight a range of important issues and provide a set of interesting interpretations and intriguing hypotheses, which help us to expand our current perception of the early social complexity of Eurasian steppe cultures.

References

Blanton, R. T., G. M. Feinman, S. A. Kowalevsky, and P. N. Perigrine. 1996. A Dual-Processual Theory for the Evolution of Mesoamerican Civilization. *Current Anthropology* 37: 1–14.

Bondsrenko, D. M., and A. V. Korotayev (eds.). 2002. *Tsivilizatsionnye modely polito-geneza*. Moscow: Institute of African Studies.

Bondarenko, D. M., L. E. Grinin, and A. V. Korotayev. 2006. Altrnativy sotsial'noi evolutsiyi. In L. E. Grinin, D. M. Bondarenko, N. N. Kradin, and A. V. Korotayev (eds.), *Ranneye gosudarstvo, ego alternativy i analogiyi*. Volgograd: Uchitel Press, pp. 15–36.

Chase-Dunn, C., and T. Hall. 1997. *The Rise and Demise: Comparing World-Systems*. Boulder, CO: Westview Press.

Crumley, C. L. 1995. Heterarchy and the Analysis of Complex Societies. In R. M. Ehrereich, C. L. Crumley, and J. E. Levy (eds.), *Heterarchy and the Analysis of Complex Societies*. Archaeological Papers of the American Anthropological Association 6. Washington, DC, pp. 1–6.

Epimakhov, A. V., B. Hanks, and C. Renfrew. 2005. Radiouglerodnaya khronologiya pamyatnikov bronzovogo veka Zauralya. *Rossiyskaya Arkheologiya* 4: 92–102.

Feinman, G. M. 2001. Mesoamerican Political Complexity. The Corporate-Network Dimension. In J. Haas (ed.), *From Leaders to Rulers*. New York: Kluwer Academic/Plenum Publishers, pp. 151–175.

Harding, A. 2000. *European Societies in the Bronze Age*. Cambridge World Archaeology Series. Cambridge: Cambridge University Press.

Klimenko, V. V. 1998. Klimat i istoriya v epokhu pervykh vysokikh kultur (3500–500 gg. do n.e.). *Oriens* 4: 5–41.

Kohl, P. L. 2007. *The Making of the Bronze Age of Eurasia*. Cambridge: Cambridge University Press.

Korotayev, A. V. 1997. *Faktory sotsialnoi evolitsiyi*. Moscow: Nauka.

Koryakova, L., and A. V. Epimakhov. 2007. *The Urals and Western Siberia in the Bronze and Iron Ages*. Cambridge: Cambridge University Press.

Kradin, N. N. 2001. *Politicheskaya antropologiya*. Moscow: Ladomir.

Kradin, N. N., and V. A. Lynsha (eds.). 1995. *Alternativnyje puti k rannei gosudarst-vennosti*. Vladivostok: Dalnauka.

Kristiansen, K., and T. Larsson. 2005. *The Rise of Bronze Age Society: Travels, Transmissions and Transformations*. Cambridge: Cambridge University Press.

Kuz'mina, E. E. 2007. *The Origin of the Indo-Iranians*. Leiden: Brill.

Popov, V. A. (ed.). 1993. *Ranniye formy sotsialnoi stratifikatsiyi*. Moscow: Nauka.

Popov, V. A. (ed.). 1995. *Ranniye formy politicheskoi organizatsiyi: ot pervobytnosti k gosudarstvennosti*. Moscow: Nauka.

Tairov, A. D. 2003. *Izmemeniya klimata stepei i lesostepei Tsentralnoi Evraziyi vo II–I tys. do n.e.: meterialy k istoricheskim rekonstruktsiyam*. Chelyabinsk: Rifei.

Trigger, B. G. 1998. *Sociocultural Evolution*. Oxford: Blackwell.

Vandkilde, H. 2004. Archaeology, Anthropology and Globalization. Inaugural lecture, October 22nd, 2004. In H. Vandkilde (ed.), *Archaeology and Anthropology: Inaugural Lectures of the Institute of Anthropology, Archaeology, and Linguistics at Aarhus University, 1996–2004*. Aarhus: Aarhus University Press.

Zdanovich, G. B., and I. M. Batanina. 2002. Planography of the Fortified Centers of the Middle Bronze Age in the Southern Trans-Urals according to Aerial Photography Data. In K. Jones-Bley and D. Zdanovich (eds.), *Complex Societies of Central Eurasia from the 3rd to the 1st Millennium BC*, vol.1. Washington, DC: Institute for the Study of Man, pp. 120–138.

2007. *Arkaim-Strana Gorodov: Prostranstvo i obrazy*. Chelyabinsk: Krokus.

CHAPTER 3

Differentiated Landscapes and Non-uniform Complexity among Bronze Age Societies of the Eurasian Steppe

MICHAEL D. FRACHETTI

Differentiated Eurasian Landscapes

ARCHAEOLOGICAL RESEARCH increasingly illustrates that Bronze Age societies of the Eurasian steppe were inherently more diverse in their ways of life than their related material culture might imply (D. Zdanovich 1997; Kosintsev 2001; Nelin 2000; Epimakhov 2003). Bronze Age steppe communities illustrate comparatively different scales of social, economic, and political organization, as well as local variability in their extents of mobility and geographic ranges of interaction (e.g., Anthony et al. 2005; Chernykh 2004; Shishlina 2003; Frachetti 2008a). Even in local settings, evidence suggests that Bronze Age steppe communities were organizationally heterogeneous – meaning they were not politically or economically centralized under a shared corpus of functional institutions. Yet widespread distribution of related forms of material culture has prompted archaeologists to define an expansive cultural community through the broad lens of culture history and economic interaction throughout the second millennium BCE. One may observe that the emergence of a seemingly extensive socio-economic landscape throughout the Bronze Age stands at odds with the organizationally small-scale and locally rooted societies that occupied this vast territory. Current archaeological models of social complexity to date do not adequately fit the Bronze Age conditions evident across the Eurasian steppe zone (see Koryakova 2002). The apparent disjuncture between the scale of socio-political institutions of steppe populations and the geographic extent of their functional economic arena provides a unique case for the investigation of alternative models of interaction and social complexity among regional communities.

Models of Mesopotamian organizational dynamics define complexity as "the degree of functional differentiation among societal units or sub-systems" (Rothman 1994: 4). This definition is fitting for Mesopotamia, because there is an ascertainable "societal" framework from which "units" or "sub-systems" may be derived (Stein 1994: 13). Archaeological signatures of a diverse yet integrated society were generated through an emerging world system of economic and social transactions, such as surplus exchange, specialized production, resource management, and bureaucratic management, which drew the region's populations into institutionally structured chiefdoms and states through time (Algaze 2001; *generally*, Stein and Rothman 1994). Rothman (1994: 4) has suggested that in this setting complexity is rooted in societal organization, which he defines as "the arrangements that structure the functioning of individuals and groups as they attempt to meet social, economic, political, or ideological needs." This definition reflects what new-institutional economists like Douglass North call "institutions."[1] Merging these terminologies suggests that greater complexity comes with a wider applicable scale of institutional forces; a larger diversity of "functional units" can be organized in relation to one another through a common set of institutional constraints. Such a metric for qualifying and quantifying complexity is useful in Mesopotamian contexts and has been adapted in various forms to describe myriad other archaeological examples, ranging from the emergence of archaic states to the structuring of complex economic and social frameworks among chiefdoms of various forms (Carneiro 1970; Blanton et al. 1996; Earle 1977; for discussion, Chapman 2003).

But what if one cannot easily circumscribe the geographic boundaries of the participant communities or locate the growth of a shared or consistent institutional framework that applies to different populations intersecting across a shared geography? Socio-political or economic complexity cannot be charted as easily on a "functional scale of differentiation" if the societies that co-generate it subscribe to independent institutional parameters or exhibit non-uniform definitions of general institutions to begin with. In place of such models, I propose a theory of non-uniform complexity that builds from North's definition of institutions in order to chart institutional heterogeneity, consolidation, and fragmentation. This model supplies a useful metric for describing complexity among differentiated populations that nonetheless interact across a common geography.

For the purpose of this study, I employ North's definition of institutions as "the humanly defined constraints that shape human interaction" (1990: 3). These constraints can be informal or formal, but for the scope of this discussion we are primarily concerned with informal institutional constraints.[2] From the perspective of Bronze Age steppe archaeology, we may further identify specific and general institutions. Specific institutions refer to culturally particular practices, such as specific taboos or certain practices that constrain behavior among discrete in-groups. For example, David Anthony describes a case of Bronze Age dog sacrifice at Krasnosamarskoe, which, although potentially part of a wider ideological treatment of dogs, is evidence of a specific institution that conditioned how that particular community behaved in a particular contextual setting (Anthony et al. 2005: 412–413). General institutions are conceived here as categorically broader constraints, such as trade parameters, building conventions, ideological symbology, or even human burial. These institutions reflect common currents that crosscut communities but are uniquely transformed by diverse groups given their individual motivations or vantage point (e.g., Freidel 1983). The rules of engagement for trade, for example, may be considerably differentiated between two participant groups, even though, functionally speaking, economic transactions may be carried out with mutually perceived success. Anyone who has shopped in a foreign market has experienced the often disparate expectations and assessments of relative value and client-agent responsibility that surface from the non-uniform institutional frame of reference that shapes such encounters. It is important to distinguish specific and general institutions because one might expect that different societies will have different specific institutions, but, as in the Mesopotamian examples cited previously, societal boundaries are commonly defined by the degree of identifiable homology in general institutions. Although complexity is typically predicated upon the establishment of durable and codified institutional frames,[3] I suggest that the range of institutional constraints that shaped Bronze Age interactions on the steppe reflects a non-uniform degree of parity among related or neighboring groups. Non-uniformity is the result of some general institutional codes being homogenized between diverse groups or re-shaped among them for strategic purposes, while other institutions remain individually or specifically defined (Fig. 3.1). Thus, for each participant community, its degree of organizational consolidation or fragmentation vis-à-vis its neighbors depends on the scalar cohesion of various institutional structures and the periodic willingness of those communities to

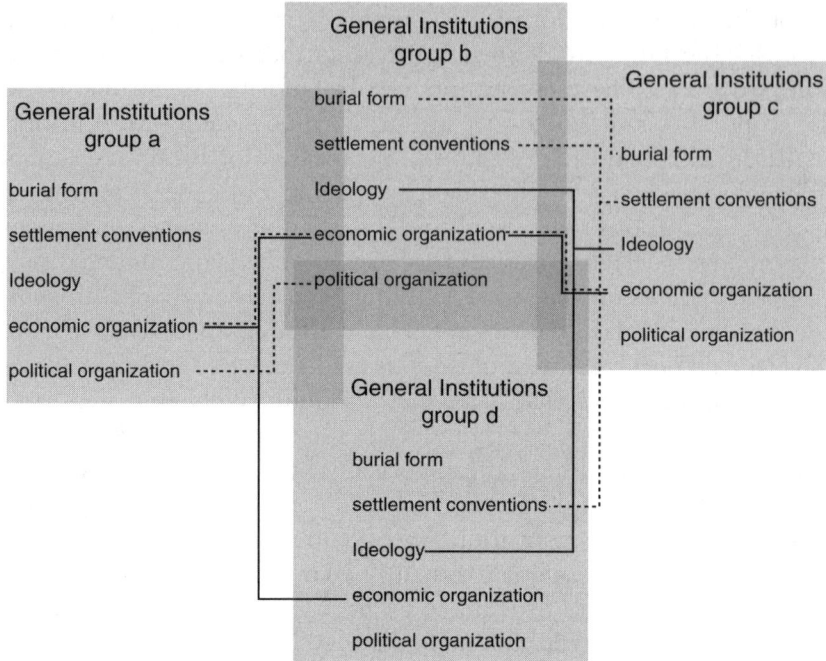

Figure 3.1. Conceptual model illustrating non-uniform complexity and periodic extent of geographic overlap among four hypothetical neighboring communities. Lines indicate institutional cohesion at period 1 (*solid*) and period 2 (dashed).

adopt or develop similar constraints to their modes of interaction. As such, the degree of "complexity" for different Bronze Age steppe communities cannot be assessed as a societal whole, because their degree of institutional cohesion may be temporarily connected and organized in some aspects, while diverse and at odds in others. The social complexity of settled prehistoric agricultural societies of the Near East, Asia, and the Mediterranean is often assessed in terms of the durability of institutions over temporal and geographic scales. Complexity among steppe communities is better evaluated in terms of institutional integration or fragmentation at the interstices of diverse populations whose economic and political interests co-exist geographically but are not necessarily bound by a shared sense of society.

In the case studies presented here, I suggest that institutions were transmutable for strategically flexible pastoralists, meaning they were periodically reformed according to a number of changing contextual factors (e.g., economic prosperity, productive alliances, ideological cohesion, environmental change). According to this paradigm, various populations can be traced along different trajectories of structural complexity. From a relational perspective, the scope of various institutional constraints considered by different groups, such as genealogical ties, trade agreements, or political alliances, may expand or contract according to particular social, economic, and political relationships. For example, a group's economic institutions may reflect a long-lived and codified set of wide-scale relationships, while their functional political institutions are comparatively local and organizationally simple. Thus, the overall level of complexity for these communities – in terms of social organization – cannot reflect a uniform trajectory of growth or decline.

Recognizing the multitude of factors conditioning steppe social interaction, Ludmila Koryakova has proposed a flexible world system model to explain the long-term structural change in the western steppe region and Trans-Urals in the Middle Bronze Age (Koryakova 2002; Chapter 2 in this volume). She suggests that uneven economic and technological developments between centralized populations at fortified sites like Sintashta and clan-tribal groups living in surrounding territories led to the establishment of a variety of scales of core-periphery relationships (Koryakova 2002: 103). Koryakova argues that, around 2200 BCE, the structure of regional interaction was reflective of a complex chiefdom, though she also notes that the trajectory of social complexity in the Trans-Urals at this time did not reflect progressive evolution to an arbitrarily greater level of complexity – at least not until the start of the first millennium BCE. Rather, she argues for cycles in the rise and collapse of local and regional relationships throughout the Bronze Age, until the start of the Iron Age.

Although her characterization of the dynamic nature of inter-group relationships is accurate, a world system–type model may be only partly supported by the archaeological evidence, especially considering the conditions typically cited for functioning world systems (see Kohl 1996). Starting with the paradigm of core-periphery relationships, large fortified sites, like Sintashta and Arkaim, may indeed have defined focal areas of greater social and political cohesion when compared to other sites within the broader regional landscape (Epimakhov 2003). A world system core (e.g., Algaze 1993), however, should also demonstrate the capability to maintain economic and political influence over both the materials and means of production in the periphery (Wallerstein 2000: 90–92). The fortified sites in the "country of towns" seem to demonstrate only localized influence, although they participated in the wider circulation of a diversity of materials or technologies (Anthony, Chapter 4 in this volume). Rather than these larger centers dictating productive control over the periphery, it instead appears that decentralized pastoralist groups had more control over the source side and distribution side of exchange, especially in terms of one of the most precious technologies of the time – metallurgy.[4] Koryakova herself notes other ways in which the chiefdom–world system model may less exactly describe the diversity of ideological, political, and economic organization evident in the steppe in the Middle Bronze Age – specifically in terms of codified burial ritual and social hierarchy (Koryakova 2002: 107). Although a world system model is debatable, Koryakova's discussion of the disproportional scales of economic and demographic cohesion in Middle Bronze Age Trans-Uralia nonetheless provides valuable theoretical building blocks for the theory of non-uniformity presented here.

Scale and Non-uniform Complexity

A KEY variable in assessing complexity is scale. When systems are viewed at the widest geo-political scale (i.e., globally), vast differences in institutional structure among different societies are not surprising. Such global perspective has sparked powerful theoretical paradigms, such as Wallerstein's (1974) conception of the modern world system. Non-uniform complexity, however, describes a condition of scalar transformation of various general institutional forces that shape social interactions among differentiated local communities. At the most local scale, we might imagine groups whose institutional constraints at first serve to organize the immediate community, while enabling them to strategize among their neighbors. At this scale, non-uniform complexity may be comparable to better-known models such as heterarchy, discussed by Crumley (1995). In heterarchical systems, political power and social institutions may be expressed among related groups according to different measures. However, heterarchy and heterarchical processes, such as peer-polity interaction (Renfrew and Cherry 1986), more often describe coherent socio-cultural systems wherein regional "peers" participate in a shared socio-political process by manipulating common elements of cultural and material currency. Here the variability of scalar consolidation and fragmentation is a result of strategic institutional change, rather than the systemic variation of a shared societal fabric through different stages of organization (Wiessner 2002). Institutional change, as North

(1990: 8) argues, "comes from the perception of entrepreneurs in political and economic organizations that they could do better by altering the existing institutional framework at some margin." It may be that some strategic alterations to a particular institution will expand its scalar resonance among interacting groups, while other institutions remain locally contextualized (i.e., specific). If we envision such transformations on the widest scale, some institutional alignments may help to shape the boundaries of a world system or facilitate the envelopment of populations into larger economic or political bodies. However, if such scalar transformations are formulated from groups with heterogeneous motivations, they are also prone to fragmentation or reform by subsequent transformations in other, parallel institutional arenas. Non-uniformity characterizes how independent communities find strategic mechanisms to shape their various modes of interaction using dynamic institutional parameters. Institutionalized practices, like the relation of metallurgy to prestige goods, may illustrate a relatively long-lived and wide-scale resonance, while other institutions, like burial form and ritual, may be comparatively more diverse and responsive to local social and political dynamics. Yet, for a period of time, burial institutions and metallurgical circulation reinforced one another because metal objects were commonly used to aggrandize buried individuals. As a result, contingent social motivations reciprocally impacted how each institution was experienced and translated among communities.

Scalar reorganization is well documented ethnographically among pastoralist groups (e.g., Irons 1974; Beck 1991). Pastoralists strategically negotiate their political and environmental landscape, periodically causing aspects of institutional parity to collapse, or *fraction*. In such periods, groups commonly regress to smaller-scale units, and their shared institutions and practices may be regionally reformulated to be relevant at the extant scale of political integration (e.g., Shahrani 1979). *Fractioning* is reflected archaeologically in periods of material diversity and landscape reform (changes in settlement geography, burial diversity, etc.). The inherent variability of mobile pastoral strategies often precludes long-term stability at a given state of organization; throughout the second millennium BCE, various steppe groups teetered on the edge of more highly institutional forms of social and political integration. These periods of social consolidation and fractioning, fundamental to the model proposed here, confound a conventional picture of progress from simpler to categorically more complex organization in a linear socio-evolutionary sense.

Viewed over the long-term, some social, political, and economic relationships across the steppe do seem to reflect consistent growth in terms of greater geographic distribution and structural stability. Certain conditions, such as metallurgical production and resource-trade relations, are less responsive to the fluctuation of non-uniform political dynamics, perhaps because of their obvious social and functional benefit. Once introduced, it is likely that even societies in a state of fractioning would not disintegrate so far as to sever all ties with certain economic or social networks. So, it follows that inter-regional resource trade and metallurgy, for example, developed along a fairly stable trajectory and seemingly promoted most steppe groups to a "higher" level of institutional coherency throughout the Bronze Age (Chernykh 1992; Chernykh et al. 2004).

Nevertheless, economic or technological prosperity does not necessitate categorical growth in socio-political organization. Characteristic forms of material culture may be distributed across the boundaries of local landscapes and have various impacts on each context. Formal characteristics and materials are transportable, but their meanings can be re-shaped and re-applied to fit the current socio-political setting and array of participants (Appadurai 1986; Urban 2001). On a larger scale, the archaeological data may reflect inter-regional interactions such as those identified across the steppe throughout the second millennium BCE, yet the overall state of complexity may still be described as non-uniform.

Depending on the degree of social solidarity or the duration of particular relationships among various groups, processes of assimilation, emulation, and transplantation of economic and socio-political forms may help to formalize the structure of wider scale geo-cultural arenas and relationships through time. Such geo-cultural extensions are specific to the motivations of particular actors and their strategies for maintenance or promotion of themselves, their group, or their institutions, but they are also subject to the historical reiteration of formalized relations of power. This may explain why some populations develop core-periphery relationships in some arenas, whereas other populations may selectively disengage from such interactions. These groups may variously benefit or lose out according to their proximity or association with neighbors with greater stability or geographic extension in their socio-political institutions (Sherratt 1993). I propose that periodic variability in the local economic and socio-political strategies of Bronze Age mobile pastoralists acted as a generative force behind the diversity of institutional expressions among steppe populations throughout the second millennium BCE.

At its most basic level, non-uniform complexity describes the scalar growth or decline of differentiated relationships between groups as they fluctuate through various levels of social integration or institutional cohesion across their landscape. A group's social landscape (i.e., the geo-cultural extent of their interactions) is determined by its practices and strategies (e.g., mobility, resource exploitation, ritual, trade) set within the ambit of regional ecological settings and geo-history (Ingold 1993). From this perspective, a given social group may go through a series of stages of consolidation and disintegration, expressed by the extent of its political control and social consolidation of power (e.g., Marcus 1998). Bronze Age pastoralists of Eurasia, as well as agricultural civilizations of southern Central Asia, demonstrate this dynamic process of socio-political growth and decline throughout the second millennium BCE (Kohl et al. 2002; Hiebert 1994: 176–177).

Non-uniform Complexity in Three Archaeological Landscapes

THREE BRONZE Age landscapes can be used to illustrate the scalar transformation of institutional organization among regional steppe populations and between the diverse societies that co-occupied the territories between the "steppe and the sown" circa 1900 BCE. Here, we individually consider the conditions of organizational complexity within the Trans-Urals region, Semirech'ye (eastern Kazakhstan), and the Margiana Oasis (Fig. 3.2).

NON-UNIFORM COMPLEXITY IN THE TRANS-URALS REGION

The economy and political structure of the Trans-Urals region in the Middle Bronze Age (ca. 2100–1700 BCE) appears to have been generated through a variety of relationships between differently organized communities. Some of the communities exhibit characteristic traits of "chiefdom-like" polities, with centralized settlements (e.g., Sintashta, Arkaim) and burials that indicate the existence of institutionalized structures of social and political hierarchy. Yet other contemporary groups in the neighboring steppe and forest-steppe regions were succeeding as mobile pastoralists and metallurgists and do not illustrate institutional coherency with these more centralized communities (Epimakhov

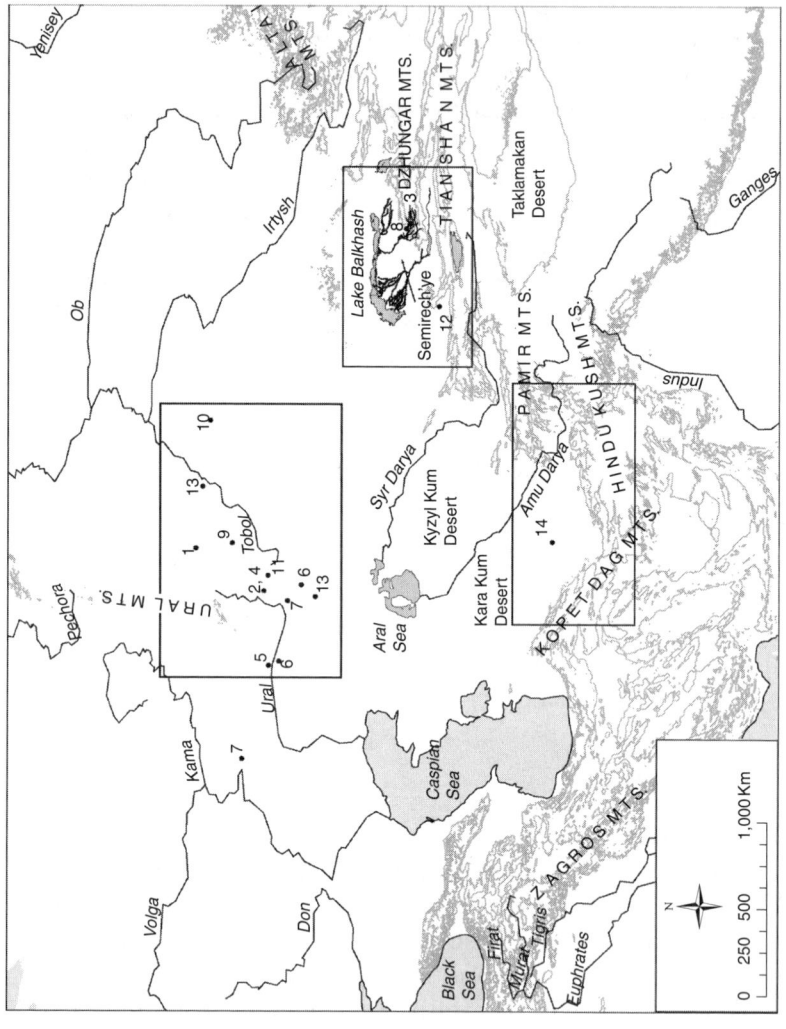

Figure 3.2. Central Eurasia, the Eurasian Steppe, and principal sites mentioned in the text: (1) Alakul'; (2) Arkaim; (3) Begash; (4) Bol'shekaraganskii; (5) Gorny; (6) Kargaly; (7) Krasnosamarskoe; (8) Kuigan; (9) Kulevchi III; (10) Petrovka, Petrovka-2; (11) Sintashta; (12) Tamgaly; (13) Tasty-Butak; (14) Margiana Oasis, Gonur. Contours at 1,000 and 2,000 meters above sea level (Map data source: ESRI Digital Chart of the World).

2003: 87–88; Koryakova 2002; Anthony et al. 2005). Epimakhov (chapter 5 in this volume) provides a well-conceived reconstruction of social dynamics in the Sintashta cultural landscape at the start of the second millennium BCE, reflected in intensified interaction among pastoralists, hunter-herders, and other groups across the Circumpontic and Trans-Urals regions (also G. Zdanovich and Zdanovich 2002: 251). Differentiated settlement and burial scales provide a starting point from which we may trace the development of non-uniform institutional structures in the Trans-Urals region in the Middle and Late Bronze Age. Then, we focus on a few institutional trajectories related to exchange and production, and ritual expression, as well as the implications that various scales of settlement and burial have for institutional cohesion among neighboring communities.

At the start of the second millennium BCE, the Sintashta-Arkaim settlement type dominated the riparian valleys between the Ural and Tobol rivers, defining the "country of towns" (G. Zdanovich and Zdanovich 2002). These settlements were large, fortified, and centralized; the outer fortifications at Sintashta enclose roughly 6.2 hectares (Gening et al. 1992: 390). Trans-Uralian communities were primarily herding groups at this time, taking advantage of the rich ecological niche along the many rivers and tributaries in the region. Stable isotope analyses from bodies of the Bol'shekaraganskii cemetery (2000–1800 cal. BCE) illustrate a diet rich in protein from sheep and cattle, with limited proportions of plant foods (K. Privat in D. Zdanovich 2002: 168–170). Differentiated burial structures and burial treatment in the form of diverse grave materials from Bol'shekaraganskii reflect an emergent social hierarchy in the "country of towns," although the reach of this political system was likely limited to a fairly localized landscape (G. Zdanovich and Zdanovich 2002: 249).

Articulated with the Sintashta cultural landscape were a number of smaller-scale groups that were likely less socio-politically centralized but were more extensive in their geographic scale of interaction, represented by Petrovka and Abashevo cultural materials (Tkachev 1998). Petrovka settlements such as Kulevchi III (Vinogradov 1982) are located within the territory of the "country of towns," but are substantially different in their construction and permanence of habitation (G. Zdanovich and Zdanovich 2002: 252). The disparities in scale and density of settlements of the Petrovka type indicate that a diversity of groups contributed to the settlement landscape of the "country of towns" with different economic strategies and demographic organization from those of Sintashta folk (Tkachev 1998). Further points of institutional difference between these contemporary communities are evident in the diversity of burial traditions across the Trans-Urals. Tkachev (1998: 41) observes that for communities associated with the Petrovka type "a specific burial planigraphy [sic] was developed – one to two central graves containing adult remains under a kurgan and children's burials typically laid-out in a ring on the periphery."[5] Sintashta-type burials are sometimes surmounted by kurgans as well, but reflect a greater focus on individual status of the person interred through elaborate tomb staging, animal sacrifice, and inclusion of high-status goods (e.g., cemetery "CM" at Sintashta; Gening et al. 1992). Sintashta-type burials illustrate a clear institutional departure from the family structure or extended lineage ritual of burial associated with the Petrovka typology. From a comparative perspective, we may infer that ideological institutions – at least as they pertain to burial and potentially to social organizational structures – were differentiated among these populations. However, a link between these populations is evident in their shared material forms (Potemkina 1995a), which may be explained through the strategic growth of mutual economic institutions related to the elaboration of metallurgy and emerging trade networks in the first centuries of the second millennium BCE (Kuz'mina 2004). Chapters 4, 5, and 9 in this volume, which deal directly with the growth of metallurgical and material exchanges throughout the Trans-Urals at the start of the second millennium BCE, demonstrate that a wide array of material and technological overlap was essential to the functional success of a variety of communities occupying the western steppe region. The success of such interaction was likely generated through the negotiation of institutionalized economic mechanisms that, at first, simply spurred a wider distribution of technology and material forms and later set the groundwork for more formal relationships between specialized communities of metallurgists, pastoralists, and others.

The geographic extent of shared economic relationships among various Trans-Uralian communities grew for more than 200 years, evident in the widening distribution of common ceramics, metals, and technologies (e.g., chariots) across the western steppe around 1900 BCE (Kuz'mina 2004; Anthony, Chapter 4 in this volume). Yet, by 1800–1700 BCE the institutional organization of the "country of towns" began to decline, demonstrated in the discontinuity of the Sintashta-Arkaim settlement type in the Trans-Urals and its ultimate complete replacement by smaller pastoralist encampments of the "late Petrovka" and the "Alakul" types (Kuz'mina 1994) (Fig. 3.3). Epimakhov cites environmental change and

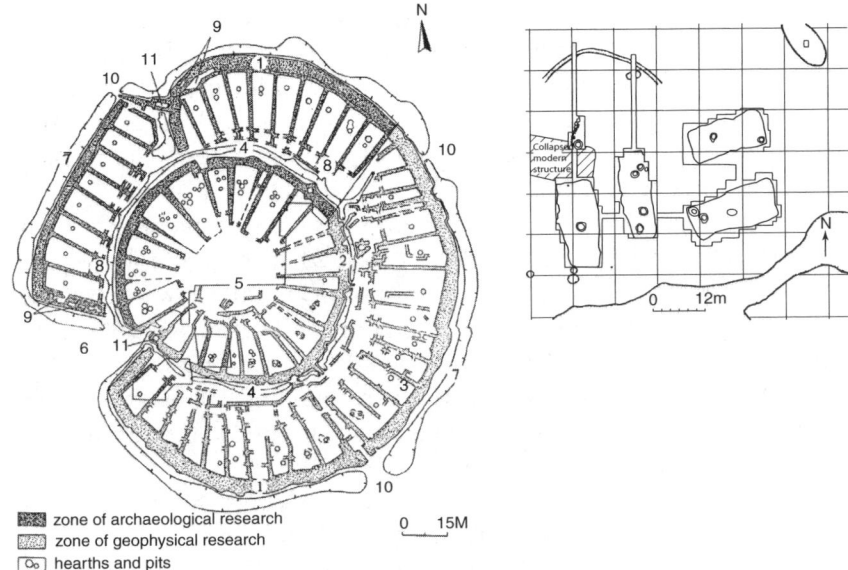

zone of archaeological research
zone of geophysical research
hearths and pits

0   15M

Figure 3.3. Comparative scale of Arkaim (after Zdanovich 2002) and Tasty-Butak (after Sorokin 1966). Both settlements are depicted at a common drawing scale.

ethnic pluralism as the cause of decentralization in the Trans-Urals (Epimakhov 2003: 88). Whatever the reason, diversity in late Petrovka and Alakul cultural assemblages and settlements may index a process of institutional fractioning of the more centralized polity of the "country of towns" into a geographically dispersed social landscape of mobile pastoralists across the region (G. Zdanovich and Zdanovich 2002: 252).

Comparing the settlement and burial records from a wider temporal perspective shows that only some groups flourished at the transition to the Late Bronze Age. Although consolidation was evident in the settlement scale of the "country of towns" at the beginning of the second millennium BCE, it would seem that extant institutional mechanisms of power did not resonate for long among those living in these larger centralized settlements. This point is illustrated poignantly by the period after 1800 BCE, when the Sintashta-type settlements were in decline and expanding technologies, like bronze metallurgy, became even more pivotal to the political economy of the Trans-Urals region (Hanks, Chapter 9 in this volume). The scalar transformation of settlement geography indexes a process of fractioning in the shift to the Alakul' phase around 1700–1600 BCE in the Trans-Urals region. In fact, the distribution of late Petrovka-Alakul' ceramics extends both west and east, and the diversity of ceramic forms at sites such as Bol'shekaraganskii may indicate an increasing expansion of economic networks while the centralization of the "country of towns" had all but collapsed.

Viewing socio-political organization from the perspective of mobile and strategically flexible pastoralists reveals that there simply may have been an institutional reorientation of political and economic relationships and a restructuring of the geographic and social scale of interacting communities. The adaptation of institutional relationships, such as a new focus on production and resource exchange, is evident around 1700 BCE at sites such as Gorny and Kargaly, located at the southwestern boundary of the Sintashta territory.

The settlement site of Gorny is located beyond the main regional extent of Sintashta-type settlements (Chernykh 2004). Gorny dates to the period when sites like Arkaim had been abandoned, roughly 1700–1400 cal. BCE. The scale of domestic structures in this period is comparably smaller than the Sintashta period, although the region can still be categorized as regionally diverse. Gorny itself encompasses an activity zone of nearly 4 hectares (Chernykh 2004: 235), which reflects a considerable social investment in the ores available in the territory around Kargaly. Gorny's significance to the development of metallurgical trade relationships between populations in the north Caspian steppe region and Trans-Urals is evident in the scale of its metallurgical production and corresponding distribution of metal artifacts (Chernykh et al. 2004), but also in the corollary scale of consumption of domestic fauna recorded at the site, apparently fostered through exchange relations with regional pastoralists (Antipina 1999).

The regional exchange relationship evident at Gorny is important for two reasons. First, the population of Gorny was likely a specialized community of miners and artisans (Chernykh 2004) who, by and large, did not directly generate its own subsistence. Thus, its scale of consumption reflects an equally specialized production of domestic animals by neighboring populations. In comparison with earlier Arkaim, Chernykh rightly notes that Gorny was analogous in terms of its scale of impact in the local political economy and in shaping power relations between variously specialized groups. Thus, in comparison with earlier exchange relationships in the Sintashta phase, the amplification of a commodity-based trade institution at Gorny illustrates how comparably less-consolidated pastoral populations were able to adapt their economic institutions in spite of relative fractioning in the social organization of the region.

These brief examples illustrate how non-uniform institutional growth and decline describe the varied expressions of social and economic organization among different groups in the Trans-Urals region from the period 2100 to approximately 1600 BCE. Specifically, we may look to diversity in the scale and construction of settlements and differentiation in burial structure to illustrate heterogeneity in relevant social and ritual institutions between interlaced populations in these periods. I define this context as institutionally non-uniform, rather than simply diverse, because the populations organized by different practical constraints and opportunities generated a mutually functional institutional relationship in some realms, evident in the distribution of metal and ceramics through flexible networks of trade and material exchange, yet they were seemingly autonomous in other institutional arenas. Thus, the political economic complexity of this period may best be described from a paradigm of non-parallel but related trajectories of institutional cohesion and fragmentation through time.

NON-UNIFORM COMPLEXITY IN SEMIRECH'YE

The Bronze Age of Semirech'ye (southeastern Kazakhstan) is less comprehensively studied than that of the Trans-Urals but nonetheless provides us with another view of non-uniform complexity in a Bronze Age steppe context. As recently as five years ago, archaeologists believed the earliest Bronze Age settlements in the region dated to about 1600 BCE (Kuz'mina 1986; Mar'yashev and Goryachev 1993; Frachetti 2002; Goryachev 2004). New radiocarbon dates from the settlement site of Begash illustrate that Semirech'ye was inhabited by pastoralists at least as early as 2450 cal. BCE (Frachetti and Mar'yashev 2007), which demands a re-conceptualization of our frameworks for understanding the development of pastoral societies in Semirech'ye and the nature of their political and economic organization throughout the Bronze Age.

Unlike the rapid consolidation of population that corresponds with the centralized settlements of the Sintashta-Arkaim "country of towns," the archaeological landscape of Semirech'ye illustrates a long duration of localized fluctuations in the scale of interaction and institutional cohesion among Bronze Age mobile pastoralists. Archaeological evidence from the Koksu Valley illustrates the establishment of large (~1200 square meter) stone-foundation villages and their associated quasi-megalithic Bronze Age cemeteries, which may reflect periods of institutional stability and group cohesion among seasonally transhumant pastoralists. However, fluctuations in the geographic reach and social coherency of various institutional parameters are indexed by periodic shifts in the construction scale of pastoralist winter villages and in the size and elaboration of Bronze Age cemeteries, most notably in the Koksu Valley (Frachetti 2008a).

Organizational consolidation may have been spurred by favorable environmental conditions (i.e., ample pastureland), excellent leadership, or positive trade relationships – to propose a few likely factors. On the basis of environmental reconstruction and range capacity studies, I have suggested that growth at settlements such as Begash indexes a concentration in the extent of pastoral mobility, because elaborate winter settlements are located less than 20 kilometers from the rich upland pastures that supported domesticated cattle and sheep (Frachetti 2006). Unfortunately, it is difficult to document archaeologically exactly how long such periods of consolidation may have lasted. Chronologically detailed reconstructions of stratigraphic accumulation at Begash illustrate periods possibly as long as 500 years with no evidence of disruption in the local patterns of mobility and settlement (Frachetti and Mar'yashev 2007). Burials at large cemeteries such as Kuigan and Begash-2 contain prestige goods and exotic grave goods (Goryachev 2004), which may be evidence for emerging social differentiation within local political institutions. The structure and materials of these burials are not, however, suggestive of a wide-reaching institutionalized form of political hierarchy.

The periods of scalar consolidation at the settlement and burial grounds of Begash seem to alternate with phases of more stochastic habitation at the site. Specifically, phase 2 (1650–1000 cal. BCE) at Begash reflects higher mobility and less substantial occupation on the part of local pastoralists. Although the site never appears to go completely out of use, these periods also last as long as 300–400 years. Smaller settlements in more marginal ecological settings, however, are also evident in the valley at this time, and the occurrence of smaller, unassociated burial grounds, such as Begash-1 (Mar'yashev and Karabaspakova 1988), may also be indicators of periodic fractioning, perhaps when larger settlements were ecologically unappealing or when smaller groups sought social or political separation from dominant groups. As with periods of cohesion, social fractioning also may have resulted from a variety of conditions. Because there are very few settlements excavated in the Koksu Valley, it is unclear what caused structural changes in this region of Semirech'ye. Elsewhere I have suggested that regular patterns of ecological variation coupled with wider regional developments in technology and production may have led mobile pastoralists to periodically expand or contract their scope of interaction in the mountain corridors of Semirech'ye (Frachetti 2002).

Geographic and scalar variation in contemporaneous settlement and burial contexts in the Koksu Valley serves as a proxy for the non-uniform trajectories of socio-political and economic relationships within the local pastoralist landscape. Yet, comparing the archaeology of the Dzhungar Mountains with that of neighboring regions throughout Semirech'ye illustrates more broadly how the scalar expansion of select social and ideological institutions generated a landscape of non-uniform complexity.

Because intensive archaeological survey was not a major thrust of research in Semirech'ye during the late Soviet era,[6] there are few detailed archaeological studies that provide comparable archaeological documentation of local Bronze Age settlement geography. Alexeii Rogozhinskii's work (and others') on the Bronze Age archaeology of Tamgaly is a notable exception that provides a well-documented case study of political differentiation to compare with the Koksu Valley.

Tamgaly is a site most noted for its spectacular Bronze Age rock art. Its main "sanctuary" boasts more than 10,000 skillfully carved rock engravings, which is why the site has been promoted recently to Unesco's world heritage list. The area around the heaviest concentration of rock art also provides a rich archaeological record of domestic and burial contexts, and Rogozhinskii's detailed excavations supply a distinct characterization of socio-political organization in this part of Semirech'ye (Rogozhinskii 1999). Tamgaly functioned as a social and ideological center for pastoralists whose practical scale of interaction may have been more extensive than other Late Bronze Age populations across Semirech'ye.

As Tamgaly's most abundant archaeological data, the site's rock art documents wide-scale regional communication, as many motifs are commonly found throughout the mountains from the Altai to the Pamirs (Samashev 1993; Mar'yashev 1994). Specifically, figures known as "sun dieties" (Fig. 3.4) have been regularly recorded throughout Semirech'ye, most notably at Byan-Zherek (Mar'yashev 2002), Eshkiolmes (Koksu Valley), and Anrakhai (Mar'yashev and Goryachev 2002). These images also show direct stylistic associations with sun dieties known from the Tianshan Mountains at the sites of Saimaly-Tash and Chopan-Ata (Kyrgyzstan). "Sun-head" motifs are also well known in the shamanistic imagery of the Altai Mountains and southern Siberia (Devlet 1980). The

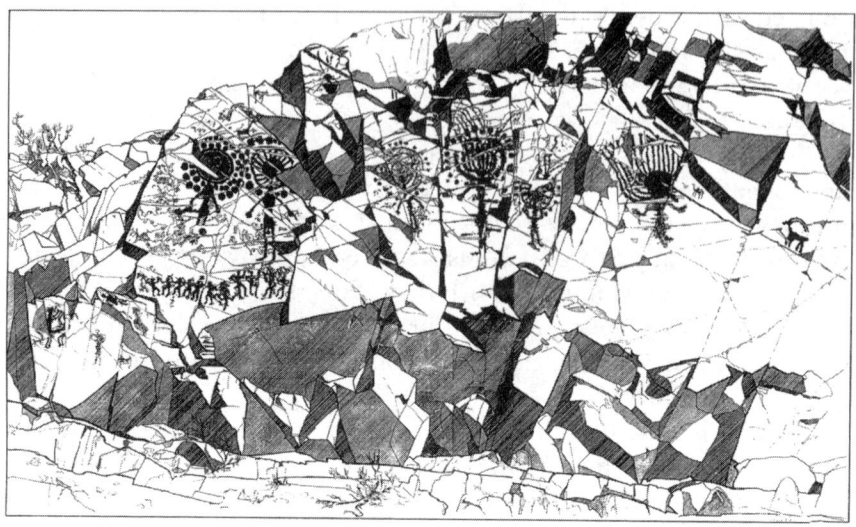

Figure 3.4. Sun-head "deities" at Tamgaly (group V) (after Rogozhinskii 2001).

identifiable cords and hanging elements of the costumes of Bronze Age sun-head images in the Dzhungar Mountains, such as at Eshkiolmes and Byan-Zherek, have been convincingly associated with Eurasian shaman costumes of the Altai, suggesting that shamanism played a role in the ritual and ideological practices of Bronze Age peoples in the Koksu Valley and at Tamgaly (Samashev 1998).

Rogozhinskii argues that Tamgaly functioned primarily as a ritual center, while practical and economic life was carried out on the site's periphery. The first period of ritual centralization is represented by the burial ground Tamgaly-1, which chronologically correlates with the Alakul'-Atasu period (1600–1400 BCE), whereas the second period of intense construction (at Tamgaly-2 and 4) reflects a later, "mixed" cultural phase in Semirech'ye corresponding to the Final Bronze Age (1400–1000 BCE).

Throughout the second millennium BCE, the geography and environment of the Tamgaly region was semi-arid, much as it is today. Thus, the regional ecology would not easily sustain pastoral populations on a year-round basis. Given the seasonality of the landscape, the site's social and ritual significance likely accumulated over time as annual rites and burial events were artistically memorialized in the canyons, serving as centralized ritual locales in the cultural geography of regional mobile pastoral groups. The wide geographic currency of Tamgaly's most prominent motifs, the sun deities, suggests the growth of a shared ideological expression among numerous groups across Semirech'ye, such as those around Begash. The scalar resonance of these semiotic forms indexes a growth in ideological semantics around 1600 BCE, which may be correlated with the fractioning of social and political geography at this time.

The two major cemeteries at Tamgaly provide some indication of the scale of socio-economic interactions associated with those interred at this important ritual center. The stone cist burials at Tamgaly-1, which are arranged in a burial group with a central cist and associated cists around it, date as early as 1600 BCE (uncalibrated C14 – Rogozhinskii 1999). On the basis of Rogozhinskii's comprehensive comparative analysis, the ceramics and metallurgy from Tamgaly-1 illustrate close stylistic similarities with Alakul' and Atasu materials from as far as the southern Urals and central Kazakhstan (Rogozhinskii 1999: 17). In fact, the ceramic assemblage collected from the 14 burials at Tamgaly-1 is fairly consistent in its decoration, which may indicate that the dominant population was engaged in more highly institutionalized relations of exchange across an extensive but formally organized socio-economic arena. Given the ecological constraints of the region, the population was likely engaged in long migratory orbits, perhaps far north into the steppes of central Kazakhstan or beyond. Ethnographic studies of seasonal movements of nineteenth-century Kazakhs lend support to this hypothesis (Abramzon 1971), at least from an ecological point of view. What is clear, however, is that the rock art of Tamgaly represents a discrete set of semiotic forms whose wide-scale distribution alludes to the formalization of certain ideological and economic relationships among regional populations and those who controlled this important location.

Although the Koksu Valley also boasts impressive rock art, the sanctuary at Tamgaly presents a more substantially centralized and formalized ritual context. The control of this site, as well as the environmental demand to migrate longer distances throughout the year, may have prompted those groups to seek more highly institutionalized interactions than those living in more productive regions of Semirech'ye. Again, a comparison of Begash to Tamgaly shows that the Bronze Age ceramic assemblage from Begash in the Koksu Valley is highly differentiated and inconsistent when compared with that of other Bronze Age settlements of the region (Frachetti 2008b: 168). Of all the diagnostic sherds recovered in excavations at Begash, 25% are decorated vessels, and of these, more than half can be associated stylistically with forms provenienced beyond Semirech'ye. The array of exotic ceramics at Begash includes sherds directly comparable with Fedorovo and Atasu ceramics known from central and eastern Kazakhstan, as well as the Altai and southern Siberia. Taken together, the ceramic assemblage at Begash reflects a variable expansion of interactive relationships in the Koksu Valley throughout the Bronze Age, which likely correlates with the periods of fractioning of socio-political organization of pastoral groups.[7]

Comparing the inventories of Begash with other Bronze Age settlements across Semirech'ye, we see that the stylistic details in the ceramic assemblages are different for each site, which suggests a relative variability in the institutional structure of each population's regional interactions. Although Bronze Age settlements across the Dzhungar Mountains and Semirech'ye, such as Talapty, Kuigan, Acy, Kyzylbulak, and Tasbas, do share a common set of coarseware jar forms with Begash (likely domestically produced quotidian ceramics), each of these sites also has a unique and rich set of decorated ceramics, which are associated with stylistic analogues from beyond the region, ranging in provenience from the Tianshan Mountains, central Kazakhstan, southern Siberia, and the Altai Mountains (Goryachev 2001). There is surprisingly limited overlap in the occurrence of particular decorative motifs at settlements across Semirech'ye. This inter-site variation suggests that, rather than participating in a common, mutually organized trade economy, each population in Semirech'ye established suitable trade relations according to its own institutions within a regionally delimited context.

Ultimately, the pastoral population living in the Koksu Valley enjoyed a rich ecological setting, and only periodically or stochastically expanded the scale of its migrations and regional interactions. The settlement patterns in the valley and their associated ceramics illustrate that internal variation in local socio-political organization may have fueled occasional bursts of regional interconnection and diffusion, but institutional coherency among populations across Semirech'ye seems to have remained locally strategic and non-uniform. Communities in the Koksu Valley experienced scalar expansions and contractions, but geographic expansion in this case more likely reflected a reframing of socio-political institutions, rather than centralization or cohesion. Other Bronze Age sites in Semirech'ye seem to mirror the case seen for the Koksu Valley, wherein material diversity was likely a result of locally resonant institutional reform rather than wide-scale organizational consolidation. In fact, the development of a centralized ideological center at Tamgaly and the distribution and coherence of ideological forms is an indicator of exceptional scalar growth in a particular institutional realm among pastoralist populations of Semirech'ye.

Non-uniform Complexity in the Margiana Oasis

The two steppe contexts examined here illustrate how regional socio-political landscapes developed under the condition of non-uniform complexity. Both of these cases have presented predominantly pastoralist scenarios, for which we may expect a higher degree of organizational variability and scalar fluctuation in socio-political institutions. Yet non-uniform complexity is useful to describe how social interactions affect societies at a variety of structural scales. Because the Margiana Oasis is outside the geographic focus of the current volume, I explore only a few of the ways in which non-uniform institutional trajectories are evident in this context around the same period of prehistory.

Socio-political organization of the Margiana Oasis during the early and middle second millennium BCE can be fruitfully re-examined through the paradigm of non-uniform complexity. C. C. Lamberg-Karlovsky (2003) has rightly argued that the political organization of the Bactria-Margiana Archaeological Complex (BMAC) does not match well with our current attributions of chiefdoms, states, or simple tribes. Qala architecture at a number of settlement sites has been related to a khanate system for the BMAC, especially during the final phases (Hiebert 1994: 177). Ceramic standardization and stamp seals are two clear indicators of institutionalized production, social hierarchy, and bureaucratic organization in mature BMAC contexts (Hiebert 1994, 2002). The recovery of BMAC materials across the Iranian plateau and beyond provides convincing evidence of BMAC participation in a larger-scale emerging world system at the start of the second millennium BCE (Kohl 2007). There are considerable traces of steppe ceramics in BMAC contexts (Hiebert 2002) but not enough to argue for stable institutionalized relationships between the two regions. Archaeologically, the BMAC lasted a little more than 200 years, after which, in the "Takhirbai period" (1800–1500 BCE), the settlement and socio-political organization of the region become obscured by the decline in large-scale architecture and a fractioning of relationships amongst the major BMAC oasis centers.

Documented relationships between steppe pastoralists and agriculturalists in the Takhirbai period are far from clear, but simple inductive reasoning begs the question: when the BMAC fractioned, what happened to the "complex" institutional relationships that promoted the proto-state-like system that enabled the BMAC to flourish? Perhaps the relationships among populations of the BMAC were less coherently institutionalized and hierarchically formulated than the material culture suggests? Although currently the archaeological record is not rich enough to illustrate why there was a political and social shift around 1800 BCE, we must conclude that the decline of sites such as Gonur, Togolok, and Altyn-Depe' must have been coincident with a shift in the political economy of the region. In this sense, the processes that introduced both consolidation and fractioning into the socio-political organization of the Trans-Urals region and Semirech'ye may also be usefully applied to those in Margiana. These include the alternating scalar expansion and decline of ideological, economic, and political institutions that drew differentiated regional populations into complex interactions.

If we look across the steppe, it is difficult to identify a singular process or trajectory of economic, social, and political development during the Bronze Age, yet we are stuck with archaeological materials as the primary proxies for interaction between societies of the Eurasian steppe and beyond. Non-uniform complexity provides a model that allows local-scale variation to carry a wider-scale impact on the re-shaping of the cultural landscape of central Eurasia throughout the second millennium BCE.

Conclusion

IN THIS chapter, I have suggested a condition of non-uniform complexity to explain the dynamic scales and character of social and political interactions across the steppe in the second millennium BCE. The cultural landscapes of the Trans-Urals and Semirech'ye were used to highlight some of the archaeological correlates of fluctuating institutional settings for pastoral populations, whereas the Bactria Margiana Archaeological Complex case suggests its application in contexts typically considered outside the steppe pastoralist world. Each of these regional landscapes illustrates that Bronze Age societies exhibited varied trajectories of social, economic, and political organization at the start of the second millennium BCE.

Steppe archaeology further illustrates that pastoralist strategies contributed substantially to the formation of distinct yet interlocking political economies across central Eurasia during the second millennium BCE (Koryakova 2002). The distribution of shared material forms and metallurgy across Eurasia, traditionally used as landmark evidence for the Andronovo culture historical model (Kuz'mina 1986, 1994; Potemkina 1995a, 1995b), can be best understood through the conditions of non-uniform complexity. Under this paradigm, opportunism and contingency facilitate material distribution within socially and ecologically structured patterns of regional interaction between pastoralists, metallurgists, and others. The Bronze Age archaeology of southeastern Kazakhstan illustrates how distant regional systems were articulated, albeit tenuously, through a network of pastoralist societies, whose own social and political organizations are inconsistent with linear, progressive models of social and political evolution.

From this viewpoint, the Bronze Age landscape of the steppe may be depicted as a "jigsaw puzzle" of fluctuating socio-economic arenas that served to link otherwise discrete and localized pastoral populations (Frachetti 2008b: 174). Pastoralist strategies, by definition, contribute to a heightened degree of variation in mobility and subsistence strategies, in settlement ecology, and in commercial activity – both within and across regions. In eastern Kazakhstan, local Bronze Age groups were practicing vertically transhumant pastoralism throughout the Tianshan and Dzhungar mountains as early as 2450 BCE; their cultural landscape shows incredible longevity because of the variability and adaptability of their economic and political practices. Differing degrees of mobility, productivity, and interaction, as well as environmental factors, are essential to the way pastoralists practically define and change the landscape within which they live, and this variation structures the venues and geographic extent of their interaction and assimilation with their neighbors (see Frachetti 2008b for discussion). In terms of complex political organization, then, these variable contexts engender a flexible system wherein power relationships can be revamped in response to periods of greater social, environmental, and economic stability or volatility, either within a local territory or at broader geo-political scales.

Acknowledgments

I WOULD like to thank Bryan Hanks and Katheryn Linduff for their efforts in organizing this volume and their helpful comments on earlier drafts of this contribution. Furthermore, I would like to thank the organizers and participants of the 2006 symposium at the University of Pittsburgh for bridging national archaeological traditions and fostering new explorations of social complexity in the steppe region.

Notes

1. Interestingly, this is not what North (1990: 7) calls an organization. Organizations, for North, are the bodies or structures that result from the implementation of institutions – or put in Rothman's terms, the resulting "functional units" of societal organization.
2. Informal constraints are those institutional boundaries that are not necessarily formalized through legal or bureaucratic mechanisms. North (1990) distinguishes informal institutions such as ideology and norms of behavior that are codified and enforced through common consensus from formal institutions such as written laws, constitutions, or religious dogma that are explicit and can be held to the letter.
3. North (1981: 94) discusses this most notably in the form of the state.
4. Chernykh and others have argued that the Seima-Turbino metallurgical phenomenon was controlled by a mobile population (Chernykh et al. 2004). Although the social details of a such a group are not clear, the wide distribution of prestige-oriented artifacts suggests that metallurgical trade and production was brokered by groups operating across wide geographic ranges.
5. Author's translation from the Russian.
6. Archaeological research in Semirech'ye during the late 1980s and early 1990s was mainly focused on single-site excavations.
7. In addition to steppe ceramics, Begash also revealed one of the few documented cases in the steppe zone with wheel-spun ceramics similar in form and production to those known in Period 1A of the Bactria and Margiana Archaeological Complex (Frachetti 2004: 347–349). However, one should not overstate this evidence because only one diagnostic sherd was found, and it awaits spectrographic analysis to confirm its provenience.

References

Abramzon, S. M. 1971. *Kirgizy i ikh etnogeneticheskie i istoriko kul'turnye sviazi.* Leningrad: Nauka.

Algaze, G. 1993. *The Uruk World System: The Dynamics of Expansion of Early Mesopotamian Civilization.* Chicago: University of Chicago Press.

———. 2001. Initial Complexity in Southwestern Asia. *Current Anthropology* 42(2): 199–233.

Anthony, D. W., D. Brown, E. Brown, A. Goodman, A. Kokhlov, P. Kosintsev, P. Kuznetsov, O. Mochalov, E. Murphy, D. Peterson, A. Pike-Tay, L. Popova, A. Rosen, N. Russell, and A. Weisskopf. 2005. The Samara Valley Project. Late

Bronze Age Economy and Ritual in the Russian Steppes. *Eurasia Antiqua* 2005: 395–417.

Antipina, Elena E. 1999. Kostnye ostatki zhivotnykh s poseleniya Gornyi (biologicheskie i arkheologicheskie aspekty issledovaniya). *Rossiiskaya Arkheologiya* 1: 103–116.

Appadurai, A. 1986. *The Social Life of Things: Commodities in Cultural Perspective.* Cambridge: Cambridge University Press.

Beck, L. 1991. *Nomad: A Year in the Life of a Qashga'i Tribesman in Iran.* Berkeley: University of California Press.

Blanton, Richard E., Gary M. Feinman, Stephen A. Kowalewski, and Peter N. Peregrine. 1996. A Dual-Processual Theory for the Evolution of Mesoamerican Civilization. *Current Anthropology* 37(1): 1–14.

Carneiro, R. 1970. A Theory of the Origin of the State. *Science* 169(3947): 733–738.

Chapman, R. 2003. *Archaeologies of Complexity.* London: Routledge.

Chernykh, E. N. 1992. *Ancient Metallurgy in the USSR: The Early Metal Age.* New Studies in Archaeology. Cambridge: Cambridge University Press.

2004. Kargaly: The Largest and Most Ancient Metallurgical Complex on the Border of Europe and Asia. In K. Linduff (ed.), *Metallurgy in Ancient Eastern Eurasia from the Urals to the Yellow River.* Lewiston: Edwin Mellen Press, pp. 223–238.

Chernykh, E. N., Evgenii V. Kuz'minykh, and L. B. Orlovskaia. 2004. Ancient Metallurgy of Northeast Asia: From the Urals to the Saiano-Altai. In K. Linduff (ed.), *Metallurgy in Ancient Eastern Eurasia from the Urals to the Yellow River.* Lewiston: Edwin Mellen Press, pp. 15–36.

Crumley, C. 1995. Heterarchy and the Analysis of Complex Societies. In R. Ehrenreich, C. Crumley, and J. Levy (eds.), *Heterarchy and the Analysis of Complex Society.* American Anthropological Association 6. Washington, DC, pp. 1–6.

Devlet, M. A. 1980. *Sibirskie poiasnye azhurnye plastiny II v. do n.e.–I v. n.e.* Arkheologiia SSSR: svod arkheologicheskikh istochnikov: Arkheologiia SSSR Izd-vo. Moscow: Nauka.

Earle, Timothy. 1977. A Reappraisal of Redistribution: Complex Hawaiian Chiefdoms. In T. K. Earle and J. E. Ericson (eds.), *Exchange Systems in Prehistory.* New York: Academic Press, pp. 213–229.

Epimakhov, A. V. 2003. Analiz tendentsii sotsial'no-ekonomicheskogo razvitiya naseleniya Urala epokhi bronzy. *Rossiskaya Arkheologiya* 1: 83–90.

Frachetti, M. D. 2002. Bronze Age Exploitation and Political Dynamics of the Eastern Eurasian Steppe Zone. In K. Boyle, C. Renfrew, and M. Levine (eds.), *Interaction: East and West in Eurasia.* Cambridge: McDonald Institute, pp. 87–96.

2004. Bronze Age Pastoral Landscapes of Eurasia and the Nature of Social Interaction in the Mountain Steppe Zone of Eastern Kazakhstan. Ph.D. dissertation, University of Pennsylvania.

2006. Digital Archaeology and the Scalar Structure of Pastoral Landscapes. In T. Evans and P. T. Daly (eds.), *Digital Archaeology.* London: Routledge, pp. 128–147.

2008a. Variability and Dynamic Landscapes of Mobile Pastoralism in Ethnography and Prehistory. In H. Barnard and W. Wendrich (eds.), *The Archaeology of*

Mobility: Nomads in the Old and in the New World. Monograph of the Cotsen Institute Advanced Seminar, UCLA. Los Angeles, pp. 366–396.

2008b. *Pastoralist Landscapes and Social Interaction in Bronze Age Eurasia*. Berkeley: University of California Press.

Frachetti, Michael D., and Alexei N. Mar'yashev. 2007. Long-Term Settlement, Mobility, and Landscape Formation of Eastern Eurasian Pastoralists from 2500 CAL B.C. *Journal of Field Archaeology* 32(3): 221–242.

Freidel, D. 1983. Political Systems in Lowland Yucatan: Dynamics and Structure in Maya Settlement. In E. Z. Vogt, and R. M. Leventhal (eds.), *Prehistoric Settlement Patterns: Essays in Honor of Gordon R. Willey*. Albuquerque: University of New Mexico Press, pp. 375–386.

Gening, V. F., G. V. Zdanovich, and V. V. Gening. 1992. *Sintashta*. Chelyabinsk: Yuzhnoe-Uralskoe Knizhkoe Izdatel'stvo.

Goryachev, A. A. 2001. O Pegrebal'nom Obryade v Pamyatnikakh Kulsaiskogo Tipa. In A. N. Mar'yashev, Yu. A. Motov, T. A. Egorova, A. A. Goryachev (eds.), *Istoriya i Arkheologiya Semirech'ya*, vol. 2, "XXI vek." Almaty, pp. 45–61.

2004. The Bronze Age Archaeological Memorials in Semirechie. In K. M. Linduff (ed.), *Metallurgy in Ancient Eastern Eurasia from the Urals to the Yellow River*. Lewiston: Edwin Mellen Press, pp. 109–138.

Hiebert, F. T. 1994. *Origins of the Bronze Age Oasis Civilization in Central Asia*. Bulletin of the American School of Prehistoric Research, Peabody Museum of Archaeology and Ethnology. Cambridge, MA: Harvard University Press.

2002. Bronze Age Interaction between the Eurasian Steppe and Central Asia. In K. Boyle, C. Renfrew, and M. Levine (eds.), *Interaction: East and West in Eurasia*. Cambridge: McDonald Institute, pp. 237–248.

Ingold, T. 1993. The Temporality of the Landscape. *World Archaeology* 25: 152–174.

Irons, W. 1974. Nomadism as a Political Adaptation: The Case of the Yomut Turkmen. *American Ethnologist* 1: 635–657.

Kohl, P. 1996. The Ancient Economy, Transferable Technologies and the Bronze Age World-System: A View from the Northeastern Frontier of the Ancient Near East. In R. Pruecel and I. Hodder (eds.), *Contemporary Theory in Archaeology*. Oxford: Blackwell, pp. 143–165.

2007. *The Making of Bronze Age Eurasia*. Cambridge: Cambridge University Press.

Kohl, P., M. G. Gadzhiev, and R. G. Magomedov. 2002. Between the Steppe and the Sown: Cultural Developments on the Caspian Littoral Plain of Southern Dagestan, c. 3600–1900 BC. In K. Boyle, C. Renfrew, and M. Levine (eds.), *Interaction: East and West in Eurasia*. Cambridge: McDonald Institute, pp. 113-134.

Koryakova, L. N., 2002. The Social Landscape of Central Eurasia in the Bronze and Iron Ages: Tendencies, Factors, and Limits of Transformation. In K. Jones-Bley and G. B. Zdanovich (eds.), *Complex Societies of Central Eurasia from the Third to the First Millennia BC*, vol. 1. Washington, DC: Institute for the Study of Man, pp. 97–117.

Kosintsev, P. A. 2001. Khozyaistvo naseleniya Urala I Povolzh'ya v III–II tys. do n. e. In *XV Ural'sckoe arkheologischeskoe soveshchanie*. Orenburg: "Drenburgskaya guberniya," pp. 85–86.

Kuz'mina, E. E. 1986. *Drevneishie skotovody ot Urala do Tian'-Shania.* Frunze: Ilim Press.

1988. Kulturnaya i etnicheskaya atributsya pastusheskikh plemen Kazakhstana i srednii Azii epokhii bronzy. *Vestnik Drevnei Istorii* 185(2): 35–59.

1994. *Otkuda prishli indoarii? Material'naia kul'tura plemen andronovskoi obshchnosti i proiskhozhdenie indoirantsev.* Moscow: MGP "Kalina."

2004. Historical Perspectives on the Andronovo and Early Metal Use in Eastern Asia. In K. Linduff (ed.), *Metallurgy in Ancient Eastern Eurasia from the Urals to the Yellow River.* Lewiston: Edwin Mellen Press, pp. 37–84.

Lamberg-Karlovsky, C. C. 2002. Archaeology and Language: The Indo-Iranians. *Current Anthropology* 43(1): 63–88.

2003. Civilization, State, or Tribe? Bactria and Margiana in the Bronze Age. *Review of Archaeology* 24(1): 11–19.

Marcus, Joyce. 1998. The Peaks and Valleys of Ancient States: An Extension of the Dynamic Model. In G. Feinman and J. Marcus (eds.), *Archaic States.* Santa Fe: School of American Research Press, pp. 59–94.

Mar'yashev, A. N. 1994. *Petroglyphs of South Kazakhstan and Semirechye.* Almaty: Akademiia Nauk.

2002. Novye Materialy o Poseleniyakh Epokhi Bronzy v Gorakh Bayan-Zhuryk. *Izvestiya* (Inst. Arkheologii Kazakhstana) 1: 23–30.

Mar'yashev, A. N., and A. A. Goryachev. 1993. Voprosi tipologii i khronologii pamyatnikov epokhi bronzi semirech'ya. *Rossiiskaya Arkheologiya* 1993(1): 5–20.

2002. *Naskal'niye Izobrazheniya Semirech'ya,* vol. 2. Almaty: "Fond" XXI c.

Mar'yashev, A. N., and K. M. Karabaspakova. 1988. Noviye Pamyatniki Epokhi Bronzi Vostochnogo Semirech'ya. In V. Medvedev and Yu. S. Khudyakov, *Drevniye Pamyatniki Severnoi Azii i ikh Okhpanniye raskopki.* Novosibirsk: Nauka, pp. 24–39.

Nelin, D. V. 2000. Poselenie epokhi bronzy Shibaevo-I: rezul'taty issledovaniya (predvaritel'naya publikatsiya). In V. V. Tkachev (ed.), *Problemy Izucheniya eneolita i bronzovogo veka Iyuzhnogo Urala.* Orsk: Institut Evraziiskikh Issledovanii, pp. 120–125.

North, D. C. 1981. *Structure and Change in Economic History.* New York: W. W. Norton.

1990. *Institutions, Institutional Change, and Economic Performance.* Cambridge: Cambridge University Press.

Potemkina, T. M. 1995a. Problemy Svyzzei i smeny kul'tur naseleniya Zaural'ya v Epokhu Bronzy (Pozdnii i final'ny Etapy). *Rossiiskaya Arkheologiya* 2: 11–20.

1995b. Problemy Svyzzei i smeny kul'tur naseleniya Zaural'ya v Epokhu Bronzy (rannii i srednii Etapy). *Rossiiskaya Arkheologiya* 1: 14–27.

Renfrew, C., and J. F. Cherry. 1986. *Peer Polity Interaction and Socio-Political Change.* Cambridge: Cambridge University Press.

Rogozhinskii, A. E. 1999. Mogil'niki epokhi bronzy urochishcha Tamgaly. In A. N. Mar'yashev, Yu. A. Motov, A. A. Goryachev, and A. E. Rogozhinskii (eds.), *Istoriya i Arkheologiya Semirech'ya.* XXI vek. Almaty, pp. 4–43.

2001. Izobrazitel'nyi ryad petroglifov epokhi bronzy svyatilishcha Tamgaly. *Istoriya i Arkheologiya Semirechiya* 2: 7–44.

Rothman, M. S. 1994. Introduction: Evolutionary Typologies and Cultural Complexity. In G. Stein and M. Rothman (eds.), *Chiefdoms and Early States in the Near East: The Organizational Dynamics of Complexity*. Madison, WI: Prehistory Press, pp. 1–10.

Samashev, Z. S. 1993. *Petroglyphs of East Kazakhstan as Historical Sources*. Almaty: Rakurs.

1998 Shamanskie syuzhety petroglifov Kazakhstana. In *Voprosy Arkheologii Kazakhstana*, vol. 2. Almaty: Gylym, pp. 197–207.

Shahrani, M. Nazif. 1979. *The Kirghiz and Wakhi of Afghanistan Adaptation to Closed Frontiers*. Seattle: University of Washington Press.

Sherratt, A. 1993. What Would a Bronze Age World-System Look Like? Relationships between Temperate Europe and the Mediterranean in Later Prehistory. *Journal of European Archaeology* 1(2): 1–58.

Shishlina, N. 2003. Yamnaya Culture Pastoral Exploitation: A Local Sequence. In M. Levine, C. Renfrew, and K. Boyle (eds.), *Late Prehistoric Steppe Adaptation and the Horse*. Cambridge: McDonald Institute, pp. 253–266.

Sorokin, V. S., and M. P. Gryaznov. 1966. *Andronovskaya Kultura*. Moscow: Nauka.

Stein, G. 1994. Introduction Part II: The Organizational Dynamics of Complexity in Greater Mesopotamia. In G. Stein and M. Rothman (eds.), *Chiefdoms and Early States in the Near East: The Organizational Dynamics of Complexity*. Madison, WI: Prehistory Press, pp. 11–22.

Stein, G., and M. Rothman (eds.). 1994. *Chiefdoms and Early States in the Near East: The Organizational Dynamics of Complexity*. Madison, WI: Prehistory Press.

Tkachev, V. V. 1998. K probleme proiskhozheniya Petrovskoi kul'tury. In *Arkhelogicheskie pamyatniki Orenburzh'ya – vyp. II*. Orenburg: Orenburgskii Gosudarstvennyi Pedagogicheskii Universitet, pp. 38–56.

Urban, G. 2001. *Metaculture: How Culture Moves through the World*. Public Worlds. Minneapolis: University of Minnesota Press.

Vinogradov, N. 1982. Kulevchi III pamyatnik petrovskogo tipav Yuzhnom Zauralye. *Kratkie Soobshcheniya Instituta arkheologii* 169: 94–99.

Wallerstein, I. M. 1974. *The Modern World System*. New York: Academic Press.

2000. *The Essential Wallerstein*. New York: New Press.

Wiessner, P. 2002. The Vines of Complexity. *Current Anthropology* 43(2): 233–269.

Zdanovich, D. G. 1997. Arkaim – kul'turnii kompleks epokhi srednei bronzi Yuzhnogo Zaural'ya. *Rossiiskaya Arkheologiya*, no. 2: 47–62.

Zdanovich, G. B., and D. G. Zdanovich, 2002. The "Country of Towns" of the Southern Trans-Urals and Some Aspects of Steppe Assimilation in the Bronze Age. In K. Boyle, C. Renfrew, and M. Levine (eds.), *Interaction: East and West in Eurasia*. Cambridge: McDonald Institute, pp. 249–264.

2002. *Arkaim: Nekropol' (po materialam kurgana 25 Bol'shekaraganskogo mogil'nika)*. Chelyabinsk: Yuzhno-Ural'skoe knizhmoe izdatel'stvo.

CHAPTER 4

The Sintashta Genesis
The Roles of Climate Change, Warfare, and Long-Distance Trade

DAVID W. ANTHONY

Introduction

RECENT STUDIES by Di Cosmo (1999, 2002) and Vehik (2002) have emphasized the transformational political effects of inter-tribal warfare in arid grasslands on two continents. Intensified warfare in both places encouraged greater political complexity, hierarchy, and elite-centered, distance-trading activities. This chapter argues that intensified warfare and long-distance trade played powerful roles in the origins of the Sintashta culture. Sintashta is defined by a group of forti-fied settlements and cemeteries dated about 2100–1800 BCE (calibrated) in the northern Eurasian steppe between the upper Ural and upper Tobol rivers southeast of the Ural Mountains. Outside the settlements were cemeteries that yielded whole-horse sacrifices, chariots, and many weapons. Inside the settlements, almost every excavated house yielded copper slag and remains of furnaces or intensely burned hearths. The metal was copper or arsenical bronze, usually in alloys of 1–2.5% arse-nic. Pieces of crucibles were placed in two graves at Krivoe Ozero (Vinogradov 2003: 172), and broken casting molds were recovered from the Arkaim settlement. An estimated 6,000 tons of quartzitic rock bear-ing 2–3% copper was mined from the single documented mining site of Vorovskaya Yama east of the upper Ural River (Grigoriev 2002: 84; Zaikov, Zdanovich, and Yuminov 1995). The surprising evidence for metallurgical production inside every excavated structure suggests that the Sintashta settlements were the focus of intense metalworking activi-ties, although the scale and organization of metal production is not well understood either within or between them (see Hanks, Chapter 9 in this volume). Recent publications by Kuz'mina (2008), Anthony (2007), Koryakova and Epimakhov (2007), Kohl (2007), Vinogradov (2003), Grigoriev (2002), Kovaleva and Zdanovich (2002), and the volumes issuing from the Arkaim conference of 1999 (Jones-Bley and Zdanovich 2002) and the Cambridge conference of 2000 (Boyle et al. 2002; Levine et al. 2003) have outlined many of the descriptive characteristics of the Sintashta culture. Yet the causes and meanings of its novel settlements, weapons, chariots, and ritual and metallurgical extravagance remain hotly debated.

I believe that the evidence can be arranged in a causal sequence. The key transformative agent in the causal chain is the intensification of conflict and warfare. In the second part of this chapter, I describe the archaeological evidence for warfare in the Sintashta culture. But I begin by asking two contextualizing questions about conflict: what might have caused an increase in the intensity and frequency of conflict in the northern steppes just before 2100–2000 BCE, and what might have been the social and political effects of an increase in violence?

CLIMATE CHANGE AND SEDENTISM

A markedly cooler, more arid climate began in the Eurasian steppe after about 2500 BCE, reaching its coldest and driest peak around 2200–2000 BCE. This event is recorded in pollen cores across the Eurasian continent, from western Russia (Klimento, et al. 2000) through the Pontic steppes (Kremenetski 1997, 2003) to the Samara Valley (Kremenetski et al. 1999), to the southern Urals (Lopez et al. 2003), across Kazakhstan (Kremenetski et al. 1997), into the Altai (Blyakharchuk et al. 2004), and even as far as the Amur River in northeastern Siberia (Kuzmin et al. 1997). The same episode is documented in botanical studies in Anatolia, the Caucasus, and Mesopotamia, and in East Africa and India (Weiss 2000). Climate scientists Perry and Hsu (2000: 12436) called it the most influential Little Ice Age in recorded history. Forests declined, steppes and deserts expanded, and winters became colder.

In the Pontic-Caspian steppes, the period between 2500 and 2000 BCE was the late Middle Bronze Age (MBA). The steppe cultures of the Early Bronze Age (EBA; 3300–2700 BCE) and earlier MBA (2800–2400 BCE) between the Don and Ural rivers had a mobile pastoral economy that left no visible settlements, but they buried their dead in highly visible kurgan cemeteries (Figs. 4.1 and 4.2). Increased competition

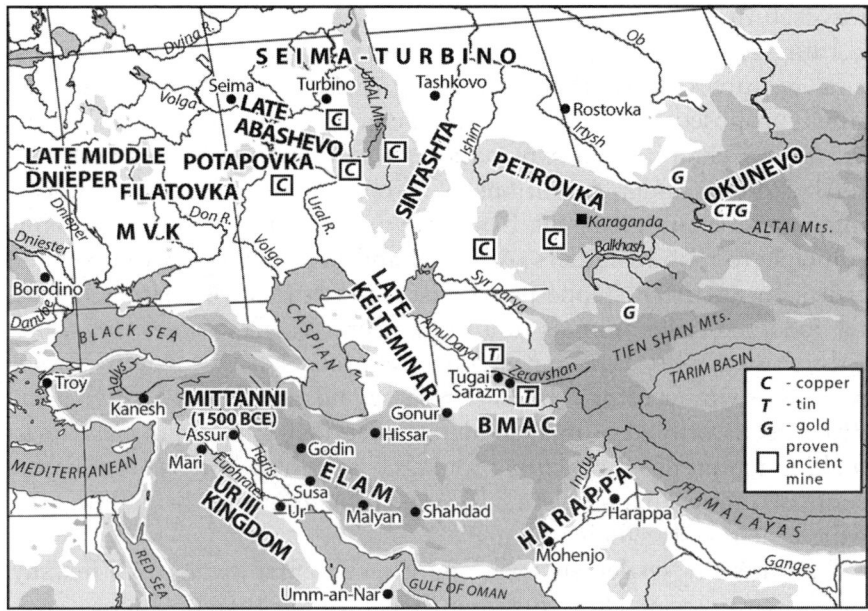

Figure 4.1. Late Middle Bronze Age culture areas of the steppe and forest-steppe zones and of West, Central, and South Asia about 2200–1800 BCE.

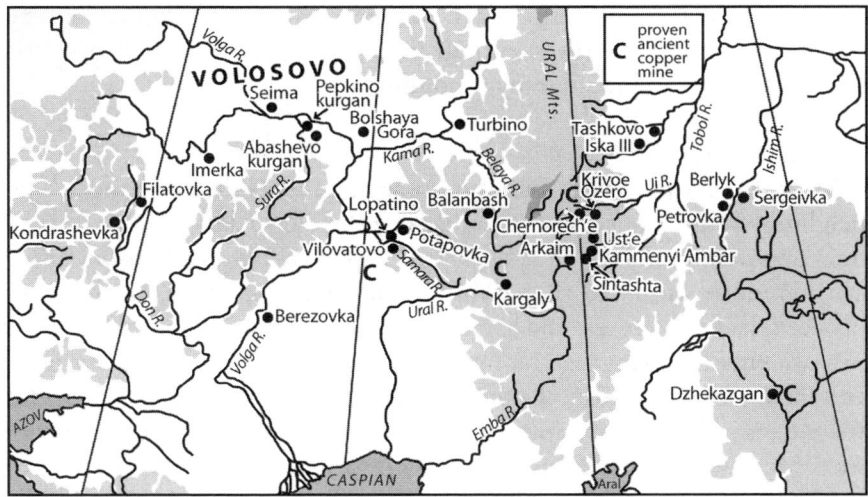

Figure 4.2. Sites in the steppe and forest-steppe zones of the late MBA referred to in the text.

for declining critical resources might have encouraged these mobile herding groups to settle in the key locations that contained the most vital resources rather than risk losing control over them. Rosenberg (1998) proposed this model, which he compared to the children's game of musical chairs, as an explanation for why mobile hunter-gatherers settled down under conditions of declining resources and increasing competition during the Neolithic period. I have applied the same principle to mobile herders (Anthony and Brown 2008; Anthony et al. 2005). In northern steppe pastoralism, the critical resource was winter fodder for the animals in a place protected from snow and wind. Ethnohistorically, Eurasian steppe pastoralists have favored marshy regions as winter refuges because 3-meter-tall stands of *Phragmites* and *Typha* reeds offered animal fodder, protection, and roofing and mat-making materials. *Phragmites* roots also are edible and sweet, like the roots of another marsh plant, *Althaea officinalis* or marsh-mallow, after which the marshmallow is named.

All Sintashta settlements were built on the first terrace of a marshy, meandering stream close to winter fodder, rather than on a more easily defended elevation, in spite of the clear concern with defense implied by fortifications (Fig. 4.3). Fortified settlements appeared during a documented decline of marsh and bog pollen and the conversion of isolated marshes and bogs to drier swamp and forest botanical communities. A "musical chairs" defensive reaction to the shrinking of large marshlands could explain both the sedentary settlement pattern adopted by the Sintashta culture and the seemingly contradictory placement of fortified settlements in low, marshy topographic locations.

Warfare and Power

Warfare is a powerful stimulus to social and political change. The heightened threat of conflict can dissolve the old social order and create new opportunities for the acquisition of power. James Madison (1865: 491) famously observed that, "War is the parent of armies; from these proceed debts and taxes; and armies, and debts, and taxes are the known instruments for bringing the many under the domination of the few." In tribal wars like those of the Bronze Age, debt was not a factor, but tribute – taxes – took the form of men and matériel, and leaders in tribal wars often were expected to exhibit generosity in rough proportion to the scale of their victories. Di Cosmo (1999, 2002) argued that statelike political structures arose among steppe nomads in the Iron Age largely because intensified warfare led to the establishment of permanent

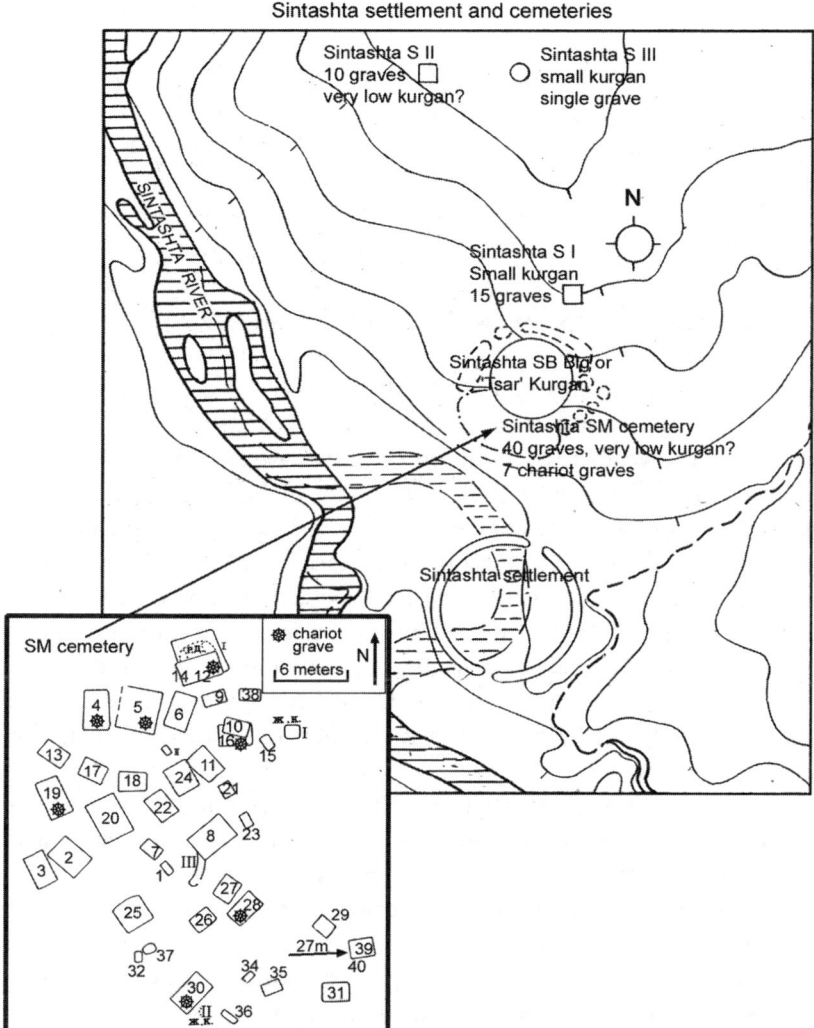

Figure 4.3. The topographic location of the Sintashta settlement and cemeteries (after Gening et al. 1992: fig. 2 and 42).

bodyguards around rival chiefs, and these grew in size until they became armies, which engendered statelike institutions designed to organize, feed, reward, and control them. Susan Vehik (2002) studied political change in the deserts and grasslands of the North American Southwest after 1200 CE, during a period of increased aridity, climatic volatility, and human conflict. Populations in some regions aggregated in larger, nucleated settlements. Increased warfare was accompanied by an increase in long-distance trade to acquire novel ornaments and valuables. Trade after 1350 CE was more than 40 times greater than it had been before. Settlement clusters appeared near high-quality flint outcrops (Vehik 2002: 49), the trade value of which was perhaps roughly analogous to the copper minerals in the Eurasian steppe. Warfare stimulated long-distance trade because exotic trade goods were useful as gifts in alliance-making negotiations between chiefs and as rewards to followers.

WARFARE AND ORNAMENTS IN THE ABASHEVO CULTURE

Warfare began to increase in the southern Urals region before the appearance of the Sintashta culture. It seems to have intensified first in the southern forest-steppe zone. At the MBA Abashevo-culture site of Pepkino, southwest of the Urals in the forest-steppe, a single grave pit contained the bodies of 28 young men, 18 of them decapitated, others with axe wounds to the head, axe wounds on the arms, and dismembered extremities (Koryakova and Epimakhov 2007: 62–63; Chernykh 1992: 200–204). Because this mass grave, dated by radiocarbon to 2400–2200 BCE (3850±95 BP, Ki-7665), was covered by a single kurgan, it reflected a single event, clearly a battle. The absence of women or children indicates that it was not a settlement massacre. Deaths in recorded battles between forces managed as tribal levies rarely reached 10% of the fighting force, and usually were more like 5% (Keeley 1996: 88–94). Casualties of 28 dead in the Pepkino grave therefore imply a force of 280–560 warriors just on the Abashevo side, or perhaps 1,000 combatants in total. Warfare on a surprising scale occurred during the Abashevo era.

The conflicts of the Abashevo period, 2500–1900 BCE, could have encouraged an increase in the demand for wealth to secure larger alliances. There was a notable increase in the production and use of copper in the southern forest-steppe zone in Abashevo sites (Grigoriev 2002: 106–118; Koryakova and Epimakhov 2007: 60). High-status Abashevo graves contained copper and silver ornaments, semi-circular solid copper and silver bracelets, cast shaft-hole axes, small socketed spearheads, and waisted knives. High-status Abashevo women wore distinctive headbands made of flat and tubular copper and silver beads decorated with hanging copper or silver double-spiral and cast rosette pendants, unique to the Abashevo culture and probably signals of ethnic status (Bol'shov 1995). The increase in copper production, the elaboration of distinctive ornamentations of the body, and the increase in warfare that characterized the Abashevo culture could have been responses to climatic deterioration, increasing conflict, and the aggrandizing behavior of local leaders who stimulated both the production of and inter-regional trade for valuables.

Transition to the Initial Sintashta Phase

Abashevo societies occupied sites in the upper Ural River valley on the east side of the Urals (Pryakhin 1976, 1980). This southern Ural group of the Abashevo culture appears to have merged with the more pastoral herders of the late Poltavka culture in the nearby northern steppes to create the core material traits and customs of the Sintashta culture (Anthony 2007: 382–389; Kuz'mina 2008: 46; Koryakova and Epimakhov 2007: 65–66). But Sintashta chiefs placed many more weapons in graves and intensified metal production in settlements. Their massive animal sacrifices and chariot burials greatly increased the scale of animal sacrifices and vehicle burials that had long been features of Catacomb and Poltavka funerals in the northern steppes. Sintashta also introduced new customs that were unique in the steppes: permanent settlements, fortifications, and chariot warfare.

What can explain this sudden, steep escalation in the scale of metal production, public funeral sacrifices, and warfare? Declining climatic conditions and escalating conflicts could have encouraged both an increase in the demand for wealth to secure larger alliances and an increase in the scale and elaboration of public funeral ceremonies as ritual settings for feasting, praise poetry, alliance making, and aggrandizement. But a new factor that must also be considered is the initiation of contact with the civilizations of Central Asia. The walled towns of the Bactria-Margiana Archaeological Complex (BMAC) in Central Asia connected the metal miners of the northern steppes with an almost bottomless market for copper. One text from Ur dated to the reign of Rim-Sin of Larsa (1822–1763 BCE) recorded the receipt of 18,333 kilograms (40,417 pounds, or 20 tons) of copper in a single shipment, most of it earmarked for one merchant. Zimri-Lim, king of the powerful city-state of Mari in northern Syria between 1776 and 1761 BCE, distributed gifts totaling more than 410 kilograms (905 pounds) of tin – not bronze, but tin – to his allies during a single tour in his eighth year (Muhly 1995; Potts 1999: 168–171, 186). The tin almost certainly came from Central Asia (Parzinger and Boroffka 2003). Long-distance trade routes were well established across West, Central, and South Asia, from the Zeravshan to India and Mesopotamia, before 2000 BCE. This old and well-oiled Asian trade network was connected to the northern Eurasian steppe for the first time around 2000 BCE. It created a relationship that fundamentally altered metal production, warfare, and ritual competition among the steppe cultures.

This interpretation of the sequence of changes that led to the evolution of the Sintashta culture can be correct only if there is substantial archaeological evidence for a sharp increase in warfare and aggrandizing chiefly behaviors in the early Sintashta culture and for contact between early Sintashta and Central Asia.

Warfare in the Sintashta Culture: Fortifications and Weapons

INCREASED WARFARE in early Sintashta is indicated by three factors: the appearance of fortified towns; increased deposits of weapons in graves; and the development of new weapons and tactics, importantly chariots. All of the Sintashta settlements excavated to date, even relatively small ones like Chernorech'ye III, were surrounded by ditches and banks with palisade walls (Koryakova and Epimakhov 2007: 68–69). The bases of the wooden palisade posts were preserved inside the banks at Ust'ye, Arkaim, and Sintashta. Although the complex gateways reconstructed at Sintashta (Gening et al. 1992: 40) might be exaggerated, the evidence for ditch-and-bank defensive walls reinforced by internal log frames and log palisades is clear. No houses have been documented outside the walls, although more research is needed to establish the nature and extent of extra-settlement activities (Hanks, Chapter 9 in this volume). Population aggregation inside walled settlements certainly was associated with increased warfare in other archaeological contexts, importantly in prehistoric North America (LeBlanc 2000; Milner and Schroeder 1999). In the Ural steppes during a period of colder climate, one wonders if houses built with shared walls against a circular external earthen bank would not perhaps have been simply warmer than the wagon camps of the earlier Bronze Age. But the walled settlements also were associated with a sharp increase in the number of weapons deposited in graves.

Epimakhov (2002) published a catalogue of all the excavated Sintashta graves from five cemeteries: Bol'shekaraganskii, Kamennyi Ambar 5, Krivoe Ozero, Sintashta (all cemeteries), and Solntse II. The catalogue listed 242 individuals (2002: 109–117). Of the 181 graves that contained any artifact inventory (2002: 124–132), 65, or 36% of the graves with artifacts, contained weapons. Only 79 (33%) of the 242 individuals were adults, and 43 of these, or 54% of all adults, were buried with weapons. Most of the adults in the weapon graves were not assigned a gender, but of the 13 that were, 11 (85%) were males. Most of the 54% of adults buried with weapons probably were males. In the earlier MBA, in graves of the Poltavka, Catacomb, or Abashevo cultures, graves with weapons were unusual. They were more frequent in Abashevo than in the steppe graves, but the great majority of Abashevo graves did not contain weapons of any kind, and when one did, usually it was a single axe or a projectile point. My reading of reports on kurgan graves of the earlier EBA and MBA suggests that less than 10%, perhaps less than 5%, contained weapons. The frequency of weapons in adult graves of the Sintashta culture (54%) was much higher.

The Sintashta culture witnessed not just an increase in weapon numbers but also the introduction of new types of weapons. Longer, heavier projectile points appeared in Sintashta-culture graves and were deposited in greater numbers. They came in two types: lanceolate and stemmed. Short lanceolate points with flat or slightly hollow bases were deposited in groups for the first time. They might have been for arrows, because prehistoric arrow points were light in weight and usually had flat or hollow bases. Lanceolate flint points with a concave or flat base occurred in seven graves at Sintashta, with up to 10 points in one grave (Sintashta Mogila grave 39, or SM 39). Four graves at Sintashta contained sets of 6 or 7 lanceolate points (SM4, SM12, SM 24, and Sintashta kurgan I grave 9, or SI 9). A set of 5 lanceolate points was deposited in the chariot grave of Berlyk II, k10.

More interesting were flint points of a new type, with a contracting stem, defined shoulders, and a long, narrow blade with a thick medial ridge, 4–10 centimeters long (Fig. 4.4). These new stemmed points might have been for javelins. Their narrow, thick blades were ideal for javelin points because the heavier shaft of a javelin (compared with an arrow) causes severe torque stresses on the embedded point at the moment of impact, and a narrow, thick point could penetrate deeper than a thin point before torque stresses caused it to break (Van Buren 1974). A stemmed point by definition is mounted in a socketed shaft or foreshaft, a complex type of attachment usually found on spears or javelins rather than arrows. Smaller stemmed points had existed earlier in Fatyanovo tool kits and were occasionally included in graves, as at the Fatyanovo cemetery of Volosovo-Danilovskii, grave 59. But this was the only grave out of 107 in this cemetery that contained a projectile point. Stemmed points had also appeared in forager sites of the Volosovo and

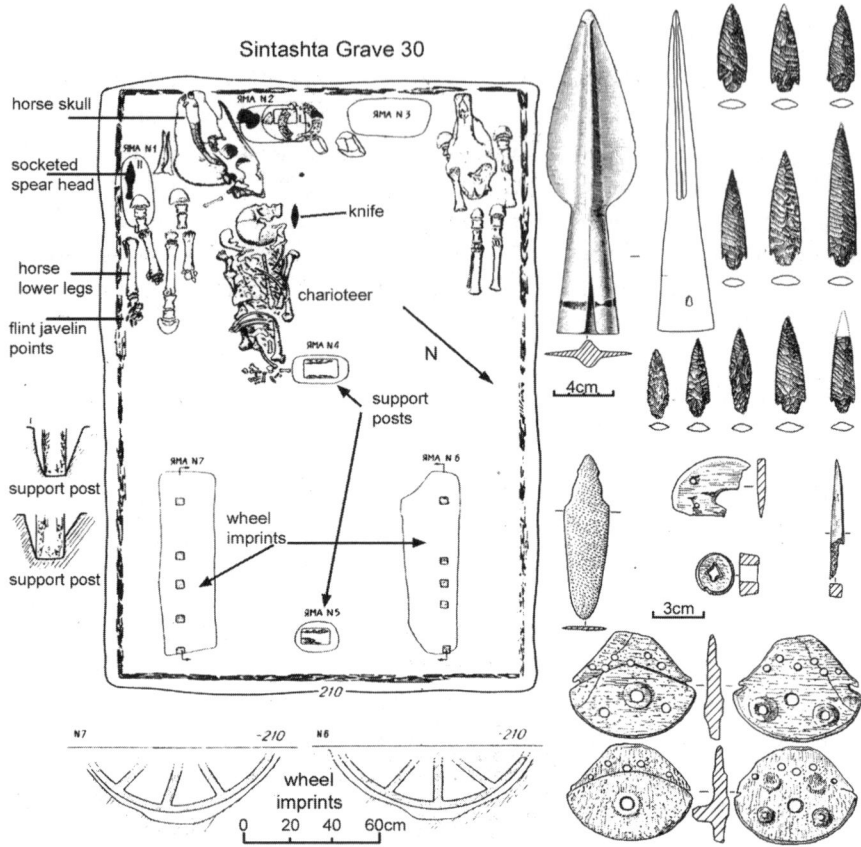

Sintashta Grave 30

horse skull

socketed
spear head

horse
lower legs

flint javelin
points

support post

support post

knife

charioteer

N

support
posts

wheel
imprints

4cm

3cm

wheel
imprints

0   20   40   60cm

Figure 4.4. Grave 30 in the Sintashta SM cemetery, with the remains of two 10-spoked chariot wheels, a charioteer, two horses, and cheek-pieces for the horse team. Weapons included 11 flint stemmed points and one bone stemmed point, possibly for a sheath of light javelins; and a copper socketed spearhead 20 centimeters long, possibly for a heavier spear (after Gening et al. 1992: fig. 111, 113, and 114).

Garin-Bor cultures in the forest-steppe zone west of the Urals, but again they were shorter than the Sintashta examples and were unusual in graves. The Seima-Turbino horizon of elite weapon types, dated about 2000–1900 BCE, featured large, long, stemmed points just like the Sintashta ones. This type was shared across the Sintashta–Seima-Turbino cultural border at the northern edge of the steppes. Long stemmed points appeared in sets of 20 points in Sintashta SM grave 5,

11 points in SM grave 30, 6 points in SM grave 24, and 9 points in SII 7. Stemmed points made of cast bronze, perhaps imitations of the flint stemmed ones, occurred in one chariot grave at Sintashta (SM 16) and in two other graves (Sintashta kurgan II grave 1, or SII 1, one example; and SII 7, five examples).

Most adult Sintashta males were buried with weapons, unlike the male graves of earlier cultures; new kinds of weapons appeared, notably long points possibly intended for javelins; and they were deposited in sets of up to 20 that appear to represent warriors' equipment. We should not expect to see much increased trauma in human skeletons because projectile wounds, even fatal ones, rarely leave evidence in the human skeleton (Milner 2005). But another indicator of increased conflict is the most hotly debated artifact of this period in the steppes – the light, horse-drawn chariot.

Warfare and Sintashta Chariots

The chariot was the first wheeled vehicle built for speed. Spoked wheels were wonders of carpentry and joinery, much more difficult to craft than solid wheels, but they made a light vehicle possible. The chariot had two spoked wheels, a standing driver, and a team of horses guided with bits and reins. Chariots have been found in graves of the Sintashta culture (2100–1800 BCE) and in some graves of its later eastern offshoot, the Petrovka culture (1900–1700 BCE). The vehicles can be identified as chariots because the lower thirds of the wheels were placed in slots in the grave floors, where stains show an outer circle of bent wood with 10 to 12 square-sectioned spokes. Depending on whether all graves with two wheel-slots are counted or just those with clear imprints of spokes, the number of chariot graves varies from 21 (Kuz'mina 2003) to 16 (Epimakhov 2002). The Sintashta-Petrovka chariots have been interpreted as effective instruments of war (Anthony and Vinogradov 1995; Nefedkin 2001) or as merely symbolic vehicles designed only for parade or ritual use, made in barbaric imitation of superior Near Eastern originals (Littauer and Crouwel 1996; Jones-Bley 2000; Vinogradov 2003: 264, 274). Arguments over their functionality have focused, surprisingly, on the distance between their wheels.

Near Eastern war chariots had crews of two or even three – a driver and an archer, and occasionally a shield bearer to protect the other two from incoming missiles. The gauge or track width of Egyptian chariots of ca. 1400–1300 BCE built for crews of two or three was 1.54–1.80 meters. The hub or nave of the wheel, a necessary part that gave it stability, projected at least 20 centimeters along the axle on each side of the wheel (often it was longer; see Sandor 2004 for a technical analysis of Egyptian chariots). A gauge around 1.4–1.5 meters seems to be the minimum to provide enough room between the wheels for the two inner hubs or naves (20 + 20 centimeters) and a car at least 1 meter wide to carry two men. Sintashta and Petrovka culture chariots with less than 1.4–1.5 meters between their wheels have been interpreted as parade or ritual vehicles unfit for battlefield use (Littauer and Crouwel 1996; Jones-Bley 2000; Vinogradov 2003).

This interpretation suffers from six flaws. First, steppe chariots were made in many sizes, including two at Kamennyi Ambar 5, two at Sintashta (SM 4, 28), and two at Berlyk (Petrovka culture) with a gauge between 1.4 and 1.6 metres, big enough for a crew of two. The first examples published in English (Gening 1979; Anthony and Vinogradov 1995; Anthony 1995) from Sintashta (SM 19) and Krivoe Ozero (KO 9:1) had gauges of only about 1.2–1.3 meters, as did three other Sintashta chariots (SM 5, 12, 30) and one other Krivoe Ozero chariot. The debate has focused on these narrower vehicles, most of which, in spite of their narrow gauges, also contained weapons (Fig. 4.4). But it has ignored the six wider vehicles. One of them (Sintashta SM 28) with a gauge of about 1.5 meters occupied a grave that also contained the partial remains of two adults – possibly a two-man crew. Even if we assume that war chariots needed a crew of two, many steppe chariots were big enough.

Second, a war chariot did not need a two-man crew. A single warrior-driver could hold the reins in one hand and hurl a javelin with the other. A driver-warrior could use his entire body to throw, whereas a man on horseback without stirrups (not invented until the post-Scythian era) could use only his arm and shoulder. This meant that a javelin-hurling charioteer could strike a man on horseback before the rider could strike him. Archers of the steppe Bronze Age seem to have used bows 1.2–1.5 meters long, on the basis of bow remains at Berezovka kurgan 3, grave 2, and Svatove kurgan 12, grave 12 (Grigoriev 2002: 59–60; Shishlina 1990; Bratchenko 2003). Bows this long would have made mounted archery difficult and clumsy (see bow measurements in Zutterman 2003), so riders gained no advantage using long bows. The stemmed projectile points of Sintashta were longer and heavier than those of the earlier MBA, more like javelin points. If javelins were used from steppe chariots, the narrow cars could have been driven by a single warrior.

Third, even if all chariot warfare did rely exclusively on archery, it is possible for a single driver-warrior to shoot a bow and drive by wrapping the reins around his hips. The Egyptian pharaoh was shown driving with his hips and shooting a bow with his hands in tomb paintings. Littauer (1968) noted that a royal scribe was painted in the same hip-driving, bow-shooting pose; and in tomb paintings of Ramses III, two-man chariot teams are shown fighting the Libyans with the Egyptian driver-archer guiding his team with his hips and the second man just holding a shield. The narrow-gauge steppe chariots with a single driver-warrior, like Krivoe Ozero 9: 1, could have been used for archery.

Fourth, most steppe chariots, including the narrow-gauge ones, were buried with weapons. I have seen complete inventories for 12 Sintashta and Petrovka chariot graves, and 10 of them contained weapons. The weapons included projectile points, metal waisted daggers, flat metal axes, metal shaft-hole axes, polished stone mace heads, and one metal socketed spearhead 20 centimeters long (Sintashta SM 30; see Fig. 4.4). When the skeleton can be sexed, it is always a male. If steppe chariots were not designed for war, it is odd that they were buried with male drivers and weapons.

Fifth, a new kind of bridle cheekpiece appeared in the steppes just when chariots did (Hüttel 1992; Pryakhin and Besedin 1999; Usachuk 2002; Kuz'mina 2003). These were disk-shaped cheekpieces made of antler or bone with a central perforation for the end of the bit and various other holes for noseband and bridle attachments, and with pointed studs or prongs on the inner face. The pointed studs pressed into the soft flesh at the corners of the horse's mouth when the reins were pulled on the opposite side. The development of a new, more severe form of driving control suggests that rapid, precise maneuvers by the driving team were necessary. Chariot cheekpieces of the same general design, a disk with sharp prongs on its inner face, appeared later in the Near East (as at Tel Haror) but were made of metal (Littauer and Crouwel 2001).

Finally, the sixth flaw in the argument that steppe chariots were ritual imitations of Near Eastern originals is that the oldest steppe chariot graves are older than any of the dated chariot images in the Near East (Kuznetsov 2006). Eight radiocarbon dates are now available from five Sintashta-culture graves containing the impressions of spoked wheels, including three chariot graves at Sintashta (SM cemetery, graves 5, 19, 28), one at Krivoe Ozero (KO k9:1), and one at Kamennyi Ambar 5 (k2:8). These dates show that chariots *certainly* appeared in the Ural-Tobol steppes by 1900–1800 BCE, and because three of the dates (3760 ± 120 BP, 3740 ± 50 BP, and 3700 ± 60 BP) have probability distributions that fall predominantly before 2000 BCE, the earliest chariots *probably* appeared in the steppes before 2000 BCE (Table 4.1).

Table 4.1. Selected radiocarbon dates for the Sintashta (S) and Potapovka (P) cultures in the south Ural steppes and middle Volga steppes, respectively

Lab number	BP date	Sample source	Grave goods[a]	Calibrated date
Sintashta SB Big Kurgan (S)				
GIN-6186	3670 ± 40	Birch log		2140–1970 BCE
GIN-6187	3510 ± 40	Birch log		1890–1740 BCE
GIN-6188	3510 ± 40	Birch log		1890–1740 BCE
GIN-6189	3260 ± 40	Birch log		1610–1450 BCE
Sintashta SM cemetery (S) ±				
Ki-653	4200 ± 100	Grave 11, wood	K	2900–2620 BCE
Ki-658	4100 ± 170	Grave 39, wood	K	2900–2450 BCE
Ki-657	3760 ± 120	Grave 28, wood	C	2400–1970 BCE
Ki-862	3360 ± 70	Grave 5, wood	C, K	1740–1520 BCE
Krivoe Ozero cemetery, kurgan 9, grave 1 (S)				
AA-9874b	3740 ± 50	Horse 1 bone	C, K	2270–2030 BCE
AA-9875a	3700 ± 60	Horse 2 bone	C, K	2200–1970 BCE
AA-9874a	3580 ± 50	Horse 1 bone	C, K	2030–1780 BCE
AA-9875b	3525 ± 50	Horse 2 bone	C, K	1920–1750 BCE
Kamennyi Ambar 5 (S)				
OxA-12532	3604 ± 31	k2:grave 12, human bone		2020–1890 BCE
OxA-12530	3572 ± 29	k2:grave 6, human bone	K	1950–1830 BCE
OxA-12533	3555 ± 31	k2:grave 15, human bone		1950–1780 BCE
OxA-12531	3549 ± 49	k2:grave 8, human bone	C, K	1950–1770 BCE
OxA-12534	3529 ± 31	k4:grave 3, human bone		1920–1770 BCE
OxA-12560	3521 ± 28	k4:grave 1, human bone		1890–1770 BCE
OxA-12535	3498 ± 35	k4:grave 15, human bone		1880–1740 BCE
Utyevka cemetery VI (P)				
AA-12568	3760 ± 100	k6:grave 4, human bone	K	2340–1980 BCE

Lab number	BP date	Sample source	Grave goods[a]	Calibrated date
OxA-4264	3585 ± 80	k6:grave 6, human bone		2110–1770 BCE
OxA-4306	3510 ± 80	k6:grave 4, human bone	K	1940–1690 BCE
OxA-4263	3470 ± 80	k6:grave 6, human bone	K	1890–1680 BCE
Potapovka cemetery I (P)				
OxA-4265	3710 ± 80	k5:grave 13, human bone		2270–1960 BCE
OxA-4266	3510 ± 80	k5:grave 3, human bone		1940–1690 BCE
AA-47802	3536 ± 57	k3:grave 1, horse skull		1950–1770 BCE
Grachevka II k5:3 (P)				
AA-53806	3752 ± 52	Human bone		2280–2030 BCE

[a] Graves that contained chariots are marked C; graves that contained studded disk cheekpieces are marked K.

In the Near East, in contrast, the oldest images of true chariots – vehicles with *two spoked* wheels, pulled by *horses* rather than native asses or onagers, controlled with *bits* rather than lip or nose rings, and guided by a *standing warrior*, not a seated driver – first appeared about 1800 BCE, on Old Syrian seals (Moorey 1986; Littauer and Crouwel 1979). The oldest vehicles with two spoked wheels appeared on seals from Karum Kanesh II, dated about 1900 BCE, but the equids were of uncertain type (possibly native asses or onagers), and they were controlled by nose rings. Excavations at Tell Brak in northern Syria recovered 102 cart models and 191 equid figurines from levels dated to the late Akkadian and Ur III periods, 2350–2000 BCE by the standard middle chronology (Oates 2003) or 2230–1900 BCE by the increasingly persuasive low chronology (Reade 2001). None of the equid figurines was clearly a horse. Two-wheeled carts were common among the vehicle models, but they had built-in seats for a sitting driver and solid wheels. No chariot models were found, which suggests that chariots were unknown here as they were elsewhere in the Near East through the end of the Ur III period. Chariots first appeared in Iran and the Near East about 1900–1800 BCE.

Chariots were invented in the southern Ural steppes. They were effective in tribal wars in the steppes because they provided an elevated platform from which a skilled driver could hurl a sheath full of javelins, and because they were noisy, fast, and intimidating. The whole point of the chariot was lightness and speed; in later Egyptian chariots, the floor was woven of leather straps and the walls were wickerwork, with just a few struts providing a rigid frame. As the open-backed car advanced over uneven ground at high speed, the driver's legs absorbed the shock. To turn the vehicle, the inside horse had to be pulled in while the outside horse was given rein. Doing this well and hurling a javelin at the same time required a lot of practice. Chariots were supreme advertisements of wealth: difficult to make (Sandor 2004) and requiring great athletic skill *and* a team of specially trained horses to drive, they were available only to those who could delegate much of their daily labor to hired herders. A chariot was material proof that the driver had the means to fund a substantial alliance or was supported by someone who did. Taken together, the evidence from fortifications, weapon types and numbers, and the tactical innovation of chariot warfare indicate that conflict increased in both scale and intensity in the northern steppes during the early Sintashta period, after about 2100 BCE.

AGGRANDIZING BEHAVIOR: CHIEFS AND FEASTS

The scale of animal sacrifices in Sintashta cemeteries implies very large funerals. At the northern edge of the Sintashta SM cemetery, Sacrificial Complex 1 was placed in a 50 centimetre-deep pit; it contained the skulls and leg bones of six horses, four cattle, and two rams, laid in two rows facing each other around an overturned pot – clearly a single event. This single sacrifice provided about 6,000 pounds (2,700 kilograms) of meat, or enough to supply each of 3,000 participants with 2 pounds (.9 kilogram). The Bolshoi Kurgan (BK), built just a few meters to the north, required, by one estimate, 3,000 man-days (Gening et al. 1992: 234–235 for Sacrificial Complex 1 and p. 370 for the man-days). The work force needed to build the kurgan matched the amount of food provided by Sacrificial Complex 1. But the Bolshoi Kurgan (looted, unfortunately) was unique. The other burial mounds at Sintashta were small and low. If the horse, cattle, and sheep-goat sacrifices that accompanied the other graves at Sintashta were meant to feed work parties, we cannot see what it was that they built. It seems likely that most of the animals were instead sacrificed as food for the funeral guests. With up to eight horses sacrificed for a single ceremony (SM grave 5), Sintashta funerals would have fed hundreds or even thousands of guests. Feast-hosting behavior is among the most common and consistent avenues open to aggrandizing individuals in tribal societies (Hayden 2001).

The central role of horses in Sintashta funeral sacrifices was unprecedented. Horse bones had appeared in earlier steppe graves, but not in great numbers, and usually less frequently than sheep or cattle. The animal bones from the Sintashta and Arkaim settlement middens were about 60% cattle, 26% sheep-goat, and 13% horse (Kosintsev 2001). Cattle and sheep supplied the great majority of the meat diet. But the funeral sacrifices in the cemeteries reversed those frequencies: 23% cattle, 37% sheep-goat, and 39% horse. Horses were the principal animal sacrificed at these two cemeteries and were three times more frequent in funeral sacrifices than in kitchen middens. Zoologist L. Gaiduchenko (1995) suggested that the Arkaim citadel specialized in horse breeding for export, and this hypothesis might be supported by stable isotope data suggesting that cattle and sheep, not horses, supplied most of the protein in the human diet at Arkaim (Privat 2002: 170). The sacrifice of whole horses, with no share for the human celebrants, was an entirely new Sintashta practice. Whole-horse sacrifice probably was a potlatch-type excess in a setting of rapidly escalating political competition. Horse and chariot sacrifices were central parts of a mortuary ritual performance that emphasized exclusivity, hierarchy, and power – what Appadurai (1986) called "tournaments of value," ceremonies meant to define membership in the elite and to channel political competition within clear boundaries.

Several investigators have questioned whether the Sintashta culture had an elite chiefly social stratum (Epimakhov 2002; Hanks, Chapter 9 in this volume). Little jewelry occurs in Sintashta graves, and no large houses or storage facilities in settlements. The signs of craft specialization, a signal of social hierarchy, are very weak in all crafts except metallurgy, and even there every household in every settlement seems to have worked metal (Epimakhov 2002: 57). Sintashta cemeteries contained the graves of a cross section of the entire age and sex spectrum, including many children, a more inclusive funeral ritual than had been normal in EBA and earlier MBA kurgans, which were raised over just one to three adult graves.

On the other hand, Sintashta cemeteries contain just a fraction of the individuals who must have lived in the associated walled settlements. The Sintashta citadel contained about 50–60 structures, and its associated cemeteries contained just 66 graves. If the settlement maintained a steady population of 300 people for six generations (150 years) there should have been more than 1,800 graves. Most Sintashta cemeteries were even smaller. Just 31 individuals were identified in the Bol'shekaraganskii cemetery for the settlement at Arkaim, which was comparable in size to Sintashta. Only a few exceptional families were given funerals in Sintashta cemeteries, but the entire family, including children, was honored this way.

The best evidence for hierarchy in Sintashta cemeteries is the extravagant public sacrifice of animals, particularly horses, and of vehicles and weapons. Horses and weapons were strongly associated in these sacrificial acts. Epimakhov's (2002: 97–132) catalogue of five Sintashta cemeteries shows that chariots were buried in 16 graves, or just 11% of the 181 graves catalogued. Horses were sacrificed in 48 graves, or 27% of graves; multiple horses were sacrificed in just 13% of graves. Weapons were deposited in 36% of graves. But 66% of graves with horse sacrifices contained weapons, and 83% of graves with multiple horse sacrifices contained weapons. Only a quarter of Sintashta graves contained horse sacrifices, but those that did also contained weapons, a symbolic association between the ownership of large horse herds, feast hosting, and the identity of the warrior.

CONTACT WITH CENTRAL ASIA

Central Asian decorative motifs (stepped pyramids) and raw materials (lead and lapis lazuli) appeared in the northern steppes (in Sintashta sites), and lost-wax metal-casting techniques appeared in sites in the forest-steppe (Seima-Turbino horizon) just when steppe pottery, Sintashta-type disk cheekpieces for chariots, horse bones with teeth showing wear from bits, and actual steppe settlements appeared in Central Asia and/or Iran. Evidence for contact can be divided into two phases.

The first phase of contact occurred in 2100–2000 BCE. Steppe pottery appeared in a stratified, radiocarbon-dated midden at Gonur North in Margiana in Central Asia (Hiebert 1994: 61–62). The bones of one horse and a horse-image pommel on a staff appeared in brick tombs in the associated Gonur cemetery (Kohl 2007: 197–198; Salvatore 2003; Sarianidi 2002). Horses were not native to Central Asia; the local wild equid was the onager (*Equus hemionus*). The horse in the Gonur tomb was an exotic animal, decapitated and deposited with a dog and a camel in a sacrificial deposit with 10 humans. In Mesopotamia, in Ur III contexts dated 2100–2000 BCE by the middle chronology, horses for the first time became subjects in royal and aristocratic art and propaganda. They were fed to lions, and King Shulgi compared himself in one inscription to "a horse of the highway that swishes his tail" (Owen 1991). Where did these newly imported horses come from? They were probably driven across the steppes and deserts of Kazakhstan into Central Asia by Sintashta or Petrovka traders. In the sixteenth century the Bukhara Khanate in Central Asia, drawing on horse-breeding grounds in the Ferghana Valley, exported 100,000 horses *annually* to the Mughal rulers of South Asia (Levi 2002). While I am not suggesting trade on that scale, horses were desirable and expensive, and the annual demand for steppe horses could easily have totaled many thousands.

The commodities that were traded from south to north during the first century of contact probably included woven textiles, the prime export commodity of Mesopotamia (Kohl 2007: 221–223). Almost all decorated objects of Elam, the BMAC, and Sarazm were decorated with the stepped pyramid, or, if it was repeated in four cardinal directions, the stepped cross (Fig. 4.5). This classic BMAC motif was absent from the steppes. It appeared for the first time in any steppe context on Sintashta and Potapovka pottery, in both cases on only about 5% of pots. Perhaps it was copied from BMAC textiles. In addition, a lead wire made of two braided strands was found in the Sintashta settlement of Kuisak (Maliutina and Zdanovich 1995:103). Lead had never before appeared in the northern steppes as a pure metal but was common in the Zeravshan – an ingot of lead weighing 10 kilograms was found at Sarazm. A Bactrian handled bronze mirror was found in a Sintashta grave at Krasnoe Znamya (Kuz'mina 2001:20). A lapis lazuli bead was reported by Kuz'mina (2001: 20) from the Sintashta settlement; and another lapis bead was found in a Seima-Turbino grave at Rostovka (Matiushchenko and Sinitsyna 1988: 7). Imported faience beads, made of glaze over a clay core, occurred at Sintashta and Krivoe Ozero but not in sites of the second phase of contact – not in Petrovka or Alakul' (Vinogradov 2003: 239). Finally, the technique of lost-wax metal casting first appeared in the north during the Sintashta period in metal objects of Seima-Turbino type; the nearest source for this new technology was the Central Asian BMAC, where lost-wax metal casting was highly developed.

The second phase of contact, 2000–1700 BCE, witnessed a dramatic intrusion of steppe materials and even settlements into Central Asia south of the Zeravshan River. Steppe pottery appeared in many sites of the BMAC (Hiebert 2002) and a Petrovka-culture settlement appeared at Tugai on the Zeravshan, near the now-abandoned Central Asian city of Sarazm (Hiebert 2002; Kuz'mina 2003). A grave at Zardcha Khalifa,

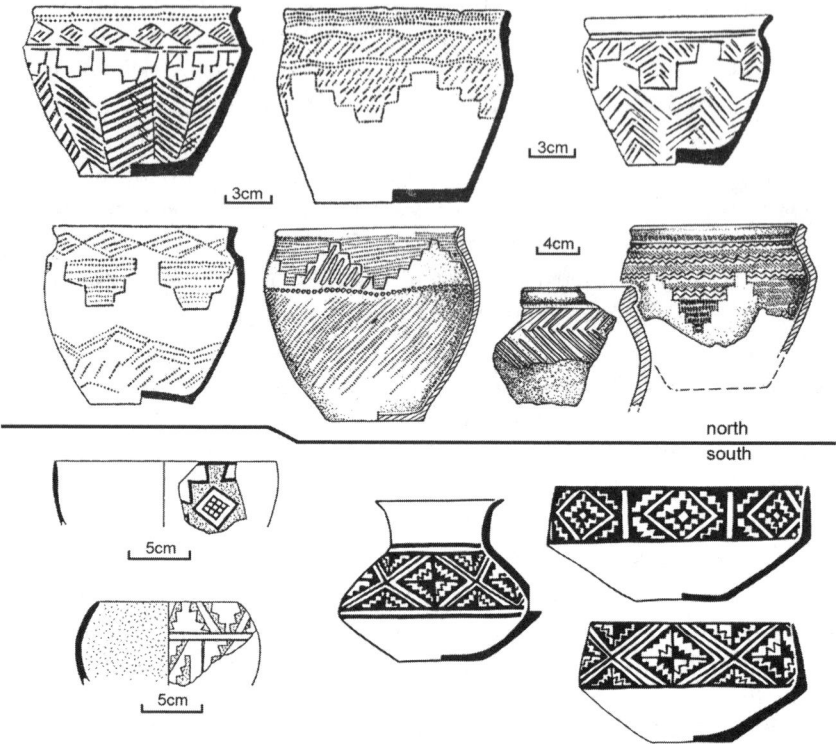

Figure 4.5. The stepped-pyramid or crenellation motif on the pottery of Sintashta and Potapovka (*top*) and Central Asia, Sarazm (*left bottom*) and Altyn-Depe (*right bottom*). This was the first appearance of this motif in the steppes; in Central Asia and Iran it was the central organizing motif in the decoration of pottery, mosaics, cast silver objects, and even a palace mural at Malyan. After Vasil'ev, et al. 1994 (top row and left pot in second row); Gening et al. 1992 (remaining pots in second row); Lyonnet 1996 (Sarazm, bottom left); and Masson 1988 (Altyn-Depe, bottom right).

near Tugai, contained four horse cheekpieces of a specific Sintashta type and bar bits for two horses, apparently a chariot team, associated with a variety of BMAC metal goods and pottery (Bobomulloev 1997). Horse bones and teeth appeared for the first time at Malyan (Zeder 1986) and Godin Tepe (Gilbert 1991) in Iran. The premolar teeth of a horse and a mule from Malyan (Kaftari phase) were worn in a way that is diagnostic of wear made by a hard (probably metal) bit, perhaps the oldest direct evidence for bitting in the ancient Near East (Anthony and Brown 1989; Anthony et al. 2006). About 1800–1700 BCE, at the end of this second phase, Andronovo miners settled the Zeravshan Valley and began to mine the rich tin deposits at Karnak (Parzinger and Boroffka 2003). An Andronovo settlement, Pavlovka, in the northern steppes near Kokchetav, Kazakhstan, contained many sherds of imported Namazga VI pots from Central Asia (Maliutina 1991). Chariot warfare was adopted during this phase by urban kingdoms throughout the Near East (Oates 2003; Littauer and Crouwel 1979), and horses probably were imported by the thousands through Central Asia and Iran. By about 1600 BCE, the cities of the BMAC were abandoned and left in ruins, replaced by pastoral populations that made a coarse incised pottery like that of Andronovo.

Summary: A Model for the Sintashta Genesis and Decline

As THE climate in the northern hemisphere became colder and more arid after 2500 BCE, previously mobile herders in the northern steppes adopted a defensive "musical chairs" reaction and settled in permanent settlements near the shrinking marshes that were vital for wintering their herds. Conflict and competition between rival tribal groups in the northern steppe and southern forest-steppe zones encouraged trade for novel insignia and valuables that could be used as gifts in public ceremonial and feasting events where alliances were made. Feasting and ceremonies became more elaborate and their scale intensified as competition between rival hosts led to potlatch-type excesses. Intensified fighting also encouraged tactical innovations in warfare, the most important being the light war chariot, used as a platform for javelin-hurling warriors.

This developing system of tribal competition was transformed when the steppe chiefs made contact with Central Asia. Central Asia and its trade system provided a new source of imported wealth and prestige goods, probably including textiles, but it also represented an enormous market for horses and metal, greatly increasing the value of copper and horses as exports. The societies of the northern steppes probably had direct access to Central Asian markets only for a short time between 2100 and 1900 BCE. By about 1900–1800 BCE, a contact zone developed in the Zeravshan Valley, and the Petrovka immigrants who filtered southward and occupied that contact zone (Tugai, Zardcha Khalifa, Karnab) could have intercepted most Central Asian commodities and supplied both metal and horses to the urban south. Many of the defining aspects of the Sintashta culture – fortified settlements, chariot burials, and extravagant animal sacrifices – disappeared from the northern steppes after direct contact with the south was blocked. But that brief time was enough to transform northern steppe economies. Once the metallurgical pump was primed, so to speak, it continued to flow. The priming happened partly because of contact with urban markets, but the flow after that raised the usage of metal in the steppes and in the forest zone to the north, both starting an internal cycle of exchange that would lead to a metal boom in the Eurasian steppe after 2000 BCE and laying the foundation for the Andronovo and Srubnaya societies of the Late Bronze Age.

References

Anthony, D. W. 1995. Horse, Wagon, and Chariot: Indo-European Languages and Archaeology. *Antiquity* 69(264): 554–565.

2007. *The Horse, the Wheel, and Language*. Princeton: Princeton University Press.

Anthony, D. W., and D. R. Brown. 1989. Looking a Gift Horse in the Mouth: Identification of the Earliest Bitted Equids and the Microscopic Analysis of Bit Wear. In P. Crabtree, D. Campana, and K. Ryan (eds.), *Animal Domestication and Its Cultural Context: Essays in Honor of Dexter Perkins and Pat Daly*. University of Pennsylvania, MASCA Research Papers in Science and Archaeology, Special Volume (6). Philadelphia, pp. 99–116.

2008. Herding and Gathering during the Late Bronze Age at Krasnosamarskoe, Russia, and the End of the Dependency Model of Steppe Pastoralism. In L. Popova, C. Hartley, and A. T. Smith (eds.), *Social Orders and Social Landscapes*. Newcastle: Cambridge Scholars Publishing, pp. 393–415.

Anthony, D. W., D. Brown, E. Brown, A. Goodman, A. Kokhlov, P. Kosintsev, P. Kuznetsov, O. Mochalov, E. Murphy, D. Peterson, A. Pike-Tay, L. Popova, A. Rosen, N. Russell, and A. Weisskopf. 2005. The Samara Valley Project: Late Bronze Age Economy and Ritual in the Russian Steppes. *Eurasia Antiqua* 11: 395–417.

Anthony, D. W., D. Brown, and C. George. 2006. Early Horseback Riding and Warfare: The Importance of the Magpie around the Neck. In S. Olsen, S. Grant, A. Choyke, and L. Bartosiewicz (eds.), *Horses and Humans: The Evolution of the Equine-Human Relationship*. BAR International Series 1560. Oxford: Archaeo Press, pp. 137–156.

Anthony, D. W., and N. B. Vinogradov. 1995. The Birth of the Chariot. *Archaeology* 48(2): 36–41.

Appadurai, A. 1986. Introduction: Commodities and the Politics of Value. In A. Appadurai (ed.), *The Social Life of Things: Commodities in Cultural Perspective*. Cambridge: Cambridge University Press, pp. 3–63.

Blyakharchuk, T. A., H. E. Wright, P. S. Borodavko, W. O. van der Knaap, and B. Ammann. 2004. Late Glacial and Holocene Vegetational Changes on the Ulagan High-Mountain Plateau, Altai Mts., Southern Siberia. *Palaeogeography, Paleoclimatology and Paleoecology* 209: 259–279.

Bobomulloev, S. 1997. Ein bronzezeitliches Grab aus Zardča Chalifa bei Pendžikent (Zeravšan-Tal). *Archäologische Mitteilungen aus Iran und Turan* 29: 122–134.

Bol'shov, S. V. 1995. Problemy kulturogeneza v lesnoi polose srednego povolzh'ya v Abashevskoe vremya. In I. B. Vasil'ev and O.V. Kuz'mina (eds.), *Drevnie IndoIranskie Kul'tury Volgo-Ural'ya*. Samara: Samara Gosudarstvennogo Pedagogicheskogo Universiteta, pp. 141–156.

Boyle, Katie, Colin Renfrew, and Marsha Levine (eds.). 2002. *Ancient Interactions: East and West in Eurasia*. Cambridge: McDonald Institute.

Bratchenko, S. N. 2003. Radiocarbon Chronology of the Early Bronze Age of the Middle Don, Svatove, Luhansk Region. *Baltic-Pontic Studies* 12: 185–208.

Chernykh, E. N. 1992. *Ancient Metallurgy in the USSR*. Cambridge: Cambridge University Press.

Di Cosmo, N. 1999. State Formation and Periodization in Inner Asian Prehistory. *Journal of World History* 10(1): 1–40.

  2002. *Ancient China and Its Enemies: The Rise of Nomadic Power in East Asian History*. Cambridge: Cambridge University Press.

Epimakhov, A. V. 2002. *Iuzhnoe Zaural'e v Epokhu Srednei Bronzy*. Chelyabinsk: YUrGU.

Gaiduchenko, L. L. 1995. Mesto i znachenie Iuzhnogo Urala v eksportno-importnikh operatsiyakh po napravleniu vostok-zapad v eopkhu Bronzy. In *Rossiya i Vostok: Problemy Vzaimodeistviya*. Chast V, Kniga 1: *Kul'tury Eneolita-Bronzy Stepnoi Evrazii*. Chelyabinsk: 3-ya Mezdunarodnaya Nauchnaya Konferentsiya, pp. 110–115.

Gening, V. F. 1979. The Cemetery at Sintashta and the Early Indo-Iranian Peoples. *Journal of Indo-European Studies* 7:1–29.

Gening, V. F., G. B. Zdanovich, and V. V. Gening. 1992. *Sintashta*. Chelyabinsk: Iuzhno-Ural'skoe Knizhnoe Izdatel'stvo.

Gilbert, Allan S. 1991. Equid Remains from Godin Tepe, Western Iran: An Interim Summary and Interpretation, with Notes on the Introduction of the horse into Southwest Asia. In R. H. Meadow and H.-P. Uerpmann (eds.), *Equids in the Ancient World*, vol. 2. Wiesbaden: Ludwig Reichert, pp. 75–122.

Grigoriev, S. A. 2002. *Ancient Indo-Europeans*. Chelyabinsk: RIFEI.

Hayden, B. 2001. Fabulous Feasts: A Prolegomenon to the Importance of Feasting. In M. Dietler and Brian Hayden (eds.), *Feasts*. Washington, DC: Smithsonian Press, pp. 23–64.

  2002. Bronze Age Interaction between the Eurasian Steppe and Central Asia. In K. Boyle, C. Renfrew, and M. Levine (eds.), *Ancient Interactions: East and West in Eurasia*. Cambridge: McDonald Institute, pp. 237–248.

Hiebert, F. 1994. *Origins of the Bronze Age Oasis Civilizations of Central Asia*. Bulletin of the American School of Prehistoric Research 42. Cambridge, MA.

Hüttel, Hans-Georg. 1992. Zur archäologischen Evidenz der Pfredenutzung in der Kupfer- und Bronzezeit. In B. Hänsel and S. Zimmer (eds.), *Die Indogermanen und das Pferd: Festschrift für Bernfried Schlerath*. Budapest: Archaeolingua, 4:197–215.

Jones-Bley, K. 2000. The Sintashta "Chariots." In J. Davis-Kimball, E. Murphy, L. Koryakova, and L. Yablonsky (eds.), *Kurgans, Ritual Sites, and Settlements – Eurasian Bronze and Iron Age*. BAR International Series 890. Oxford: Arechaeo Press, pp. 135–140.

Jones-Bley, K., and D. G. Zdanovich (eds.). 2002. *Complex Societies of Central Eurasia from the 3rd to the 1st Millennium BC*, vol. 1 and 2. Washington, DC: Institute for the Study of Man.

Keeley, L. H. 1996. *War before Civilization*. Oxford: Oxford University Press.

Klimenko, V. V., V. A. Klimanov, A. A. Sirin, and A. M. Sleptsov. 2001. Climate Changes in Western European Russia in the Late Holocene. *Doklady Earth Sciences* 377(2): 190–194.

Kohl, P. L. 2007. *The Making of Bronze Age Eurasia*. Cambridge: Cambridge University Press.

Kosintsev, P. 2001. Kompleks kostnykh ostatkov domashnikh zhivotnykh iz poselenii i mogilnikov epokhi Bronzy Volgo-Ural'ya i ZaUral'ya. In Yu. I. Kolev (ed.), *Bronzovyi Vek Vostochnoi Evropy: Kharakteristika Kul'tur, Khronologiya i Periodizatsiya*. Samara: Samarskii Gosudarstvennyi Pedagogicheskii Universitet, pp. 363–367.

Koryakova, L., and A. V. Epimakhov. 2007. *The Urals and Western Siberia in the Bronze and Iron Ages*. Cambridge: Cambridge University Press.

Kovaleva, V. T., and G. B. Zdanovich, (eds.), 2002. *Arkaim: Nekropol (po materialam kurgana 25 Bol'shekaraganskoe Mogil'nika)*. Chelyabinsk: Yuzhno-Ural'skoe Knizhnoe Izdatel'stvo.

Kremenetski, C. 1997. The Late Holocene Environment and Climate Shift in Russia and Surrounding Lands. In H. Dalfes, G. Kukla, and H. Weiss (eds.), *Climate Change in the Third Millennium BC*. New York: Springer, pp. 351–370.

2003. Steppe and Forest-Steppe Belt of Eurasia: Holocene Environmental History. In M. Levine, C. Renfrew, and K. Boyle (eds.), *Prehistoric Steppe Adaptation and the Horse*. Cambridge: Cambridge University Press, pp. 11–27.

Kremenetski, C. V., T. Böttger, F. W. Junge, and A. G. Tarasov. 1999. Late- and Postglacial Environment of the Buzuluk Area, Middle Volga Region, Russia. *Quaternary Science Reviews* 18: 1185–1203.

Kremenetski, C. V., P. E. Tarasov, and A. E. Cherkinsky. 1997. Postglacial Development of Kazakhstan Pine Forests. *Geographie Physique et Quaternaire* 51: 391–404.

Kuzmin, Y. V., G. S. Burr, J. M. O'Malley, and A. J. T. Jull. 1997. Radiocarbon Dating of Climatic and Cultural Changes on the Russian Far East during the Late Glacial and Holocene. Paper presented at the Seventh Annual V. M. Goldschmidt Conference, Tucson, AZ [online abstract]. http://www.lpi.usra.edu/meetings/gold/pdf/2198.pdf.

Kuz'mina, E. E. 2001. The First Migration Wave of Indo-Iranians to the South. *Journal of Indo-European Studies* 29(1–2): 1–40.

2003. Origins of Pastoralism in the Eurasian Steppes. In Marsha Levine, Colin Renfrew, and Katie Boyle (eds.), *Prehistoric Steppe Adaptation and the Horse*. Cambridge: McDonald Institute, pp. 203–232.

2008. *The Prehistory of the Silk Road*. Ed. Victor H. Mair. Philadelphia: University of Pennsylvania Press.

Kuznetsov, P. F. 2006. The Emergence of Bronze Age Chariots in Eastern Europe. *Antiquity* 80: 638–645.

LeBlanc, Steven A. 1999. *Prehistoric Warfare in the American Southwest*. Salt Lake City: University of Utah Press.

Levi, S. C. 2002. *The Indian Diaspora in Central Asia and Its Trade, 1550–1900*. Leiden: Brill.

Levine, M., C. Renfrew, and K. Boyle (eds.). 2003. *Prehistoric Steppe Adaptation and the Horse*. Cambridge: McDonald Institute.

Littauer, M. A. 1968. A 19th and 20th Dynasty Heroic Motif on Attic Black-Figured Vases? *American Journal of Archaeology* 72: 150–152.

Littauer, M. A., and J. Crouwel, 1979. *Wheeled Vehicles and Ridden Animals in the Ancient Near East*. Leiden: Brill.

1996. The Origin of the True Chariot. *Antiquity* 70: 934–939.

2001. The Earliest Evidence for Metal Bridle Bits. *Oxford Journal of Archaeology* 20(4): 329–338.

Lopez, P., J. A. Lopez-Saez, E. N. Chernykh, and P. Tarasov. 2003. Late Holocene Vegetation History and Human Activity Shown by Pollen Analysis of Novenki Peat Bog (Kargaly region, Orenburg Oblast, Russia). *Vegetation History and Archaeobotany* 12(1): 75–82.

Lyonnet, B. (ed.). 1996. *Sarazm (Tajikistan). Céramiques (Chalcolithiques et Bronze Ancien)*. Mémoire de la Mission Archéologique Française en Asie Centrale 7. Paris.

Madison, J. 1865 [1795]. *Letters and Other Writings of James Madison, Fourth President of the United States*. Vol. IV. Ed. William C. Rives and Philip R. Fendall. Philadelphia: Lippincott.

Maliutina, T. S. 1991. Stratigraficheskaya pozitsiya materilaov Fedeorovskoi kul'tury na mnogosloinikh poseleniyakh Kazakhstanskikh stepei. In V. V. Nikitin (ed.), *Drevnosti Vostochno-Evropeiskoi Lesostepi*. Samara: Samarskii Gosudarstvennyi Pedagogicheskii Institut, pp. 141–162.

Maliutina, T. S., and G. B. Zdanovich. 1995. Kuisak – ukreplennoe poselenie protogorodskoi tsivilizatsii iuzhnogo zaUral'ya. In *Rossiya i Vostok: Problemy Vzaimodeistviya*, Chast V, Kniga 1: *Kul'tury Eneolita-Bronzy Stepnoi Evrazii*. Chelyabinsk: 3-ya Mezdunarodnaya Nauchnaya Konferentsiya, pp. 100–106.

Masson, V. M. 1988. *Altyn-Depe'*. Trans. Henry N. Michael. University Museum Monograph 55. Philadelphia.

Matiushchenko, V. I., and G. V. Sinitsyna. 1988. *Mogil'nik u d. Rostovka Vblizi Omska*. Tomsk: Tomskogo Universiteta.

Milner, G. R. 2005. Nineteenth-Century Arrow Wounds and Perceptions of Prehistoric Warfare. *American Antiquity* 70(1): 144–156.

Milner, G. R., and S. Schroeder. 1999. Mississippian Sociopolitical Systems. In J. E. Neitzel (ed.), *Great Towns and Regional Polities*. Dragoon, AZ: Amerind Foundation, pp. 95–107.

Moorey, P. R. S. 1986. The Emergence of the Light, Horse-Drawn Chariot in the Near East, ca. 2000–1500 BC. *World Archaeology* 18(2): 196–215.

Muhly, J. D. 1995. Mining and Metalwork in Ancient Western Asia. In J. Sasson, J. Baines, G. Beckman, and K. Rubinson (eds.), *Civilizations of the Ancient Near East*, vol. 3. New York: Charles Scribner's Sons, pp. 1501–1519.

Nefedkin, A. K. 2001. *Boevye Kolesnitsy i Kolesnichie Drevnikh Grekov (XVI-I vv do n.e.)*. St. Petersburg: IIMK.

Oates, J. 2003. A Note on the Early Evidence for Horse and the Riding of Equids in Western Asia. In M. Levine, C. Renfrew, and K. Boyle (eds.), *Prehistoric Steppe Adaptation and the Horse*. Cambridge: McDonald Institute, pp. 115–125.

Owen, D. 1991. The First Equestrian: An Ur III Glyptic Scene. *Acta Sumerologica* 13: 259–273.

Parzinger, H., and N. Boroffka. 2003. *Das Zinn der Bronzezeit in Mittelasien I: Die siedlungsarchäologischen Forschgungen im Umfeld der Zinnlagerstätten*. Archäologie in Iran und Turan, Band 5. Mainz am Rhein: Philipp von Zabern.

Perry, C. A., and K. J. Hsu. 2000. Geophysical, Archaeological, and Historical Evidence Support a Solar-Output Model for Climate Change. *Proceedings of the National Academy of Sciences* 7(23): 12433–12438.

Potts, D. T. 1999. *The Archaeology of Elam*. Cambridge: Cambridge University Press.

Privat, K. 2002. Preliminary Report of Paleodietary Analysis of Human and Faunal Remains from Bolshekaragansky Kurgan 25. In V. T. Kovaleva and G. B. Zdanovich (eds.), *Arkaim: Nekropol (po materialam kurgana 25 Bol'shekaraganskoe Mogil'nika)*. Chelyabinsk: Yuzhno-Ural'skoe Knizhnoe Izdatel'stvo, pp. 166–171.

Pryakhin, A. D. 1976. *Poseleniya Abashevskoi Obshchnosti*. Voronezh: Voronezhskogo Universiteta.

——— 1980. Abashevskaya kul'turno-istoricheskaya obshchnost' epokhi bronzy i lesostepe. In A. D. Pryakhin (ed.), *Arkheologiya Vostochno-Evropeiskoi Lesostepi*. Voronezh: Voronezhskogo Universiteta, pp. 7–32.

Pryakhin, A. D., and V. I. Besedin. 1999. The Horse Bridle of the Middle Bronze Age in the East European Forest-Steppe and the Steppe. *Anthropology and Archaeology of Eurasia* 38(1): 39–59.

Reade, J. 2001. Assyrian King-Lists, the Royal Tombs of Ur, and Indus Origins. *Journal of Near Eastern Studies* 60(1): 1–29.

Rosenberg, M. 1998. Cheating at Musical Chairs: Territoriality and Sedentism in an Evolutionary Context. *Current Anthropology* 39(5): 653–681.

Salvatori, S. 2003. Pots and Peoples: the "Pandora's Jar" of Central Asian Archaeological Research; on Two Recent Books on Gonur Graveyard Excavations. *Rivista di Archeologia* 27: 5–20.

Sandor, B. I. 2004. The Rise and Decline of the Tutankhamen-Class Chariot. *Oxford Journal of Archaeology* 23(2): 153–175.

Sarianidi, V. I. 2002. *Margush: Drevnevostochnoe tsarstvo v staroi del'te reki Murgab*. Ashgabat: Turkmendowlethebarlary.

Shishlina, N. I. 1990. O slozhnom luke Srubnoi kul'tury. In S. V. Studzitskaya (ed.), *Problemy Arkheologii Evrazii*. Moskva: Trudy Gosudarstvennogo Oedena Lenina Istoricheskogo Muzeya 74, pp. 23–37.

Usachuk, A. N. 2002. Regional Peculiarities of Technology of the Shield Cheekpiece Production (Based on the Materials of the Middle Don, Volga, and South Urals). In K. Jones-Bley and D. G. Zdanovich (eds.), *Complex Societies of Central Eurasia from the 3rd to the 1st Millennium BC*, vol. 1. Washington, DC: Institute for the Study of Man, pp. 337–343.

Van Buren, G. E. 1974. *Arrowheads and Projectile Points.* Garden Grove, CA: Arrowhead Publishing.

Vasil'ev, I. B., P. F. Kuznetsov, and A. P. Semenova. 1994. *Potapovskii Kurgannyi Mogil'nik Indoiranskikh Plemen na Volge.* Samara: Samarskii Universitet.

Vehik, S. 2002. Conflict, Trade, and Political Development on the Southern Plains. *American Antiquity* 67(1): 37–64.

Vinogradov, N. 2003. *Mogil'nik Bronzovogo Beka: Krivoe Ozero v Yuzhnom Zaural'e.* Chelyabinsk: Yuzhno-Ural'skoe Knizhnoe Izdatel'stvo.

Weiss, H. 2000. Beyond the Younger Dryas: Collapse as Adaptation to Abrupt Climate Change in Ancient West Asia and the Eastern Mediterranean. In G. Bawden and R. Reycraft (eds.), *Environmental Disaster and the Archaeology of Human Response.* Maxwell Museum of Anthropology Anthropological Papers 7. Albuquerque, pp. 75–98.

Zaikov, V. V., G. B. Zdanovich, and A. M. Yuminov. 1995. Mednyi rudnik Bronzogo veka "Vorovskaya Yama." In *Rossiya i Vostok: Problemy Vzaimodeistviya, Chast V, Kniga 1: Kul'tury Eneolita-Bronzy Stepnoi Evrazii.* Chelyabinsk: 3-ya Mezdunarodnaya Nauchnaya Konferentsiya, pp. 157–162.

Zeder, M. 1986. The Equid Remains from Tal-e Malyan, Southern Iran. In R. Meadow and H.-P. Uerpmann (eds.), *Equids in the Ancient World*, vol. 1. Wiesbaden: Ludwig Reichert, pp. 366–412.

Zutterman, C. 2003. The Bow in the Ancient Near East, a Re-evaluation of Archery from the late 2nd Millennium to the End of the Achaemenid Empire. *Iranica Antiqua* 38: 119–165.

CHAPTER 5

Settlements and Cemeteries of the Bronze Age of the Urals
The Potential for Reconstructing Early Social Dynamics

ANDREI V. EPIMAKHOV

THEORETICAL MODELS that were developed in Anglo-American social anthropology and that have subsequently been seen widely within the discipline of archaeology (Service 1962; Fried 1975; Carneiro 1981) were constructed according to a strong evolutionary principle. That is, social developmental stages moved from simple to more complex. Because the societal typologies originally used within such schemas were rigid, a considerable reaction to this form of classification has emerged within scholarship in recent years (Crumley 1995; Semenov 1999; Shanks and Tilley 1996; Yoffee 1993).

Within these discussions, some Russian scholars have suggested the actual *degree* of complexity of a society is less important than the quantity of hierarchical levels that contribute to the "complexity" (Vas'utin et al. 2005). The concept of social complexity often has been recognized as the degrees of functional differentiation that exist within the various subsystems of a society, which provide for the optimum adaptation to specific environments (Flannery 1972). However, I argue that this concept is constrained through the actual practice of archaeological analysis. Such a problem exists as a result of placing too much emphasis on the single dimension of vertical differentiation. Consequently, if a particular case study does not exhibit evidence of formal mechanisms for management and control, with concomitant attributes of stratification, the society is then perceived as *egalitarian* almost automatically. In contrast to this, approaches to social complexity should focus more on understanding *complexity* as a multivariate concept. Otherwise, scholarship loses the opportunity to recognize societies that may comprise specialized segments but have in effect an overall low level of social hierarchy. Typical examples of this include tribal societies that form wide networks of interactions, not necessarily determined by economic factors, that offer cumulative complexity and scalar relationships that may be comparable to what has often been considered "chiefdom"-level developments.

Such conceptual issues and their connections to theoretical models for societal reconstruction have been discussed in recent years for Bronze Age socio-economic developments in north central Eurasia (Koryakova 1996, 2002; Epimakhov 2002a; Koryakova and Epimakhov 2007). In spite of the various distinctions that have been illustrated for Bronze Age cultures in this region, there are no specialists vindicating the "state" character of these societies (Gorbunov 1992; Epimakhov 2002b; Zdanovich and Zdanovich 2002). However, the archaeological sites attributed to this region and period display a number of important distinctive elements that require more systematic investigation in terms of basic stages of development and long-term economic and social trends.

Before addressing these issues, it is necessary to examine several other factors: the conventional cultural-historical interpretations of the periods and territories within the region of the southern and middle Ural Mountains; questions regarding the actual patterning of social structures and processes that may be reflected by archaeological data; and problems connected with the use of incongruent analogies, such as those from ethnographic studies, for the reconstruction of Bronze Age socio-cultural dynamics.

Archaeological Patterns

THE URALS are situated at the boundary of eastern Europe and western Siberia. This territory has a greater meridian extent and is characterized by a significant variety of landscape-climatic zones – ranging from tundra in the north to arid steppe in the south. The primary focus of this chapter is on the social processes and developments in the southern region, which can be divided into three main zones: the Cis-Urals, Trans-Urals, and forested-mountain Urals (Fig. 5.1). Each of these areas has a distinct environment, which factored significantly into various prehistoric social, cultural, and economic processes.

The significance of this region cannot be examined exclusively through the model of cultural contact in the region, which has been traditionally seen in terms of a zone of interaction between two ancient

Figure 5.1. Map of the Ural Mountains region.

language groups: the Indo-Iranian (steppe) and Finno-Ugric (forest). Rather, because of significant mineral resources – in particular, concentrations of copper ores (Zaikov et al. 2005) – from the Bronze Age onward the region played an independent role in economic, social, and cultural processes within north central Eurasia. These developments are reflected through a variety of unique archaeological discoveries made in the region, which have subsequently inspired numerous questions regarding possible stimuli for social evolution.

Currently, the Bronze Age, especially in the region of the southern Urals, is the best represented of the prehistoric periods as a result of the high number of archaeological sites that have been identified and the scale of scientific investigation carried out during the Soviet and post-Soviet periods. Even though there has been a high level of research activity in this region for several decades, some areas have received more investigation than others, and much research remains to be done. Nevertheless, for the steppe and forest-steppe zones, the relative chronology for the Bronze Age is relatively well agreed upon, although its correlation with the cultural developments in the northern forest zone is fervently debated. Cultural developments in the northern forested region, contemporaneous with monuments of the Early and Middle Bronze Age of the southern part of the region, are not presently well known. Regarding this, two hypotheses have been posited: the persistence of Eneolithic traditions until the movement of the steppe population into this zone (Kuz'minykh 1993); and the existence of currently unknown materials connected with the Early and Middle Bronze Age. More systematic approaches to these issues through the use of absolute-dating methods have been implemented recently (Chernykh et al. 2002; Epimakhov et al. 2005; Hanks et al. 2007) and the calibrated values of these data are presented in Figure 5.2.

The beginning of the Bronze Age dates to between the fourth and third millennia BCE and is connected to the existence of the Yamnaya (Pit-Grave) cultures in the steppe zone of the Cis-Urals (Morgunova et al. 2003; Bogdanov 2004). Currently, the Yamnaya pattern is very well represented by *kurgan* (tumulus) burial grounds, which comprise more than 150 cemeteries and cenotaphs. Settlement sites, however, are for the most part completely absent, although recent studies have shown that the exploitation of the extensive Chernyka deposits of copper ore were initiated during this period (Chernykh 2002, 2003, 2004, 2005). As a result, an independent, regional center of metallurgy was formed that is well reflected in the material culture recovered from mortuary excavations in the region.

Yamnaya burials indicate an obvious heterogeneity in the expenditure of labor on tomb construction (i.e., size of barrow and grave pit construction), some features of ritual activity, and the deposition of animal remains. Adult male burials are strongly represented and account for

cal BC	Periodization (Eastern Europe scheme)	Metallurgical networks (provinces)		Territory				
				Cis-Urals		Trans-Urals		
				Steppe	Forest-Steppe	Mountain forest	Forest-Steppe	Steppe
6	Early Iron Age			early nomads	Ananyino	Itkul', Gamayun	Itkul', Gorokhovo etc.	early nomads
7								
8						6epe3OBO		
9	Final Bronze Age	Eurasian metallurgical network (province)	III	?	Mezhovka	Mezhovka	Late Barkhatovo	Belokluchyovka
10							Barkhatovo Mezhovka	
11								
12								
13								
14	Late Bronze Age		II	Srubnaya-Alakul'	Srubnaya	Cherkaskul'	Cerkaskul'	Alakul'-Fyodorovka-Stubnaya Petrovka
15								
16						Koptyaki	Fyodorovka-Alakul'	
17				Petrovka	Early Srubnaya			
18			I	Sintashta	Abashevo	Seima-Turbino	Odinivo, Krokhalyovka	Sintashta
19								
20	Middle Bronze Age			Vol'sk-Lbische Letest Catakombnaya		?		?
21								
22		Circumpontic metallurgical network (province)	II					
23								
24								
25	Early Bronze Age			Yamnaya (Pit-Grave)	Volosovo Garino Agidel'	Kysykul', Surtandy, Lipchino? Shapkul', Ayat etc.	Kysykul', Surtandy, Lipchino? Shapkul', Ayat etc.	Eneolitic cultures
26								
27			I					
28								
29								
30				Turganyik type				
31								
32								
33								

Figure 5.2. Table of chronology and key cultural-historical patterns of the southern and middle Ural archaeological cultures in the Bronze Age.

approximately 80% of all excavated burials. This mortality profile suggests that the kurgan ceremony was used only for a certain component of the society and that the inhumation rite was probably an attribute associated with high status connected with the construction of large barrows. For the erection of the most monumental constructions, a large-scale concentration of labor resources was carried out in a short period of time – with evidence suggesting that this was done during one warm season of activity (Bogdanov 2004). The subsistence economy for this period appears to be complex mobile stockbreeding. The question of whether this was a local development or a transfer of tradition from the outside (e.g., eastern Europe) remains a subject of active discussion. However, it is possible to state with a high degree of confidence that the Yamnaya traditions represent the first type of *productive* economy in the Urals region.

Various data (e.g., radiocarbon dating, rare combination of artifacts) support the theory that Yamnaya traditions, which are traditionally linked with the Indo-European language family before its disintegration, co-existed in the Urals with indigenous Eneolithic populations. These groups probably had local Finno-Ugric roots. The Eneolithic populations in this region had a subsistence economy based on hunting and fishing, which reflects deep local subsistence traditions. It is impossible to estimate the duration of the co-existence of these two differing lifeways; however, the contrast between their economic systems and the variety of landscapes associated with the southern Ural Mountains (from steppe to the mountain-forest zone) appears to support this proposition.

Within the limits of the eastern European chronological system, this period corresponds with the Early Bronze Age. The final phase of this archaeological pattern is relatively unclear, but attempts to coordinate the late Yamnaya and Sintashta developments, corresponding approximately with the first quarter of the second millennium BCE, have not been easily established. However, it is likely that the break of a uniform cultural tradition took place in the steppe zone at the end of the third millennium BCE, corresponding with the Middle Bronze Age of the eastern European chronology. In any case, this cultural overlap has not been substantiated, and more archaeological data are required.

The second major cultural development, that of the Sintashta archaeological pattern, has been traced on both sides of the Ural Mountains; however, on the eastern side of the Urals there are numerous settlements and cemeteries but to the west of the Urals only mortuary sites.

Because the spatial organization of the Sintashta fortified settlements has been discussed repeatedly elsewhere (Zdanovich and Batanina 2002; Koryakova and Epimakhov 2007), only a few details need to be emphasized here.

Sintashta settlements are supported by monumental systems of fortification, and the internal space has a very structured organization that is almost entirely occupied by standard rectangular shaped buildings. The total area of the settlements is from 0.8 to 3.6 hectares, and in every case traces of metalworking (slag, metal droplets, furnaces, etc.) have been recovered during excavation.

Cemeteries, typically up to 10 barrow constructions, contain from 1 to 30 burial pits with individual and collective inhumations. The total number of individuals recovered from Sintashta burial sites is now approximately 200–250. Of this sample, males, females, and nearly all age categories are represented – with the recovery of old adult individuals being rare. The structure of the mortality pattern therefore differs somewhat from what one would consider a normal mortality profile. Complex burial rituals, such as the sacrifice of animals, high-value grave goods such as bronze weaponry and occasionally stone mace heads and spoke-wheeled chariots, and elaborate tomb construction, allow one to consider that these burial grounds were places of burial for only one segment of the population, most likely individuals or families of higher status (Epimakhov 2002a, 2005).

On the western slopes of the Ural Mountains, along the border of the forest-steppe zone, there is evidence of the Abashevo culture (for the Urals, it is often named Balanbash) (Sal'nikov 1967). Some finds exhibit features that resemble Sintashta metallurgy and ceramic patterns. According to the opinion of some researchers, the Sintashta and Balanbash developments were contemporaneous. Nevertheless, Balanbash settlements differ markedly from Sintashta as there is no clear system of fortification, they are several times smaller in size, and there are no indications of standardized construction. Funeral complexes also do not contain objects that may be seen as markers of status, such as weaponry. For the Balanbash sites, however, elements of settlement hierarchy have been established – in connection with large stationary centers (e.g., Beregovskoe, Tyubyak) and smaller settlements. About 10 hoards of bronze objects are also known and their locations appear to be distributed at the borders of the Balanbash cultural zone.

In considering the Middle Bronze Age, it is also necessary to mention the Seima-Turbino phenomenon, with associated cemeteries and finds that are widespread within a large territory of western Siberia and eastern Europe. Available archaeological evidence suggests a stable and long-functioning network of communications, which provided a distribution of metal from the east to the west. Of course, this does not exclude migratory processes (Chernykh and Kuz'minykh 1989; Chernykh et al. 2003), which may have taken place along the ecological zone of the forest and forest-steppe.

The next chronological phase, the Srubnaya-Andronovo, reflects a sharp change in both the number and organization of archaeological sites. The archaeological evidence of this period is most significant with regard to known Bronze Age sites and the occurrence and long-term functioning of large cemeteries, with the number of barrows in some of them exceeding 100. Reliable examples of settlement sites are not common, and the tradition of compact settlement organization ends, as settlement excavations have routinely revealed multi-purpose buildings and structures. Although the total area of sites quite often reaches 1–2 hectares, the actual number of dwellings is reduced by four to five times from the previous period. Mortuary sites generally reflect a greater emphasis on kinship or "group-oriented" activities. In terms of burial goods and ritual practices, however, there are a number of "deviations," including complicated tomb design, changes in corpse orientation, and the deposit of rare categories of artifacts. In the opinion of some scholars, these findings are a reflection of complex social development (Gorbunov 1992; Tsimidanov 2004). As indirect support for this hypothesis, highly specialized settlements, such as Gorny in the southeastern Urals, may serve as a vivid example of the appearance of new patterns of social organization (Chernykh 2002, 2004, 2005).

Very small and modest cemeteries of the Final Bronze Age look paradoxical when compared with the large and complex elite barrows of the Early Iron Age. However, it is likely that in the steppe Trans-Urals there was a change in cultural traditions between these two periods. There is widespread evidence for a number of permanent settlements, including settlements with significant occupation areas that contain small concentrations of artifacts within cultural occupation sequences. This evidence indicates that an increase of mobility associated with animal husbandry occurred in combination with other populations that continued a more traditional settled way of life. Exceptions to this pattern are archaeological sites in the upland foothill areas and forest-steppe zones, where populations did not make a transition toward higher mobility or nomadic patterns of movement. The cemeteries of this period have not been well investigated, and their scale and organization are modest when compared with the earlier Middle Bronze Age. Alongside earthen mound mortuary constructions, flat burials have also been encountered. Artifact inventories from these sites are made up exclusively of ceramics; therefore, interpretations of the structure of the societies and their possible organization have been difficult. Overall, there is an impression among scholars that at this time there was a process of fragmentation of larger social collectives.

Variations in Complexity

THE KEY cultural-historical patterns discussed in the previous section represent a very diverse set of materials for archaeological interpretation; however, in many cases there are traces of cultural continuity through time. The level of social complexity was obviously wide-ranging during the Bronze Age; moreover, it is possible to confidently suggest that changes were varied in nature and were stimulated by a number of different internal and external factors. In the first case, it is necessary to take note of climatic changes, migrations, relationships with neighboring groups, and more complete use of local resources. Among the social and cultural processes of the Urals Bronze Age, two scenarios can be distinctly identified: *transformational* (especially in the beginning and in the last centuries of the Bronze Age) and *evolutionary*. In different landscape-climatic zones, these scenarios had their own specific characteristics, although variations of both models are represented in the forest and steppe ecological areas. Among the stimuli of radical cultural change, it is necessary to identify transitions within the economic systems as well. Important transitions relating to the introduction of a productive form of economy occurred during the Bronze Age, the most significant of which was the emergence of livestock breeding and metallurgy. Both, possibly, were introduced into the Cis-Urals steppe from outside this region.

Following this introduction, a clearer transition occurred between the Middle and Late Bronze Age during the development of the Abashevo and Sintashta archaeological cultures. Despite the absence of clarity on the problem of their genesis, we admit a connection between the eastern European Catacombnaya populations. For the Sintashta traditions, we see even more distant sources of influence. Actually, both cultures were situated in the forest-steppe and north of the steppe zone on both sides of the Ural Mountains. It should be understood that the natural geographic boundary of the Urals was permeable in both directions.

The period of the Urals Late Bronze Age became a time of inclusion into broader systems of interaction associated with the Eurasian metallurgical network (province) – including the Seima-Turbino phenomenon and the Andronovo and Srubnaya families of cultures. Certainly we cannot call this a "world system" (in Wallerstein's sense), but we must not ignore the fact of similarity of many artifacts, technologies, and rituals across a broad area of Eurasia. For example, weapons of the Seima-Turbino type have been found from northern China to Scandinavia (Chernykh 1992; Mei 2003). Convergent invention is highly unlikely in this and other cases. Forest populations were also involved in this system, and that is where synchronous cultures were formed. During the Final Bronze Age, we observe a transition toward a more mobile economic system although this process occurred over several centuries and extended into the first millennium BCE. A change in the direction of cultural communications can be distinctly recognized at this time and the possibility of an inflow of west Siberian populations (e.g., Irmen' or related to it) cannot be excluded as a part of larger social processes. In spite of some changes, the subsistence system, which was based on diversified animal husbandry, was stable. It is possible that variations in the structure of this economy centered on the percentage ratio of certain species in the domestic herd structure (Fig. 5.3). These changes were caused by distinctive ecological niches among groups and the degree of mobility of cattle breeding.

The second economic component of social and cultural processes is metallurgical production and trade. Because stockbreeding was the primary branch of subsistence, metallurgy was not an essential influence for basic subsistence production, as animal husbandry can develop successfully without any metal tools. Moreover, the majority of metal products (e.g., decorations, instruments for woodworking, weapons) were only slightly involved in the process of food production. This situation changed only during the Late Bronze Age, when metal instruments apparently became not only part of a prestige economy but also a part of the economy as a whole. Nevertheless, metal decorations and weapons were produced abundantly at this time and were technologically difficult objects to produce. It is also important to note that the production of metal became the stimulus for the establishment and maintenance of new channels of distant communication. Periodic changes in the system of the production centers rendered significant influence on the course

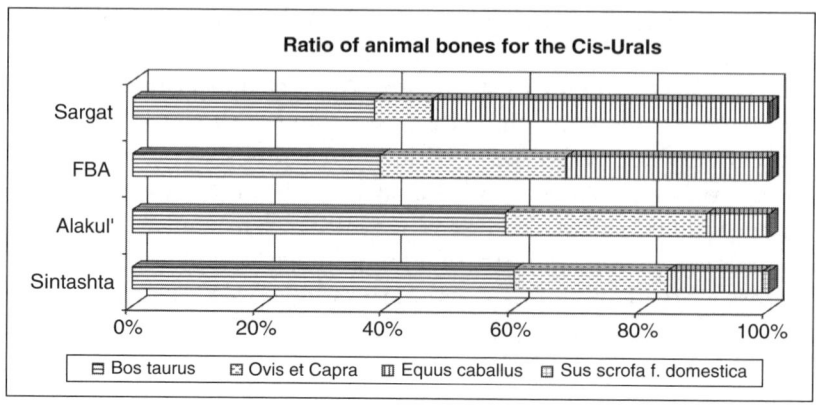

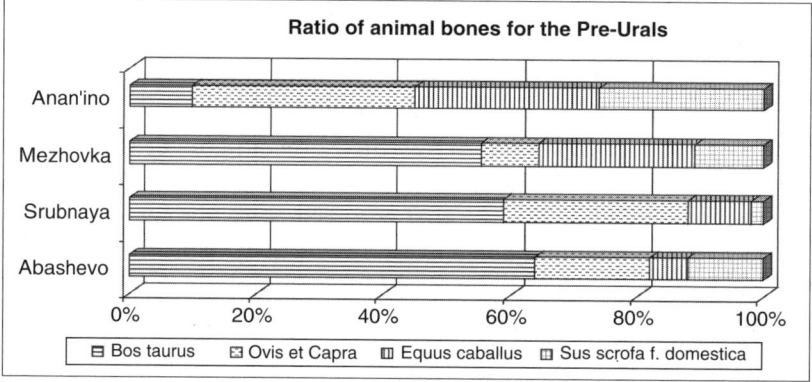

Figure 5.3. Long-term trend of cattle breeding based on faunal remains recovered from archaeological sites in the second and first millennia cal. BCE.

and direction of societal evolution. This is most obvious for the Urals (especially the eastern side) during the Late and Final Bronze Ages when central Kazakhstan and the Altai centers emerged.

The study of prehistoric social complexity in the Urals region can be greatly oversimplified if one overlooks the possibility of a system of larger regional communication and networking. The extent and stability of such systems are interpreted with a wide degree of reliability; however, on the basis of archaeological evidence, it is possible to argue that populations in the Urals region were included within different systems during the Bronze Age. Some of these networks may be considered as more global in nature, although the actual character and scale of such connections still need to be established. While much progress has been made on understanding cycles of transformation and stabilization for significant areas of northern Eurasia, the extent of contact between populations in the central steppe region and the southern urban societies of Central Asia is more difficult to ascertain (Frank 1993). Although we do not exclude such a connection, the scale of such interaction was probably quite low. Additional problems connected with achieving a better understanding of this inter-regional dynamic are created by the sharp contradiction of chronological systems used in these areas.

The last factor that must be discussed is the significance of environmental change. Climate change has traditionally been seen as one of the primary causal factors for marked socio-cultural transition in north central Eurasia. For the Urals Bronze Age, unfortunately, it is not possible to talk about the creation of a uniform conventional scale of climatic fluctuations. The conclusions for the Cis-Urals and Trans-Urals territories appear to differ greatly, a result of the degree of continental weather patterns, the direction of basic air currents, and other factors. However, it is necessary to recognize that the variety of landscapes within a limited territory, which is exactly what we have in the Urals, allows the use of various resources without serious consequences – such as the modification of the species composition of domestic livestock. In this case, it is important to generate several models that will relate to the variation of landscape zones. This kind of work has already been started by regional scholars and allows us to consider the movement of landscape borders in the Ural mountain foothills on a west-east line as well as some other unusual characteristics of this ecological zone.

Characterizing Bronze Age Social Complexity

THE RELATIVE stability of the economy of the Urals Bronze Age does not appear to correlate well with the more complex models of socio-political organization that have been suggested. I would emphasize in this case that it is difficult to propose a very complex level of social organization for this time period. The seemingly small demographic parameters of the Bronze Age communities and the absence of sharp property distinctions (i.e., ranking) within mortuary data do not appear to support models indicative of greater social stratification. A significant part of Bronze Age populations was undoubtedly affected by the specific conditions of its ecological niche and the type of economy that complicated the control of resources by certain individuals or groups. Though the achieved level of development of the economy provided a stable existence, it had no possibility for further intensification. The result was a "choice" for consistency, along with the preservation of communication with remote groups that were related through similar cultural traditions.

The second issue connected to a non-linear type of social complexity relates to the sphere of ideology and its connection to social structures and ritual activities. Most interpretations concerning social stratification within Bronze Age societies are based on materials recovered from funerary sites. For the Urals Bronze Age, kurgan inhumation was obviously used for the minority of dead, perhaps excluding the Srubnaya-Andronovo period. Therefore, because the investigation relies on mortuary data, it is possible to interpret only the social institutions that were reflected and emphasized in mortuary ritual traditions. It is, of course, difficult to believe that during the Early Bronze Age there was no age stratification or clear social structure; however, these appear not to have been emphasized during funerary practices. For example, Figure 5.4 details the various structures and degrees of differentiation that have been posited for societies during the Bronze Age. The most obvious conclusion from these data is that there is a common lack of property differentiation within mortuary rites. In fact, it is possible to recognize only Yamnaya burials as a clearer example of this, although the absence of the most "valuable" copper items indicates that copper grave goods were likely not the property of the deceased individuals.

A more detailed discussion of the mortuary data from the Bronze Age is beyond the scope of the present article; however, it is important to note that social distinctions are mirrored not only in the *quantity and composition* of artifacts but also in the *nature* of their display. Thus, attributes of rank and status can be demonstrated through the *monumental* character of kurgan constructions, the spatial organization of mortuary features and grave pits, animal sacrifice and inclusion of specific species and body elements, and features of burial manipulation (e.g., excarnation and secondary burials). In consideration of these and similar factors, it is only in rare cases that a combination of such attributes is recovered for specific burials or cultural patterns. In fact, such patterns are confined primarily to the mortuary patterns of the Sintashta culture.

In this case, by the beginning of the Late Bronze Age, it is likely that migration, territorial expansion, and conflict over resources stimulated the emergence of a military elite within independent Sintashta communities. Such a development also may be connected with settlement fortification, personal weaponry, and other related military equipment

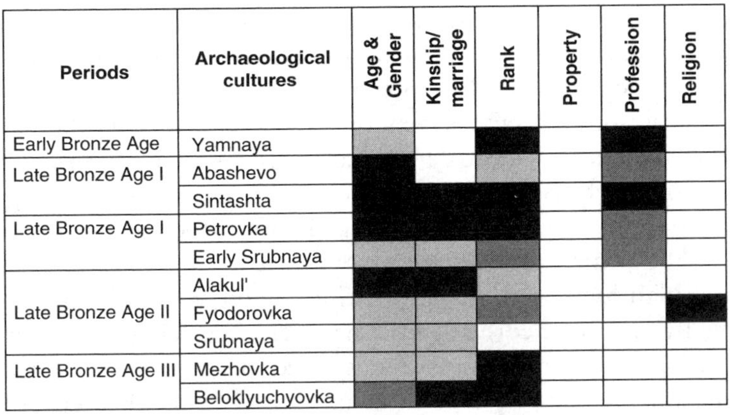

Periods	Archaeological cultures	Age & Gender	Kinship/ marriage	Rank	Property	Profession	Religion
Early Bronze Age	Yamnaya						
Late Bronze Age I	Abashevo						
	Sintashta						
Late Bronze Age I	Petrovka						
	Early Srubnaya						
Late Bronze Age II	Alakul'						
	Fyodorovka						
	Srubnaya						
Late Bronze Age III	Mezhovka						
	Beloklyuchyovka						

Figure 5.4. Manifestation of different social structures in burial rites of the Urals population in the Bronze Age.

commonly recovered from Sintashta sites. Over time, societal priorities changed with the achievement of new territorial resources and an eventual lack of inter-societal conflict. As a result, by the Late Bronze Age societies developed more along the lines of horizontal (e.g., kinship) rather than vertical (e.g., status) differentiation.

Conclusion

IN THE application of theoretical models for social complexity in the Bronze Age Urals, several important issues must be considered. For example, direct ethnographic analogies for societies of the Bronze Age of the steppe region are non-existent for a number of reasons. Between the Bronze and Iron Ages a new type of economy was developed as the territory of northern Eurasia began to play the role of a "near periphery" to that of southern state formations. In this connection, it is doubtful that "universal" concepts, such as that of chiefdom or "proto-polis," can be used to explain steppe and forest-steppe dynamics. Rather, another approach seems to be more relevant – that is, working from basic archaeological materials to social reconstruction of micro- and macro-levels and further to their comparison with other models of social change.

Following this approach, one may argue that the Bronze Age Ural archaeological cultures reflect the rise and development of an independent model of society. During the process of adaptation to natural environmental and cultural conditions, the level of complexity changed not along the lines of an increase and decline in hierarchical structures but rather through the growth of spatial specialization that was accompanied by the occurrence and maintenance of a steady network of communications with other societies (economic, ideological, and others).

The development of more detailed analysis of such long-term trends is problematic. Such issues are strongly linked to the necessity of a greater variety of archaeological and ecological data, which at present do not exist. In the future, as scholars grapple with the application of models for explaining social change in this region, it will be necesary to continually test the theories against new evidence gained from archaeological studies. Such work must focus more specifically on variations in both subsistence and productive economies, evidence that will lead to a greater understanding of the scale of inter-regional trade and exchange, and more systematic chemical and physical analyses of human remains recovered from mortuary sites. Such studies are certain to both improve and challenge current interpretations of social complexity and change in the Urals region during the Bronze Age.

Acknowledgments

I WOULD like to thank Bryan Hanks and Ekaterina Efimova for their hospitality in Pittsburgh during the symposium that led to this publication, and I gratefully acknowledge their invaluable help in the translation and editing of this chapter.

References

Bogdanov, S. V. 2004. *Epokha medi stepnogo Priuralya*. Ekaterinburg: UrO RAN.
Carneiro, R. 1981. The Chiefdom: Precursor to the State. In G. Jones and R. Kautz (eds.), *The Transition to Statehood in the New World*, Cambridge: Cambridge University Press, pp. 37–79.
Chernykh, E. N. 1992. *Ancient Metallurgy in the USSR*. Cambridge: Cambridge University Press.
    (ed.). 2002. *Kargaly*. Vol. 2. Moscow: Languages of Slavonic Cultures.
    2003. Kargaly: The Largest and Most Ancient Metallurgical Complex on the Border of Europe and Asia. In K. Linduff (ed.), *Metallurgy in Ancient Central Asia from the Urals to the Yellow River*. Lewiston: Edwin Mellen Press, pp. 223–238.
    (ed.). 2004. *Kargaly*. Vol. 3. Moscow: Languages of Slavonic Cultures.
    (ed.). 2005. *Kargaly*. Vol. 4. Moscow: Languages of Slavonic Cultures.
Chernykh, E. N., L. I. Avilova, L. B. Orlovskaya, and S. V. Kuz'minykh. 2002. Drevnyaya metallurgiya v Ztircumpontiiskom areale: ot edinstva k raspadu. *Rossiiskaya Arkheologiya* 1: 5–23.

Chernykh, E. N., and S. V. Kuz'minykh. 1989. *Drevnyaya metallurgiya Severnoi Evraziyi (seiminsko-turbinskij fenomen)*. Moscow: Nauka.

Chernykh, E. N., S. V. Kuz'minykh, L. B. Orlovskaya. 2003. Ancient Metallurgy in Northern Asia: From the Urals to the Saiano-Altai. In K. Linduff (ed.), *Metallurgy in Ancient Central Asia from the Urals to the Yellow River*. Lewiston: Edwin Mellen Press, pp. 15–36.

Crumley, C. L. 1995. Heterarchy and the Analysis of Complex Societies. In R. M. Ehrereich, C. L. Crumley, and J. E. Levy (eds.), *Heterarchy and the Analysis of Complex Societies*. Archaeological Papers of the American Anthropological Association 6. Washington, DC, pp. 1–6.

Epimakhov, A. V. 2002a. Complex Societies and the Possibilities to Diagnose Them on the Basis of Archaeological Data: Sintashta Type Sites of the Middle Bronze Age. In K. Jones-Bley and D. Zdanovich (eds.), *Complex Societies of Central Eurasia from the 3rd to the 1st Millennium BC*, vol. 1. Washington, DC: Institute for the Study of Man, pp. 139–48.

2002b. *Yuzhnoye Zauralye v period sredenij bronzy*. Chelyabinsk: YuUrGU.

2005. *Rannie komplexnye obschestva Severa Tzentral'noi Evrazii (po materialam mogil'nika Kamennyi Ambar-5)*. Vol. 1. Chelyabinsk: Chelyabinskii dom pechati.

Epimakhov, A. V., B. Hanks, and C. Renfrew. 2005. Radiouglerodnaya khronologiya pamyatnikov bronzovogo veka Zauralya. *Rossiiskaya Arkheologiya* 4: 92–102.

Flannery, K. V. 1972. The Cultural Evolution of Civilizations. *Annual Review of Ecology and Systematics* 3: 399–426.

Frank, A. G. 1993. Bronze Age World System Cycles. *Current Anthropology* 4: 383–430.

Fried, M. 1975. *The Notion of the Tribe*. Menlo Park: Cumming Publishing.

Gorbunov, V. S. 1992. *Bronzovyi vek Volgo-Uraliskoy lesostepi*. Ufa: Baskir Pedagogical Institute.

Hanks, B., A. V. Epimakhov, and C. Renfrew. 2007. Towards a Refined Chronology for the Bronze Age of the Southern Urals, Russia. *Antiquity* 81: 353–367.

Koryakova, L. N. 1996. Social Trends in Temperate Eurasia during the Second and First Millennia BC. *Journal of European Archaeology* 4: 243–280.

2002. Social Landscape of Central Eurasia in the Bronze and Iron Ages: Tendencies, Factors and Limits of Transformation. In K. Jones-Bley and D. Zdanovich (eds.), *Complex Societies of Central Eurasia from the 3rd to the 1st Millennium BC*, Washington, DC: Institute for the Study of Man, pp. 97–118.

Koryakova, L. N., and A. V. Epimakhov. 2007. *The Urals and Western Siberia in the Bronze and Iron Age*. Cambridge World Archaeology Series. Cambridge: Cambridge University Press.

Kuz'minykh, S. V. 1993. Kvazieneoliticheskie kul'tury Severnoi Evrazii: problema periodizatzii. In I. B. Vasilyev (ed.), *Arheologicheskie kul'tury Bol'shogo Urala*. Ekaterinburg: IIA UrO RAN, pp. 116–122.

Mei, J. 2003. Cultural Interaction between China and Central Asia during the Bronze Age. *Proceedings of British Academy*. 121: 1–39.

Morgunova, N. L., O. S. Khokhlova, G. I. Zaitseva, O. A. Chichagova and A. A. Goly'eva. 2003. Rezultaty radiouglerodnogo datirovaniya arkheo-logicheskikh pamyatnikov Yuzhnogo Priuralya. In N. L. Morgunova (ed.), *Shumailovskiye kurgany*. Orenburg: Orenburg Pedagogical Institute, pp. 96–104.

Sal'nikov, K. V. 1967. *Ocherki drevnei istoriyi Yuzhnogo Urala*. Moscow: Nauka.

Semenov, U. I. 1999. *Filosofiya istorii ot istokov do nashikh dnei: Osnovnye problemy*. Moscow: Staryi Sad.

Service, E. 1962. *Primitive Social Organization: An Evolutionary Perspective*. New York: Random House.

Shanks, M., and C. Tilley. 1996. *Social Theory and Archaeology*. Cambridge: Polity Press.

Tsimidanov, V. V. 2004. *Sotzial'naya struktura srubnogo obschestva*. Donetzk: IA NAN Ukrainy.

Vas'utin, S. A., N. N. Kradin, and A. A. Tishkin. 2005. Rekonstruktziya social'noi struktuty rannih kochevnikov v arheologii. In N. N. Kradin, A. A. Tishkin, and A. V. Khavrinsky (eds.), *Social'naya struktura rannih kochevnikov Evrasii*. Irkutsk: Izdatel'stvo Irkutskogo gosudarstvennogo tehnicheskogo universiteta, pp. 10–38.

Yoffee, N. 1993. Too Many Chiefs? (or, Safe Texts for the 90s). In N. Yoffee and A. Sherratt (eds.), *Archaeological Theory: Who Sets the Agenda?* Cambridge: Cambridge University Press, pp. 60–78.

Zaikov, V. V., A. V. Yuminov, A. U. Dunaev, G. B. Zdanovich, and S. A. Grigiriev. 2005. Geologo-mineralogicheskie issledovaniya mednykh rudnikov na Uzhnom Urale. *Arkheologiya, Etnographiya i Antropologiya Evrazii* 4: 101–14.

Zdanovich, G. B., and D. G. Zdanovich. 2002. The "Country of Towns" of Southern Trans-Urals. In K. Boyle, C. Renfrew, and M. Levine (eds.), *Ancient Interactions: East and West in Eurasia*. Cambridge: McDonald Institute, pp. 249–264.

Zdanovich, G. B., and I. M. Batanina. 2002. Planography of the Fortified Centers of the Middle Bronze Age in the Southern Trans-Urals according to Aereal Photography Data. In K. Jones-Bley and D. Zdanovich (eds.), *Complex Societies of Central Eurasia from the 3rd to the 1st Millennium BC*, vol. 1. Washington, DC: Institute for the Study of Man, pp. 120–138.

CHAPTER 6

The Maikop Singularity
The Unequal Accumulation of Wealth on the Bronze Age Eurasian Steppe?

PHILIP L. KOHL

I N 1897 N. I. Veselovskii excavated the very large, nearly 11 meter high Oshad kurgan or barrow in the town of Maikop in the Kuban region near the foothills of the northwestern Caucasus (the present-day capital of the Adygei Republic). The kurgan contained a spectacularly rich burial assemblage, including bronze weapons and cauldrons; scores of figured gold appliqués, which had been sewn on the clothes of the principal male burial; six silver rods (some more than a meter long) with gold and silver terminals depicting bulls (Fig. 6.1); silver, gold, stone, and ceramic vessels; and numerous turquoise and carnelian beads. This discovery stimulated the excavation of other large kurgans located in the same general region, some of which seemed royal-like in their dimensions and, when not robbed in antiquity, in their materials. This research has continued to the present day and spectacular discoveries are still being unearthed, such as hoards from the Klady kurgan necropolis near the village of Novosvobodnaya that have been excavated from 1979 on (Rezepkin 2000), containing distinct but clearly Maikop-related bronze, gold, silver, polished stone, ceramic, turquoise, and carnelian artifacts.

The Maikop materials were initially brought to the attention of Western scholars through the writings of A. M. Tallgren, M. I. Rostovtseff, and, later, V. G. Childe. The absolute dating of these large kurgans was debated for years, with some scholars (Degen-Kovalevskii 1939, cited in Munchaev 1994: 159–160) relating them to the Scythians or immediately pre-Scythians and dating them as late as the early first millennium BCE, while most (Iessen 1950) dated them back to the middle to second half of the third millennium BCE. The demonstration of

Figure 6.1. Maikop kurgan: gold and silver bulls (adapted from Markovin and Munchaev 2003: 54, fig. 10).

convincing parallels to still earlier northern Mesopotamian or Syrian remains (Andreeva 1977), the new excavation of related Maikop settlements (e.g., Korenevskii 1993, 1995, 2001), and the application of calibrated radiocarbon determinations (Trifonov 1996, 2001, 2004; Chernykh et al. 2000; Chernykh and Orlovskaya 2004a, 2004a, b) have all combined to place them on a much firmer chronological footing. As a result, this research dates their earliest appearance much farther back – beginning as early as the second quarter of the fourth millennium BCE or nearly to the transitional period between late Ubaid and early Uruk times (cf. also Lyonnet 2000).

The so-called Maikop culture, or the more inclusive Maikop cultural-historical community, has always presented difficulties of analysis. Even today, it is treated as an enigma or as a phenomenon hard to conceptualize (e.g., a symposium held in Novorossiisk in 1991 was entitled *Maikopskii fenomenon v drevnei istorii Kavkaza i Vostochnoi Evropy*). The data continue to accumulate. More settlements now have been and are being excavated, correcting for the prior emphasis on mortuary constructions and remains. The Maikop parallels with northern Mesopotamia or, more broadly, with the ancient Near East, and the seemingly consistent and growing number of calibrated radiocarbon determinations (currently more than 40 such dates; E. N. Chernykh personal communication) not only date the Maikop phenomenon more securely but also suggest some connections – albeit hard to specify – with larger historical processes, such as the northern Mesopotamian and later Uruk expansion north into eastern Anatolia. The calibrated radiocarbon dates suggest that the Maikop culture seems to have had a formative influence on kurgan burial rituals and what now appears to be the later development of the Pit-Grave (Yamnaya) culture on the Eurasian steppe (Chernykh and Orlovskaya 2004a: 97). This chapter discusses some of this new evidence and its potential significance, but its central point is to stress the unique character of the Maikop materials and to tease out the implications of its singularity for our understanding of Bronze Age Eurasia.

Munchaev (1994: 178, 174; Trifonov 1991: 25) estimates that roughly 150 Maikop burial complexes have been excavated, while there are only about 30–40 known Maikop settlements (or even fewer; cf. Korenevskii 2001; B. Lyonnet personal communication), only a handful of which have been substantially excavated. The ratio of mortuary assemblages to settlements for Maikop remains is heavily weighted toward the former, and this situation is almost the opposite of what is known for the Early Bronze Kura-Araxes cultural-historical community of Transcaucasia to the south, which probably begins slightly later toward the middle or third quarter of the fourth millennium BCE. Hundreds of Kura-Araxes settlements have been found, scores of which have been excavated, whereas very few Kura-Araxes cemeteries have been located and investigated. As Chernykh (1992: 73) is at pains to observe, it is primarily this difference in the nature of the archaeological evidence that explains the apparent greater wealth of the Maikop metals relative to that of the Kura-Araxes culture. Although both areas were working and probably producing metals on a large scale, we have more metal artifacts from the Maikop culture because more rich "royal" kurgans have been uncovered.

Munchaev (1994) divides the Maikop culture into the three successive phases of Maikop, transitional, and Novosvobodnaya on the basis of changes in the features of the construction of the kurgans and their accompanying ceramic and metal artifacts. He accepts completely the ceramic parallels first noted by Andreeva between the early Maikop ceramic vessels and those found farther south in Syria and northern Mesopotamia (Amuq F and Gawra XII-IX). A detailed comparison of their specific attributes reveals a "similarity that is simply striking" (ibid.: 169), and it has now been claimed that some of the spherical Maikop vessels may have been turned on a slow wheel, a technological development that may reflect direct borrowing from the south, though this also could either be a local innovation or even reflect diffusion from the north because the slow potter's wheel may also have been used in late Tripol'ye CI specialized ceramic workshops.

The depiction of a deer and a "tree of life" on a cylinder seal from an early Maikop burial at Krasnogvardeiskoe (Nekhaev 1986) can be paralleled to depictions on late fourth-millennium BCE seals from northern Mesopotamia (Tepe Gawra) and eastern Anatolia (Degirmentepe), while a toggle-pin with a triangular-shaped head from the Late Uruk–related Arslantepe is identical to a pin found in an early Maikop burial at the Ust'dzhegutin cemetery (for references, see Munchaev 1994: 169). Intriguingly, microlithic chipped stone tools were found interred beneath the floor of pebbles in the great Maikop kurgan, and Munchaev (ibid., 170, 189) relates their seemingly late or archaic presence there to the long-rooted Mesopotamian tradition of depositing such archaic artifacts beneath the floors of public buildings or temples (e.g., in the earlier Yarim Tepe 2 and at Uruk itself). In other words, the fact that such a symbolic Mesopotamian practice is attested in the richest known "royal," or chiefly, Maikop burial must have significance not only for the earlier dating of the Maikop culture, but also for determining aspects of its cultural affiliation and formation.

Other scholars have focused on the northern steppe component of the Maikop culture. Most fundamentally, kurgan or raised-earth burials are not characteristic of northern Mesopotamia, but at least eight Chalcolithic, possibly pre-Maikop kurgans have been excavated in central Ciscaucasia (work of S. N. Korenevskii, cited in Munchaev 1994: 178–179) and in the Kuban area (Nekhaev 1990). Slightly later early kurgans with Maikop or Maikop-related materials also appear on the middle and lower Don on sites of the so-called Konstantinovka culture, some materials of which, such as characteristic asymmetric flint arrowheads, show clear parallels with Maikop remains (Rassamakin 1999: 117–122). V. A. Trifonov (2004: 58–60), in a reappraisal and comparison of the so-called royal tomb at Arslantepe with the Novosvobodnaya-phase Maikop burials, reverses the arrow of cultural transmission and borrowing and argues for an eastern Anatolian Chalcolithic origin for the Novosvobodnaya tombs, such as documented at Korucutepe. Thus, if Trifonov is correct, and if the calibrated radiocarbon dates securely place Maikop chronologically before the emergence of the Pit-Grave (Yamnaya) horizon, then, somewhat counter-intuitively, the origins of raising large barrows or kurgans above the broad, flat expanse of the steppes may not have been indigenous but may have derived from eastern Anatolia or the northern periphery of the greater ancient Near East. There is much that is paradoxical or enigmatic about the Maikop phenomenon.

The Maikop settlements with their relatively thin cultural deposits, light-framed, clay-plastered wattle-and-daub houses, some of which were supported with wooden posts, and many of which contain numerous deep storage pits, hardly recall characteristic ancient Near Eastern or northern Mesopotamian building traditions and techniques. Similarly, the subsistence economy of the Maikop culture, as understood from the excavations of a few of the settlements, seems to have focused more on animal husbandry, such as cattle and possibly pig raising, than on agriculture. Such subsistence practices seem to bespeak more of a northern steppe connection (ultimately, to the breakup of the Tripol'ye settlements?), than of a southern-related Near Eastern heritage. It is probably futile to seek a single source from which the Maikop culture emerged. Rather, this culture or cultural-historical community has multiple origins or is syncretic, with local roots that extend naturally north onto the steppes and with surprisingly close and distinct connections with northern Mesopotamia.

Some Maikop or, better, Maikop-related settlements, such as Meshoko and Yasenova Polyana (Munchaev 1994: 174), were perched on the top of steep ravines and surrounded by stone walls, while others, such as the Galyugai series of settlements along the Middle Terek or those recently investigated by B. Lyonnet and A. Rezepkin along the eroded southern shore of the large Krasnodar reservoir, were open and easily accessible. Rock shelters, containing Maikop materials, also have been excavated. The stone-encircled settlements, if they indeed are correctly attributed to the Maikop culture and not culturally or chronologically earlier (Korenevskii 2001: 24–25), may have been occupied permanently and over a longer period of time than the other types of settlements, which possibly were occupied seasonally (Korenevskii 1995: 80–81). For example, Korenevskii's work, in particular, has been important in showing that such Maikop settlements extended at least as far east as along the Middle Terek. The cultural deposits of the Maikop-related settlements in the piedmont rarely attain 1.5 meters and never exceed 2 meters in depth, and the open settlements on the open north Caucasian plain have much thinner deposits and, in some cases, are totally buried, a fact that long impeded their recognition and excavation.

In this respect, the Maikop settlements sharply contrast with those of the Kura-Araxes culture sites south of the Great Caucasus range, particularly those with mud-brick architecture, the deposits of which can exceed 7 meters in depth (e.g., at Dzhraovit on the Ararat plain or at Garakepektepe in southeastern Azerbaijan). Maikop houses are typically light-framed structures, plastered with clay and reinforced with small wooden posts and reeds (wattle and daub). The small villages or encampments now being revealed in the Krasnodar area contain up to 20 or so circular wattle-and-daub structures, some of which exceed 6 meters in diameter and strangely reveal evidence of being partially burned or deliberately set on fire (B. Lyonnet personal communication). Semi-subterranean houses also occur, frequently cut into by deep rubbish pits, as at Galyugai I, a practice that typically occurs also at Velikent on the Caspian plain.

Direct evidence for agriculture in the form of paleobotanical remains retrieved through flotation or seed impressions on vessels are not yet available, though grinding stones, occasional flint sickle blades (Korenevskii 1995: 62), and very substantial bronze hoes (ibid.: 170) seem to attest indirectly to the practice of some form of extensive field preparation and cultivation and collection of plant remains, although it is also possible that such hoes really functioned as adzes. Consistent with the lack of direct evidence for agriculture elsewhere on the Bronze Age Eurasian steppe, the Maikop settlements have yielded very little macrobotanical remains, only about 10 grains of wheat, for example, being recovered via flotation from the recent excavations near the Krasnodar reservoir (B. Lyonnet personal communication). Relative again to the Kura-Araxes settlements in Transcaucasia, agriculture apparently played a far less significant role in the subsistence economy of the Maikop culture, and in this respect the Maikop phenomenon prefigured later developments on the Bronze Age Eurasian steppe.

Animal husbandry, probably involving at least some form of transhumance, was the more dominant activity. Interestingly, the most thoroughly investigated Maikop-related settlements (though these now may be earlier) in the foothills, which are located along tributaries of the Kuban River, such as Meshoko and Yasenova Polyana, reveal a surprisingly high concentration of pig remains with 40% at the former site and 22.2% at the latter (Chernykh et al. 1998: 245, table 2). Maikop settlements farther east along the Middle Terek, such as Galyugai I, contain far fewer pig bones (3.3% at Galyugai I) and have a much greater concentration of sheep and goats (44.6% at Galyugai I compared to 15.2% and

12.3%, respectively, at Meshoko and Yasenova Polyana). Cattle (steers and cows) were always the principal animals raised by the Maikop herders, constituting 44.5%, 65.5%, and 49.6% of the assemblages from these three sites (Meshoko, Yasenova Polyana, and Galyugai I). Cattle were the animals principally raised by the late Tripol'ye peoples, who also kept a considerable number of pigs. The adoption of such practices by the Maikop herders may not be totally coincidental. The importance of cattle in the subsistence economy is also reflected in Maikop art, such as the silver and gold long-horned bulls that capped the "royal" staffs in the original great Maikop kurgan.

It is commonly accepted that keeping pigs implies sedentism, but this assumption may rely too much on the characteristics of contemporary pigs that have been bred for centuries to produce maximum meat per animal, making them less mobile. Hittite texts refer to the neighboring Kashka peoples to their northeast as "pig-raising nomads" (R. Matthews personal communication). Mobile, wild-appearing, and, apparently, very tasty Kakhetian pigs were moved seasonally between highland and lowland areas in central and eastern Georgia in the recent past. In fact, during Soviet times, pig herders, who also farmed, drove these animals into the high wooded Georgian forests (in the Aragvi Valley, Svaneti, and eastern Georgian mountain valleys) during the summer and let the animals forage freely in the forests, driving them to more protected lowland areas during the late fall (Z. Kikodze personal communication).

The extremely low percentage of horse bones found on Maikop settlements suggests minimally that horses were not a basic component of their diet. The only indirect possible evidence for horse-riding consists of the problematic interpretation of distinctive handled circular bronze objects as cheekpieces (or *psalia* in Russian), an interpretation open to question (Trifonov 1987) as such objects have never been found directly associated with horse remains. In any event, the Maikop culture is very distinctive, not only in terms of its metals but also in terms of what current evidence reveals about its basic subsistence economy where a range of agricultural and pastoral practices are suggested. These practices indicate some distinctive forms of transhumance, not directly comparable with later, ethnographically documented practices of steppe nomads.

The wealth of the metals, including arsenical copper/bronze objects and silver and gold artifacts, found in the Maikop "royal" kurgans is truly extraordinary, leading Chernykh (1992: 142–144) to reflect on the "problem of gold" at this time. Indeed, if we trace the occurrence of gold in the area of our concern, we see a conspicuous shift from north to south that continues through Middle Bronze Age times: the early Chalcolithic florescence of gold consumption in the Balkans, particularly in the Varna cemetery; the abundance of gold (and silver) objects in the Maikop kurgans of the northwestern Caucasus during the Early Bronze period; and the spectacular discoveries of precious gold and, to a lesser extent, silver objects in the monumental early kurgans of Transcaucasia and the famous hoards of Anatolia during the late Early and Middle Bronze periods. Although, undoubtedly, accidents of discovery play a part here, the trend is unmistakable and must reflect underlying historical processes. For example, Avilova et al. (1999: 57–58) calculate that approximately 7,400 gold and 1,000 silver artifacts have been found in Maikop-related kurgans in the northwestern Caucasus. These practically disappear in this area toward the middle of the third millennium BCE, while the number of gold and silver artifacts in Anatolia and Transcaucasia (and, not incidentally, in Mesopotamia, such as at the Royal Cemetery at Ur) sharply rises (calculated at about 32,000 objects; ibid.). This shift reflects not only changes in the production and supplies of precious metals but also, most likely, the movements of peoples with their leaders or chiefs south across or around the Great Caucasus range.

The Maikop arsenical copper/bronzes artifacts include not only ceremonial prestige weapons (which were potentially also usable, such as ribbed tanged daggers and shaft-hole axes) and ornaments but also functional tools, such as the already mentioned "hoes" and chisels and awls, and bowls and large distinctive cauldrons. Characteristic objects of uncertain significance include the so-called twisted circular "cheek-pieces" (or *psalia*) (Munchaev 1994: 211; for a different interpretation, cf. Trifonov 1987) and the large, shafted hooks resembling pitchforks (or *kryuki*; Fig 6.2).

Chernykh's work has shown that the Maikop bronzes could be divided into two groups, which include copper-arsenic alloys and copper-arsenic-nickel alloys, and he has postulated that the sources for the former were ore deposits in Transcaucasia and for the latter were deposits located farther south, possibly in eastern Anatolia, which were also utilized by Mesopotamians. Chernykh (1992: 159–160) refers to the "North Caucasian Bridge," which brought metals, presumably as ingots or in semi-worked form, across the Caucasus. He explains the wealth of the Maikop chiefs as associated with their unique role as intermediaries in this south-north metals trade, supplying vast areas of the steppes to the north and east with Caucasian-derived bronzes. The Maikop culture is present and presumably influential in the very emergence of his "Circumpontic

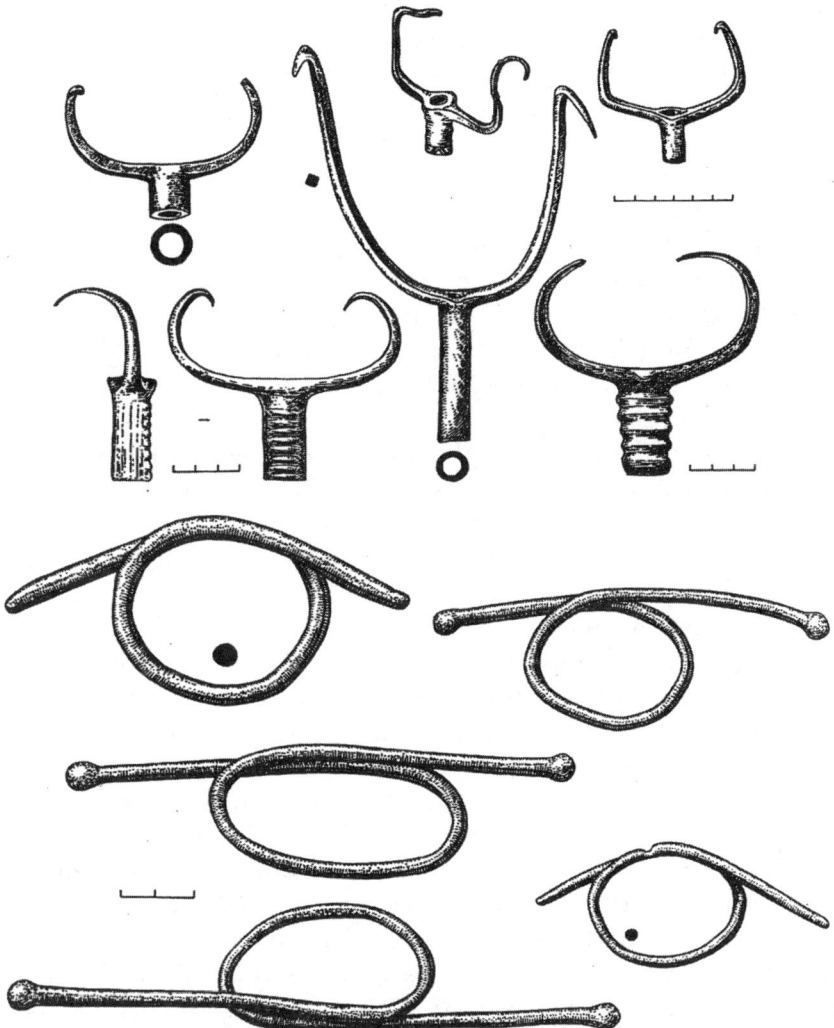

Figure 6.2. Maikop culture: bronze hooks or forks (*kryuki*) and "cheekpieces" (*psalia*) or Mesopotamian cult symbols? (adapted from Markovin and Munchaev 2003: 68, fig. 20).

metallurgical province," which dominates the production and exchange of metals on the Eurasian steppe during the Early and Middle Bronze Ages. This thesis is provocative and elegant, but it could still benefit from firmer documentation linking Maikop with the Kura-Araxes cultures of Transcaucasia and with the terminal Chalcolithic and Early Bronze cultures of eastern Anatolia, extending into northern Mesopotamia. The mining and working of ores and the exchange of metals throughout these regions require additional research.

What happened to the Maikop phenomenon? Why did it disappear or seemingly become supplanted by cultures, such as Novotitorovskaya and later regional variants of the Katakombnaya cultural-historical community that are also known to us largely from their mortuary remains? Here attention is focused not on the possible movements of Maikop peoples across the steppes or south around the Caucasus but on the virtual post-Maikop disappearance of archaeological evidence for the differential accumulation of substantial wealth. This is particularly so in the form of precious metals, such as gold and silver artifacts, on the western Eurasian steppe throughout the rest of the Bronze Age. That is, the Bronze Age cultures that subsequently develop on the western Eurasian steppe contain no such evidence for social differentiation and appear much more egalitarian, if not actually impoverished, relative to Maikop.

The Maikop phenomenon stands out for its uniqueness or singularity, particularly in terms of the precious metals buried with its presumed leaders or chiefs. From this perspective, it is not surprising that initially some scholars attempted to date the Maikop materials to immediately pre-Scythian times. In terms of the concentration of wealth, the Maikop "royal" kurgans resemble the much later royal kurgans that appear on the steppes only with the advent of real nomadic societies interacting regularly with sedentary states to their south at the beginning of the Iron Age. How does one account for Maikop's singularity? If true Eurasian nomadism finally emerged only when relations with settled state societies were firmly established – as has been convincingly argued by A. M. Khazanov (1994: 94–95) and, more recently, by L. N. Koryakova and A. V. Epimakhov (2007: 160) – then does Maikop's singularity or precocity in terms of its accumulation of wealth suggest, albeit indirectly, that it had somehow established relations by the middle of the fourth millennium BCE with a settled state(s) to south? These much later Iron Age nomadic societies, and ultimately the first steppe empires (and first appearance of "royal" kurgans), came into being in part because they were caught up in larger systems of inter-regional interaction and exchange, including regular relations with sedentary states to their south (from China to Rome, including the states of southern Central Asia, such as the Parthian and the Kushan states).

If this relationship is essentially correct, then with what settled complex state society was the Maikop culture regularly interacting? Here, admittedly, the argument becomes speculative, as convincing archaeological documentation for such relations is still largely lacking. Calibrated radiocarbon dates show that Maikop's rise and demise roughly coincide with the advent and collapse of what in the West has become known as the Uruk expansion, a complex, multi-faceted, relatively long-lived phenomenon indicating some form of southern Mesopotamian presence and/or interest in the Anatolian highlands, particularly along the Upper Euphrates drainage. The end of this southern presence, the Uruk contraction, if you will, roughly corresponds with the initial dispersal of Kura-Araxes or Early Transcaucasian peoples to the south and southwest, another extremely complex phenomenon that cannot be discussed here. As far as is known, state societies do not reappear in the eastern Anatolian highlands or in Transcaucasia until the advent of the Iron Age kingdom of Urartu at the end of the ninth century BCE. Southern Mesopotamia (including southwestern Iran) subsequently directed its primary interests first to the east, culminating in the rise of secondary states in eastern Iran, Central Asia, and western South Asia in the second half of the third and first centuries of the second millennia BCE, and then to the west, particularly to the eastern Mediterranean basin during the second millennium BCE. The western Eurasian steppe developed largely on its own during the remainder of the Bronze Age, moving and exchanging materials and ideas over vast distances and constantly developing its mobile herding economies that gradually led to the development of mounted nomadism, social differentiation, and states on the steppes during the first millennium BCE. From this perspective, the Maikop phenomenon seems remarkably singular.

Abbreviations

KSIA	*Kratkie Soobshcheniya o dokladakh i polevykh issledovaniyakh Instituta Arkheologii Akademii Nauk* [in Russian]
KSIIMK	*Kratkie Soobshcheniya o dokladakh i polevykh issledovaniyakh Instituta istorii material'noi kul'tury AN SSSR,* Moscow [in Russian]
RA	*Rossiiskaya arkheologiya, (Russian Archaeology),* Moscow [in Russian]

References

Andreeva, M. V. 1977. K voprosu o yuzhnykh svyazakh maikopskoi kul'tury. *Sovetskaya Arkheologiya* 1: 39–56.

Avilova, L. I., E. V. Antonova, and T. O. Teneishvili. 1999. Metallurgicheskoe proizvodstvo v yuzhnoi zone tsirkumpontiiskoi Metallurgicheskoi provintsii v epokhu rannei bronzy. *RA* 1: 51–65.

Cernykh, E. N., E. E. Antipina, and E. Ju. Lebedeva. 1998. Produktionsformen der Urgesellschaft in den Steppen Osteuropas (Ackerbau, Viehzucht, Erzgewinnung und Ezverhüttung'). In B. Hänsel and J. Machnik (eds.), *Das Karpatenbecken und die Osteuropäische Steppe: Nomadenbewegungen und Kulturaustausch in den vorchrhistlichen Metallzeiten (4000–500 v. Chr.)* Munich: Verlag Marie Leidorf GmbH, pp. 233–252.

Chernykh, E. N. 1992. *Ancient Metallurgy in the USSR*. Cambridge: Cambridge University Press.

Chernykh, E. N., L. I. Avilova, and L. B. Orlovskaya. 2000. *Metallurgicheskie provintsii i radiouglerodnaya khronologiya (Metallurgical Provinces and Radiocarbon Chronology)* [in Russian and English]. Moscow: Institute of Archaeology, Russian Academy of Sciences.

Chernykh, E. N., and L. B. Orlovskaya. 2004a. Radiouglerodnaya khronologiya drevneyamnoi obshchnosti i istoki kurgannykh kul'tur. *RA* 1: 84–99.

    2004b. Radiouglerodnaya khronologiya katakombnoi kultur'no-istoricheskoi obshchnosti (srednii bronzovyi vek). *RA* 2: 5–29.

Degen-Kovalevskii, B. E. 1939. Problema datirovki bol'shikh kubanskikh kurganov. *KSIIMK* 2.

Iessen, A. A. 1950. K khronologii bol'shikh kubanskikh kurganov. *Sovetskaya Arkheologiya* 12.

Khazanov, A. M. 1994. *Nomads and the Outside World*, 2nd edn. Madison: University of Wisconsin Press.

Korenevskii, S. N. 1993. *Drevneishee Osedloe Naselenie na Srednem Tereke*. Moscow: obshchestro "Znanie."

    1995. *Galyugai – poselenie makopskoi kul'tury*. Moscow.

    2001. Drevneishie Zemledel'tsy i Skotovody Predkavkaz'ya. Doctor of Historical Sciences Dissertation Abstract, Moscow.

Koryakova, L. N., and A. V. Epimakhov. 2007. *The Urals and Western Siberia in the Bronze and Iron Ages*. Cambridge: Cambridge University Press.

Lyonnet, B. 2000. La Mésopotamie et le Caucase du Nord au IVe et au debut du IIIe millnaires av. n.è.: leurs rapports et les problèmes chronologiques de la culture de Majkop. Etat de la question et nouvelles propositions. In C. Marro and H. Hauptmann (eds.), *Chronologies des pays du Caucase et de l'Euphrate aux IVe-IIIe Millenaires*. Paris: De Boccard Edition-Diffusion, pp. 299–320.

Markovin, V. I., and R. M. Munchaev. 2003. *Severnyi Kavkaz: Ocherki drevnei i sredneivekovoi istorii i kul'tury*. Moscow: RAN.

Munchaev, R. M. 1994: Maikopskaya kul'tura. In K. Kh. Kushnareva, and V. I. Markovin (eds.), *Epokha Bronzy Kavkaza i Srednei Azii: rannyaya i srednyaya bronza Kavkaza*. Moscow: Nauka, pp. 158–225.

Nekhaev, A. A. 1986. Pogrebenie maikopskoi kul'tury iz kurgana u sela Krasnog-vardeiskoe. *Sovetskaya Arkheologiya* 1.

Rassamakin, Yu. 1999. The Eneolithic of the Black Sea Steppe: Dynamics of Cultural and Economic Development, 4500–2300 BC. In M. Levine, Yu. Rassamakin, A. Kislenko, and N. Tatarintseva (eds.), *Late Prehistoric Exploitation of the Eurasian Steppe*. Cambridge: McDonald Institute, pp. 59–182.

Rezepkin, A. D. 2000. *Das frühbronzezeitliche Gräberfeld von Klady und die Majkop-Kultur in Nordwestkaukasien*. Rahden/Westf.: Verlag Marie Leidorf GmbH.

Trifonov, V. A. 1987. Nekotorye voprosy peredneaziatskikh svyazei maikopskoi kul'tury. *KSIA* 192: 18–26.

1991. Osobennosti lokal'no-khronologicheskogo razvitiya Maikopskoi kul'tury. In V. A. Trifonov (ed.), *Maikopskii fenomenon v drevnei istorii Kavkaza i Vostochnoi Evropy*. Leningrad: Leningrad Division of the Institute of Archaeology, Soviet Academy of Sciences, pp. 25–29.

1996. Popravki k absolyutnoi khronologii kul'tur epokhi eneolita – bronzy severnogo Kavkaza. In *Mezhdu Aziei i Evropoi*. St. Petersburgh: Gosudarst Vennyi Hermitage, pp. 43–49.

2001. Popravik absolyutnoi khronologii kul'tur epokhi eneolita-srednei bronzy Kavkaza, stepnoi i lesostepnoi zon vostochnoi Evropy (po dannym radiouglerodnogo datirovaniya). In Yu. I. Kolev, (ed.), *Bronzovyi vek vostochnoi evropy: kharakteristika kul'tur, khronologiya i periodizatsiya*. Samara, pp. 71–82.

2004. "Tsarskie" Grobnitsy Arslantepe i Novosvobodnoi: nekotorye aspekty sravnitel'nogo analiza. In *Problemy Arkheologii Nizhnego Povol'zh'ya/ Mezhdunardnaya Nizhne volzhskaya arkheologicheskaya konferentsiya. Tezisi Dokladov*. Volgograd: Institute of Archaeology, Russian Academy of Sciences, pp. 56–61.

PART TWO

MINING, METALLURGY, AND TRADE

CHAPTER 7

Introduction

KATHERYN M. LINDUFF

THE CHAPTERS in Part II focus on the importance of metal technology and its relationship to socio-political and economic change in the steppe zone and neighboring regions. The mining, production, and trade of bronze and other metals have been evaluated in the past through numerous analytical models, including core-periphery relationships, multiple-core developments, and the emergence of metallurgical provinces of interaction and exchange (Knauth 1974; Moorey 1985; Tylecote 1992; Chernykh 1992; Linduff 2004; Chernykh et al. 2004; Linduff et al. 2000, 2004). Such models have suggested the widespread and complex nature of early metallurgy in the steppe, and the following essays continue to refine and test our current understandings of the nature and extent of technological diffusion, the emergence of new social organization connected with mining and production communities, and inter-regional and intra-regional strategies connected with metals trade and exchange. Contributing authors were asked to discuss new ways of defining and investigating such social and technological developments in or with the steppe region. Important issues addressed within the chapters include defining the structure and organization of mining communities, theorizing elite strategies for political power foundations and their connection with trade and exchange patterns, the scale of interaction and diffusion of technology between steppe- and non-steppe-based polities, and the cycling dynamics of regional prominence associated with the rise and collapse of mining and production centers in the Eurasian region.

Perhaps most important, these chapters are based on newly available data that allow close examination of individual production centers and

regions. The data come from scientific testing of the materials themselves, from recently excavated communities, from a fresh understanding of mining and its methods based on excavated mines (Chernykh
1997; Chernykh et al. 2004), and even from environmental and human
remains in specific regions. This collection of essays adds Eurasia to the
discussion on the use and production of metal worldwide in the ancient
world and brings social and political context to the study of metallurgy
and metal production there in ways not possible even two decades ago
(Knauth 1974; Moorey 1985; Tylecote 1992; Chernykh 1992, 2004;
Chernykh et al. 2004; Linduff et al. 2000; Linduff 2004).

The development of metallurgy has been considered fundamental
to the emergence of complex societies in many regions of the ancient
world for some time. Decades ago, studies on the beginnings of metallurgy envisioned the setting in the centers of early state-level societies: Mesopotamia and Egypt in the west and China in the east and
early centers of production could be documented in West and East
Asia, but not in between (Knauth 1974; Moorey 1985; Tylecote 1992).
Both the intellectual climate and the evidence available conceptualized
the advent of metal use as a spontaneous occurrence in a single ancient
society. But, given the discrepancy in start dates and distance between
the Near East (beginning no later than the fifth millennium BCE) and
the Far East (beginning at the earliest about 3000 BCE), there has been
much speculation about what role, if any, Eurasia had in the process of
transmission. Researchers asked whether such a complex technology
could be transmitted across the vast steppe to East Asia by the peoples
of Eurasia or whether it was spontaneously generated in the Far East.
Most such speculation was based on evidence of mature metallurgical
production centers and of the movement of certain types of artifacts.
It was assumed that the technological know-how was predicated on
the existence of the division of labor to include mining and the extraction of ores, smelting, working of metals, multi-phase casting, and at
least empirical knowledge of chemical properties of the metals and the
existence of trade networks. All these tasks are, of course, part of the
process of preparing and producing metal objects, especially alloyed
ones, and earlier studies assumed that their existence required a centralized, and even hierarchical, societal organization for the communities from which the patrons of the finished products came. Recent
studies have questioned these assumptions and archaeological investigations have yielded evidence of a very complex and diverse picture
(Chernykh 2004; Linduff 2004).

Until recently, the available evidence was insufficient to explain the emergence and spread of metal technology in the third and second millennia BCE in the Eurasian steppe, where life was thought to be centered in kin-based, relatively independent pastoral or agro-pastoral communities. Now archaeologists in central and eastern Eurasia have uncovered information about early metal use and production by residents of the steppe and the significance of this development, especially in the area east and south of the Urals and west of the Yellow River (Chernykh 1992, 2004; Linduff 2004).

Debates about metallurgical technology and its consequences in the ancient world now consider several crucial factors: knowledge of the presence of ores and the corollary existence or creation of trade networks; the presence of knowledgeable local and/or itinerant artisans who knew metals and their properties; a community able to support such workers, with a degree of social and/or ritual complexity to create a demand for metal products, at least in the host society; and the ability to create high-temperature furnaces for smelting and refinement of ores and final castings.

The publication of Evgenii Chernykh's texts and bibliographies on the early metallurgy in the USSR (Chernykh et al. 2004; Chernykh and Zavialor 2005; Chernykh 1992, 2000, 2004) and many reports on individual sites have provided data on excavated materials from the territories between the Near East and the current Chinese borders. In addition, more-complete reports on copper- and bronze-using sites in Russia, especially those near the Ural Mountains, such as Arkaim and Sintashta (Gening et al. 1992; Zdanovich 1997), have been published. Chernykh and his colleagues in Moscow have collected almost 2,500 C14 dates from metal-using sites across Eurasia (Chernykh et al. 2000). The map of the earliest known metal production shows that metals were part of village life in Eurasia no later than the late fourth millennium BCE (Chernykh 2004).

With these data, Chernykh defined "metallurgical provinces" as large contiguous regions linked through shared utilization of morphologically defined ornaments, tools, and weapons; a common technology of metallurgical production; availability of, or access to, the same metallurgical resources often emerging into large trade networks; and comparable dating (Chernykh 1992: 7–16). These provinces cover distinct areas at different times and include discrete sub-areas of metallurgical knowledge and metalworking. The sub-group (or sub-province) called the Seima-Turbino chronological horizon dates from the late fourth and third millennia BCE, or Late Bronze Age of the Eurasian metallurgical province (Chernykh 1992: 7).

Although Elena Kuz'mina (2004: 37–84) has challenged the method and details of dating of this complex against the Andronovo, there seems to be little argument about the importance of this metallurgical development. The disagreements over the dating of each sub-area complex depend at least in part on the methodology used to establish a chronology. Many scholars depend largely on C14 testing with calibration (Chernykh et al. 2000; Epimakhov et al. 2005). Kuz'mina (2004), on the other hand, proposed that the most reasonable dating is derived from classification methods using evolutionary-typological analysis. Her system finds the Andronovo (and possibly the Fedorovo in Kazakhstan) and the Seima-Turbino synchronous in the seventeenth century BCE. Many authors use a combination of both methods (Konkova 2004; Legrand 2006; Goriachev 2004).

In addition, distinct centers of production have also been identified within the EMP (Chernykh 1992), and several of these have been investigated closely in the chapters here: the southern Volga (Peterson) and western Siberia (Hanks). Moreover, the area of Eurasia is extended to include western (Mei) and southwestern (Han and Li) China. For example, the easternmost region of the steppe includes the Andronovo historico-cultural community and other archaeological cultures recognized previously by archaeologists in the southern Urals and central and northern and eastern Kazakhstan to the Altai and Tianshan mountains (Kuz'mina 2004). The significance of these regional centers is evident in the study of sub-areas just east of the Ural Mountains (Konkova 2004), in eastern Kazakhstan (Goriachev 2006) and western China (Mei), where small Andronovo/Fedorovo-type metal objects have been excavated recently. Although separated in many cases by large distances, analogies in the shapes and décor of these metal objects can and have been noted. In each case, however, regional peculiarities such as pottery types or decor and/or metallurgical traditions mark important distinctions. For instance, only in the eastern region in the Altai and Tianshan do we find unique types such as socketed axes with hatched triangles and rhombuses, forked-shank spearheads, and curved knives with animal (sheep and horses) and human subjects on the pommel. Currently available materials showing these differences, however, indicate that there is interchange throughout this region no later than the third and early second millennia BCE (Linduff 2004).

Moreover, fundamental innovations in metallurgy and metalworking were made in this region at this time. Chernykh found that in the "eastern focus" area (i.e., up to the borders of present-day China) only two examples of arsenical copper were found; all other examples were tin bronze (all from Rudny Altai, Rostovka). The curved knives with horse figures on the hilts, for instance, are all high-quality tin bronze and were found by him only in the Altai. The vast quantity of tin ore in the Altai is given as the reason for such a concentration of tin-copper alloys in eastern and southern Siberia (Chernykh 1992: 224–226). The studies by Mei, Han and Li, and Chernykh himself in this volume extend the discussion of that micro-region to include parts of present-day China.

By contrast, products typical of the western sub-area were made from "pure" and arsenical copper and billon found in abundance in the Urals. And, although tin-copper and tin-arsenic-copper products were found throughout the EMP defined by Chernykh, only certain shapes such as ornately decorated Seima-Turbino socket axes were excavated in the western region, suggesting that the axes were supplied in a finished state (Chernykh 1992: 224; 2004). Finally, Chernykh proposed that the Altai was the source of particular tool types, chemical compositions of tin bronze, and depictions of animals (2000). In addition, jade and flint, bone tools, and protective armor are also exclusive to this area as well. The notion of an Eurasian metallurgical province can be extended to include Xinjiang and even probably Gansu and various northern provinces in present-day China, given the materials found and analyzed there in the past couple of decades (Han and Sun 2000, 2004; Mei 2000, 2004; Linduff 1997, 2004; Linduff et al. 2000: 1–29). In the essays in this volume, attempts are made to connect the materials with the communities and their needs, lifestyles, and socio-political organization.

When compared, these studies show that not all sites or regions were uniformly organized or participating in the same processes of production. It is clear, however, that these production areas represent societies that practiced job differentiation. In Chapter 11, Peterson has documented the recycling of metals, for instance, that suited the continuing, if changing, needs of communities. This process is, of course, very different from one where new batches of metal and alloys were being made on a regular basis, and it implies different socio-economic circumstances.

In the case of the sub-region of Xinjiang in western China presented in Chapter 12, Mei argues that socio-cultural features, including ritual, created demands for metal objects with uses determined by the local community. Consequently, the various patterns of use in the Central Plain in dynastic China, in Xinjiang and the Northern Frontier, reflect differing socio-cultural organization in each area. A uniform expectation or explanation for metal-using and metal-producing communities is unlikely, and in many cases the industry followed the already existing socio-cultural patterns as evidenced especially in ritual. On the basis of technical analysis of the metal objects themselves, Rubin Han and Xiaocen Li in Chapter 10 propose that certain technologies, for instance tinning, should be seen as signature inventions of particular areas and peoples (east of the Taihang Mountains in the Northern Frontier), and when found elsewhere, such as in Yunnan in the southwest of present-day China, they signal either technology transfer and probably trade, or the presence of itinerant workers. They go on to suggest that, once transferred, such technology was used to mark status and thereby signaled social inequality in the host society where it was introduced from outside.

In Chapter 9, Hanks outlines a micro-regional approach for the study of the relationship between early metallurgy and socio-political hierarchies. Because much of the available evidence is from mortuary settings, he lays out a manner of study that looks beyond the metals themselves to include bio-archaeological and other sorts of analyses to consider the effects of metallurgy on human life within a community. His research area is the well-known production area that includes Arkaim and its correlates south of the Urals where agro-pastoral communities have been documented, all of which have been presumed to be self-contained metal-production facilities. These communities were not the major consumers of their wares, and must have been at least interconnected and/or engaged in a larger enterprise. His preliminary study of the mortuary evidence from the region shows that social differentiation can be found, unlike what has been argued in the past on the basis of evidence of uniformity in the size of living quarters. His continuing research promises to address several vexing questions about the relationship between metal production and social hierarchies.

These chapters make clear that many regional traditions flourished in addition to the exchange of artifacts, and that the movement of ideas and people, perhaps technicians more than whole groups, was likely and was a stimulus for the transfer and invention of the variety of technologies employed in this vast area. Moreover, the community contexts examined here were shown to be diverse in technological know-how, in use of metals, and in socio-economic organization.

References

Chernykh, E. 1992. *Ancient Metallurgy in the USSR*, Trans. S. Wright. Cambridge: Cambridge University Press.

2004. Kargaly: The Largest and Most Ancient Metallurgical Complex on the Border of Europe and Asia. In K. Linduff (ed.), *Metallurgy in Ancient Eastern Eurasia from the Urals to the Yellow River*. Lewiston: Edwin Mellen Press, pp. 223–238.

Chernykh, E., E. Kuz'minykh, and L. B. Orlovskaia, 2004. Ancient Metallurgy of Northeast Asia: From the Urals to the Sayano-Altai. In K. Linduff (ed.), *Metallurgy in Ancient Eastern Eurasia from the Urals to the Yellow River*. Lewiston: Edwin Mellen Press, pp. 15–36.

Chernykh, E., L. I. Viola, and L. B. Orlovskaya. 2000. *Metallurgicheskie provintsii i radiouglerodnaia khronologiia*. Moscow: Institute of Archaeology of the Russian Academy of Sciences.

Chernykh, E., and V. I. Zavialov. 2005. *Arkeologiia i estestvennonauchnye metody*. Moscow: I Azyki slavianskoi kultory.

Epimakhov, A. B., B. K. Hanks, and C. Renfrew. 2005. Radiocarbon Dating Chronology for the Bronze Age Monuments in the Trans-Urals, Russia. *Rossiiskaia Arkheologiia* 4: 92–102.

Gening, V. F., G. B. Zdanovich, and V. V. Gening. 1992. *Sintashta: arkheologicheskie pamiatniki ariiskikh plemen Uralo-Kazakhskikh stepei*, vol. 1. Chelyabinsk: Iuzhno-Ural'skoe knzhnoe izdatel'stvo.

Goriachev, A. 2004. The Bronze Age Archaeological Memorials in Semirechie. In K. Linduff (ed.), *Metallurgy in Ancient Eastern Eurasia from the Urals to the Yellow River*. Lewiston: Edwin Mellen Press, pp. 109–138.

Han, R., and S. Sun. 2000. Preliminary Studies on the Bronzes Excavated from the Tianshanbeilu Cemetery, Hami, Xinjiang. In K. Linduff (ed.), *Metallurgy in Ancient Eastern Eurasia from the Urals to the Yellow River*. Lewiston: Edwin Mellen Press, pp. 157–172.

2004. A Study of Casting and Manufacturing Techniques of Early Copper and Bronze Artifacts Found in Gansu. In K. Linduff (ed.), *Metallurgy in Ancient Eastern Eurasia from the Urals to the Yellow River*. Lewiston: Edwin Mellen Press, pp. 175–194.

Knauth, P. 1974. *The Metalsmiths*. New York: Time-Life Books.

Konkova, L. 2000. The Emergence of the Earliest Metal; in the [Russian] Far East. In G. Kim, K. Yi and H.-T. Kang (eds.), *BUMA-V: Messages from the History of Metals to the Future Metal Age, Proceedings of the Fifth International Conference on the Beginnings of the Use of Metals and Alloys*. April 21–24, 2002, Gyonhju, Korea. Seoul: Korea Institute of Metals, BK21 Division of Materials Education and Research, Seoul National University, pp. 79–84.

Kuz'mina, Elena. 2004. Historical Perspectives on the Andronovo and Early Metal Use in Eastern Asia. In K. Linduff (ed.), *Metallurgy in Ancient Eastern Eurasia from the Urals to the Yellow River*. Lewiston: Edwin Mellen Press, pp. 37–84.

Legrand, S. 2006. The Emergence of the Krasuk Culture. *Antiquity* 80, no. 310: 843–859.

Linduff, K. 1997. Here Today and Gone Tomorrow: Bronze-Using Cultures outside the Central Plain. *Bulletin of the Institute of History and Philology.* Nankang, Taipei: Academia Sinica, 1997: 393–428.

2004. *Metallurgy in Ancient Eastern Eurasia from the Urals to the Yellow River.* Lewiston: Edwin Mellen Press.

Linduff, K., R. Han, and S. Sun. 2000. *The Beginnings of Metallurgy in China.* Lewiston: Edwin Mellen Press.

Mei, J. 2000. *Copper and Bronze Metallurgy in Late Prehistoric Xinjiang: Its Cultural Context and Relationship with Neighboring Regions.* BAR International Series 865. Oxford: Archaeopress.

2004. Metallurgy in Bronze Age Xinjiang and Its Cultural Context. In K. Linduff (ed.), *Metallurgy in Ancient Eastern Eurasia from the Urals to the Yellow River.* Lewiston: Edwin Mellen Press, pp. 173–188.

Moorey, P. 1985. *Materials and Manufacture in Ancient Mesopotamia: The Evidence of Archaeology and Art; Metals and Metalwork, Glazed Materials and Glass.* Oxford: British Archaeological Reports.

Tylecote, R. 1992. *A History of Metallurgy.* 2nd ed. London: Institute of Materials.

Zdanovich, G. 1997. Arkaim-kul'turnyi konpleks epokhi sredni bronzy Iuzhnofo Zaural'ia. *Rossiiskaia arkheologiia* 2: 47–68.

CHAPTER 8

Formation of the Eurasian Steppe Belt Cultures
Viewed through the Lens of Archaeometallurgy and Radiocarbon Dating

EVGENII N. CHERNYKH

MY PRIMARY goal in this chapter is to determine the general formative stages of Eurasian metallurgical production during the Early
Metal Age.[1] In previous publications, I have designated three such stages.
Now within the second and the third stages, it is possible to assign a number of successive phases. This research is based on extensive databases
of radiocarbon dates of the most ancient metal sites in the region, stored
and systematized in the Laboratory of the Institute of Archaeology
of the Russian Academy of Sciences in Moscow. These databases have
been used to form "metallurgical provinces" (MP) and include more
than 120,000 objects. The number of systematically and statistically
processed data of calibrated dates (C14) has reached nearly 1,700 analyses. I have attempted to give an account of the role of Eurasia in both
the emergence and transmission of metallurgical technology; artifact
typologies in western and eastern Asia; and a chronological framework
within which other studies, here and elsewhere, can be understood.

The area of concern is vast, and the contacts across it are substantial. What is not outlined in this report is the impact of what I have
described chronologically and territorially across such a distance on
societal change. This could be seen from the information on metallurgy in several ways. First, if the procurement of ores and production
of metal items were instigated by an elite represented by the richest of
the grave inventories, we could assume that social inequality was part of
the equation. On the other hand, those who provided the technological
know-how to procure and create such large quantities of metal objects
may have been part of a vast network that required supply and demand

from known communities documented here as "users." Such communities of workers existed and were unevenly organized. But many issues concerning social organization are far beyond the goals of this chapter and require very different evidence from what I discuss here and what I believe is within the boundaries of my expertise. It is hoped, however, that the chronology constructed here provides a framework that will assist others in their reconstruction of various trends in social organization and complexity in the steppe region.

The Early (First) Stage of Formation of the Steppe Belt Cultures: The Carpatho-Balkan Metallurgical Province

IN ALL likelihood, one should connect the origin of the famous Carpatho-Balkan metallurgical province (CBMP) of the Copper Age and its swift appearance with the formation of the "steppe belt" of stockbreeding cultures (Chernykh 1992: 35–53). In the period of the maximal distribution of metal and metal production in the province itself, its territory was about 1.3–1.4 million square kilometers (Fig. 8.1). With the known mining, metallurgical, and metal producing centers providing the base for

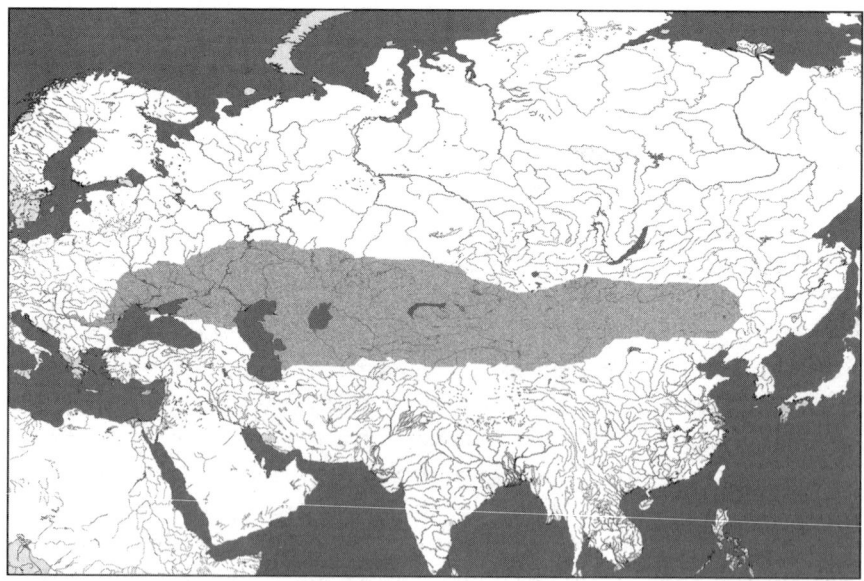

Figure 8.1. Map showing the distribution of the steppe belt domain of Eurasian stockbreeding cultures.

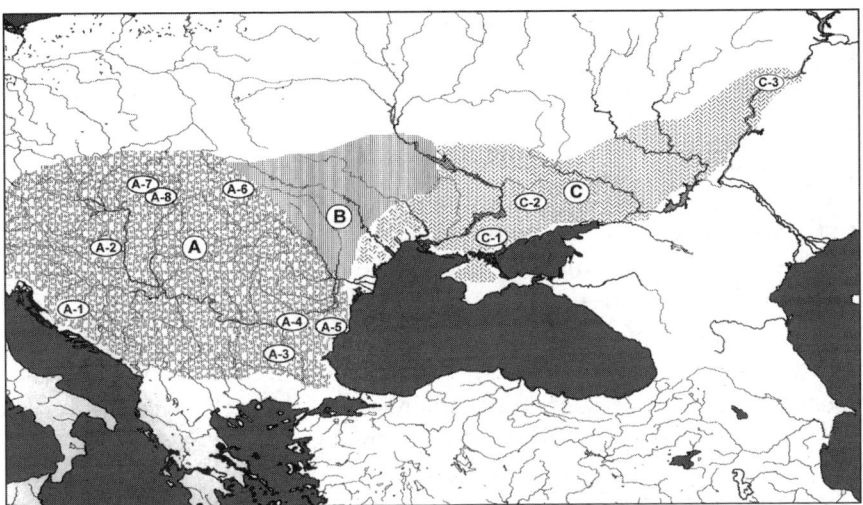

Figure 8.2. Map of the Carpatho-Balkan metallurgical province area: (A) the central block of settled farming cultures and communities; (A-1) Butmir; (A-2) Vinca C/D; (A-3) Karanovo V-Maritsa; (A-4) Karanovo VI-Gumelniţa; (A-5) Varna; (A-6) Lengyel; (A-7) Tiszapolgar; (A-8) Bodrogkresztur; (B) the cultural block Cucuteni-Tripol'ye; (C) the block of the steppe stockbreeding cultures; (C-1) Dnepr-Donets or Mariupol'; (C-2) Sredni Stog; (C-3) Khvalynsk.

the structure of the province, one can distinguish three basic blocks of cultures with a sufficient measure of reliability (Fig. 8.2).

The first (the major and central) CBMP block includes the mining and metallurgical production centers located mainly to the north of the Balkans (Jovanović 1979) and in the Carpathian basin (Fig. 8.2). Huge numbers of copper instruments and weapons as well as ornaments were produced in these centers (Todorova 1999). In this block, such unique sites as the Varna "gold" necropolis and Ai Bunar copper mine, which is the most ancient mine in the world up till now, have been surveyed in detail. The territorial scope of this specified block of cultures is equal to about 0.75–0.8 million square kilometers.

The second block is connected with the set of cultures known as the Tripol'ye or the Cucuteni-Tripol'ye community (0.16–0.18 million square kilometers). By comparison with the previous area, the Tripol'ye culture group should be regarded as peripheral (Fig. 8.2). This conclusion is based on the scale of metal production. In the Tripol'ye community, three basic types of cultures can be distinguished: Tripol'ye A, B, and C-1. In these Tripol'ye communities, the centers of metal production were minor in character in comparison to those in the CBMP. In these centers, the Tripol'ye workers produced weapons and decorations out of copper imported from the centers of the main block of the CBMP. Most likely, this block became the main point of transmission for copper ores to their east, or to the steppe populations.

The third block, which we can define as the eastern (or northeastern), was definitely marginal to the CBMP and included a territory up to 0.4–0.5 million square kilometers. It was composed entirely of cultures, or archaeological communities, of steppe stockbreeders. The schematic map in Figure 8.2 presents the distribution of these three communities. Their pinpoint presence in the area of the Danube settled farming cultures is obvious (e.g., Comşa 1991).

One must pay attention to some of the peculiarities of the steppe communities in the south of eastern Europe. Researchers of household and sepulchral sites of the steppe block separate them not only from rather remote settlements and necropolises, such as those in the Danube region, but also from adjoining Tripol'ye settlements. Here distinctions in basic cultural features are true only for *external* comparisons with this block of cultures. Many mutually exclusive conclusions can be found in the *internal* structures of these communities in previous research. All attempts to correctly distinguish separate cultures constantly rely on a "blurred" picture of the key points buried in a huge mass of stored archaeological materials. In the study of steppe cultures, we constantly come across the "pattern of cultural continuity" so characteristic for the majority of cultures of the steppe belt in Eurasia (Chernykh 2007: 35–36).[2]

In this block, it seems to me that there are three archaeological communities: the Dnepr-Donets, the Sredni Stog, and the Khvalynsk cultures (Figs. 8.2 and 8.3), even though a variety of different names have been published. For instance, certain burial grounds or settlements of the Dnepr-Donetsk community have been called "neo-Neolithic sites," the "Novo-Danilovka type" sites, or "culture of the Mariupol' necropolis type."[3]

An assessment of the basic types of metallurgical and metal-processing production in the centers of both western blocks suggests that metal processing in the third peripheral block is rather primitive by comparison (Ryndina 1998: 151–179). It fits poorly into the general morphological and technological standards of the CBMP because these cultures did not manufacture, and were not able to manufacture, magnificent metal weapons for which the central zone centers were so famous. The only basis for inclusion of these steppe centers of metal processing is their importation of copper, which was received by the steppe groups from the central CBMP block.

Problems associated with the absolute dating of cultures and communities of all three blocks were solved with the help of 470 calibrated radiocarbon dates, which included the calculation of the sum of probabilities for each set (Fig. 8.3). Actually, the total number of known dates is now much greater, and the latest ones are connected to the sites of the main block of the Carpatho-Balkan cultures. I have limited the number mentioned here because almost half of the processed chronological analyses (230) are connected to the sites of the central block of the province. A calendar range at the apogee of metallurgical activity of mining and metallurgical production in the basic centers of the CBMP is essential and is most likely a five-century time interval – between the forty-eighth and the forty-third centuries BCE. There are many fewer known dates for the sites of other blocks. For example, there are 139 dates for the three basic cultures of the Tripol'ye community and 101 calendar dates for the steppe communities.

In this case, it is also necessary to clarify the character of the distribution of sums of probabilities of the calibrated radiocarbon dates in all three blocks, as their diagrams present an uneven picture. Practically all frequency ranges of the sums of probabilities of the central block are characterized by compactness and are close to a normal distribution (Fig. 8.3). In comparison to the preceding analysis of dates of the Tripol'ye community, they are indistinct because the distribution indicates stretched ranges of a 68% probability. Second, the precise distinction of the specified ranges for the basic stages (cultures) of the Tripol'ye community, A, B, and C1, is obvious. However, it is curious that the dates for the Tripol'ye A sites correlate with a pre-metal stage or, in essence, the Neolithic period. The Tripol'ye B period in many respects coincides with an apogee of activity in the CBMP production centers, though primarily during the later centuries. The culture of the Tripol'ye C1 period is already entirely out of the limits of the mentioned range and corresponds to a period of decline for this, the most ancient Eurasian metallurgical province.

Frequency figures of the sums of probabilities of the calibrated dates of the steppe culture block differ from one another. Here, a chaotic character of distribution dominates in many respects and was maximally

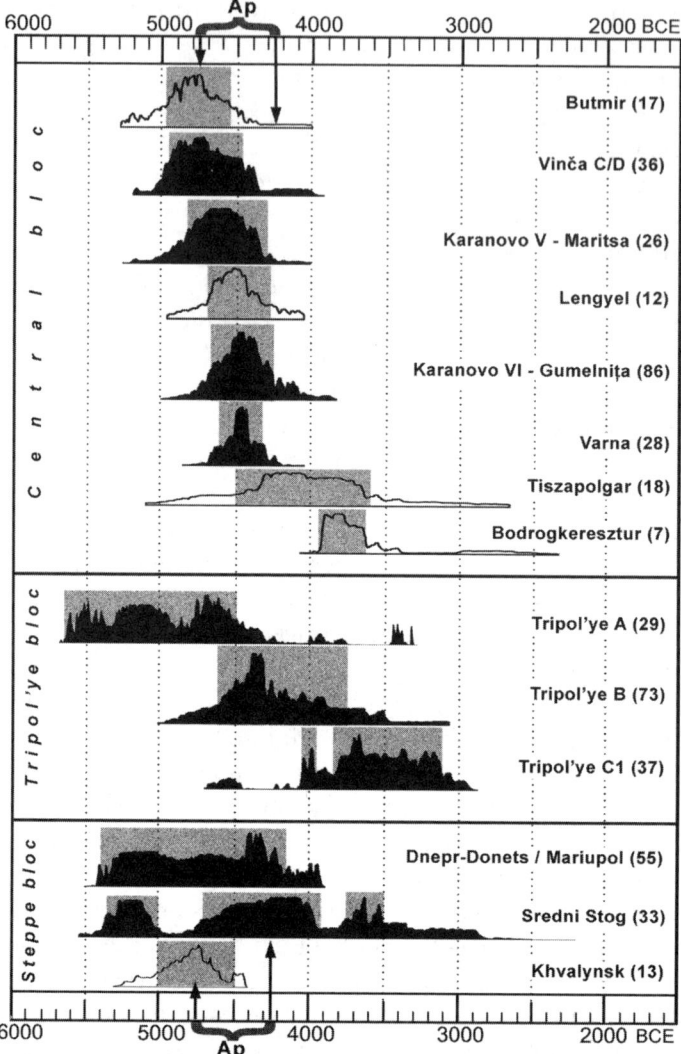

Figure 8.3. Sum probabilities of radiocarbon dates of the archaeological cultures and communities of the Carpatho-Balkan metallurgical province. *Note*: The shaded rectangle for each polygon corresponds with the probability of 68.2% (it conforms with figures 3, 5, 6, 10). Ap = the apogee of activity of the central block's productive centers of the Carpatho-Balkan metallurgical province.

expressed in the materials of the Sredni Stog culture. The distribution of the Khvalynsk culture is more compact although we have only 13 reliable dates taken from burials in the two easternmost burial grounds of the steppe block (Fig. 8.2).

The Second Stage of the Steppe Belt Formation: The Circumpontic Metallurgical Province

BETWEEN THE fifth and early fourth millennia BCE, there was a dramatic change in the settled cultural-economic systems of the Copper Age. The central change during this time was the disintegration of the Carpatho-Balkan metallurgical province and the parallel formation of a new, large Circumpontic metallurgical province (CMP), which marked the beginning of the Bronze Age. At the final stage of this province, its territory was approximately 4.5–5 million square kilometers. The system of the mining, metallurgical, and metal-processing CMP centers stretched from west to the east, from the Adriatic Sea up to the southern Ural Mountains, and also from south to north, ranging from the Levant, Mesopotamia, and Susiane up to the forest areas of the Upper Volga region.

After the accumulation and statistical processing of a large series of radiocarbon dates, I have made essential corrective amendments to our understanding of the formation of the whole Eurasian continental metallurgical province. The results of the systematic processing of 833 calibrated radiocarbon dates from numerous suites of archaeological communities, cultures, and separate sites are presented in this research.[4] At this time, the long history of both the formation and functioning of the huge CMP system has led me to propose two major chronological phases different from my earlier interpretations.

The first phase marks a formative stage of the province, to be called the *proto-CMP*. The term was formulated during the first major attempt to create a chronology based on radiocarbon dates for centers of production in the Carpatho-Balkan and Circumpontic provinces (Chernykh et al. 2000: 14–18, 37–38). The chronological range of the early phase included the whole fourth millennium BCE. The prefix *proto* and its category include the territory of productive centers of the province at the early phase and do not include all Circumpontic areas (Fig. 8.4). The northern area of the Balkan peninsula, along with the Carpathian and Danube basins and steppe zone of the northern Black Sea Coast, remained within the borders of the vanishing Carpatho-Balkan province.

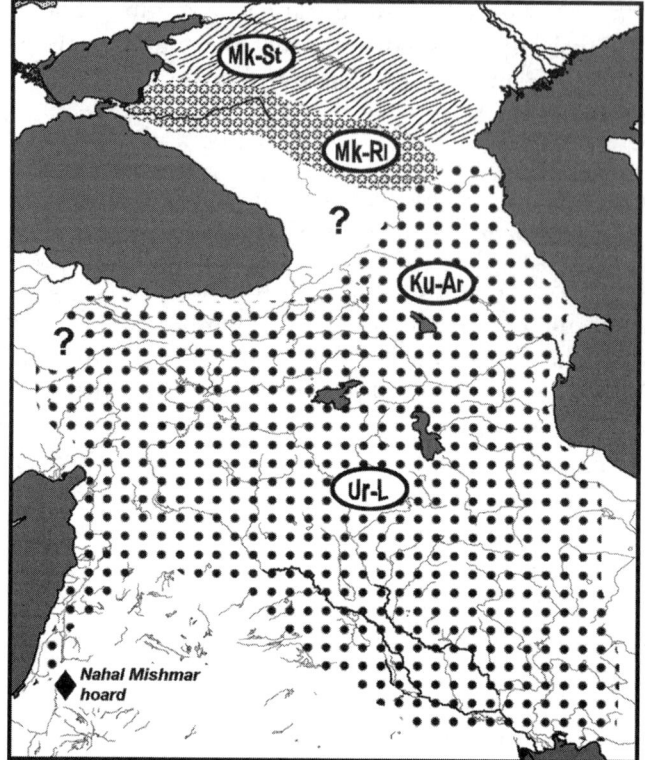

Figure 8.4. Map of the Circumpontic metallurgical province (early phase of the province formation). Mk-Rl = area of the Maikop culture/ community); Mk-St = area of so-called steppe Maikop; Ku-Ar = area of the Kura-Arax culture; Ur-L = area of the Late Northern Uruk.

The second phase included a true Circumpontic province wherein its productive centers completely encircled the Black Sea basin. By then, the CBMP had come to an end, and its former territories were occupied by metallurgical and metal-processing centers where the more morphological, technological standards of the CMP completely dominated. The chronological range of the productive centers of the second phase coincided with the third millennium BCE.

For the gigantic CMP, during both phases of its existence, a number of remarkable features are apparent and typical. The first and probably the most essential was the conformation to new technological and morphological standards, ones sharply different from those of the disintegrating CBMP system. The change included not only the main categories and forms of the tools and weapons but also the beginning use of intentional copper-arsenic alloys (arsenical bronzes). New methods of smelting and processing metals developed in the CMP centers and formed a basis for the origin of the "global" West-Eurasian model of metallurgical production. Later, by the beginning of the second millennium BCE, the differences between the West-Eurasian and East Asian models were clear.

Another important feature of the emergence of the CMP was the structure of the extensive Circumpontic world. Already from the beginning, two contrasting blocks of strikingly exclusive archaeological cultures appeared. The first, or southern block, included settled farming cultures and communities. The steppe communities known as the kurgan cultures represented the second or northern block.

The third factor is that over this long period various cultures and communities in the steppe began to play an important but dissimilar role from that which I noted for the Copper Age of the Carpatho-Balkan province. The marginal character of the steppe stockbreeding cultures of the Copper Age fell into oblivion, not only from the central block of the CBMP cultures but also from the block of the Tripol'ye communities.

THE FIRST CMP PHASE: THE MAIKOP PHENOMENON

The famous Maikop culture is at the forefront of the CMP formation (see Chapter 6 by Kohl in this volume). The phenomenal and, in some respects, paradoxical features of the Maikop development are very distinctive. Metal items from the Maikop kurgan burials have been known for nearly a century and undoubtedly are their major attribute (Munchaev 1975: 211–335; Rezepkin 2000). In the long series of Early Bronze Age (EBA) cultures and sites in the Near East we have not found anything equal to the Maikop "royal" complexes in terms of lavishness, quality, and quantity of bronze, gold, and silver objects (Chernykh et al. 2002: 5–15, fig. 3) – if we assume the limits of the first CMP phase outlined here. All attempts to see the magnificence of the Maikop metal complexes as a kind of local response to a Near Eastern impulse have been made in the absence of information on a southern zone that was either equivalent or superior to it.

Maikop grave constructions are of a type that belonged to a circle of stockbreeding kurgan cultures in the northern zone of the Circumpontic province, and because of their complexity and large size, they undoubtedly were impressive compared to other steppe kurgan communities of eastern Europe. Moreover, within the long sequence of the kurgan cultures, the Maikop community occupied a space on the boundary of the foothills of the Caucasus Mountains. Behind its ridges, communities of the other CMP zone were located, however, and these were not similar to the kurgan cultures noted previously (Fig. 8.4). Second, the Maikop culture differed markedly from the others, the more northern kurgan communities, because of its much earlier date.

Further analysis of the community introduces paradoxes about the Maikop culture. For example, there is a sharp contrast between the magnificence of the kurgan burials and the rather modest (at times poor) character of the settlements connected to these burials. Even in the most remarkable of the habitation sites that are known, such as the settlement of Meshoko situated south of the Kuban River that includes a stone defensive wall (Formozov 1965: 70–105), we are at a loss to explain the richness of the famous kurgans. Other settlements of this culture are even less impressive.

In addition, artifacts from the Maikop burials have been studied for more than a century, but the research has not located funerary or household complexes with any signs of mining, metallurgy, or even metal processing. This absence makes the amazing collections of metal objects from the kurgan burial cemeteries even more remarkable.

One surprising feature of the Maikop artifacts is that radiocarbon dates allow a more ancient chronological range for the Maikop cultures than has been suggested through comparison of types of archaeological materials (Chernykh and Orlovskayia 2007). The Maikop complexes are dated at a 68% probability range for the 37 processed dates and correspond to a chronological range of between 4050 and 3050 BCE (Fig. 8.5). The steppe sites of the so-called Maikop type ("steppe Maikop")[5] are supported by 19 dates and are also within almost the same time range of 4000–3000 BCE (Fig. 8.5).

Furthermore, the calendar age of the Maikop cultures is more ancient than many other communities, cultures, and some widely known settlements (tells) of the Early Bronze Age included in the southern block of cultures (Fig. 8.5). Only sites of the so-called late northern Uruk, or the period of well-known Uruk northern expansion, are synchronous with the Maikop complexes. However, we should remember that the Uruk-type sites are extremely metal poor. Also by comparison to the Maikop chronology, the younger date of the Kura-Araxes culture sites are conspicuous (Fig. 8.5). Paradoxically, the Maikop culture has always been considered secondary to the later communities with respect to

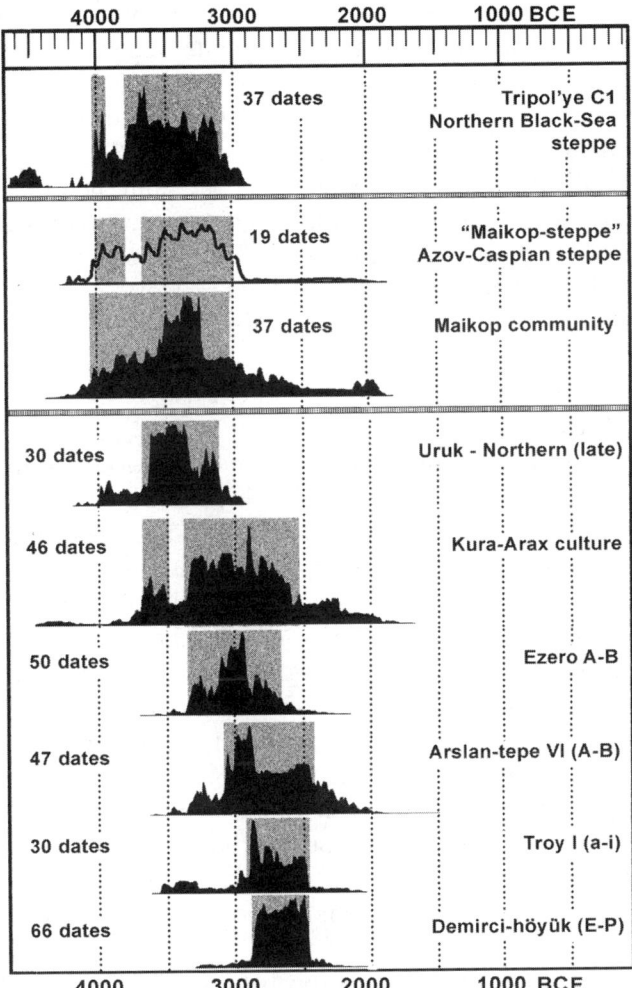

Figure 8.5. Sum probabilities of the radiocarbon dates of the archaeological cultures and communities of the Circumpontic metallurgical
province (the Tripol'ye C1 culture corresponds with the Carpatho-
Balkan metallurgical province; see Fig. 8.3).

metallurgy and metal processing. Most likely, the problem briefly outlined here for the Maikop, and its early chronology, will demand much
more detailed study in the future.[6]

We find a similar picture of dating superiority for the Maikop community when comparing it to the block of the steppe eastern European

cultures and communities (Fig. 8.6). This fact emphasizes once again the unusual anomaly of the large Maikop kurgans that were furnished with gold, silver, and bronze objects.

Last, we should pay special attention to synchronism in the range of dating of the Maikop artifacts and those from the Circumpontic provinces, especially the Tripol'ye C1 sites (Fig. 8.5), and their relation to the dates for the waning Carpatho-Balkan province. In this case, it is striking that the Maikop culture and Tripol'ye C1 were situated not only in several different territories but also without material evidence of contact between them.

THE SECOND (LATE) CMP PHASE: THE STEPPE KURGAN CULTURES

Among the centers of metallurgy and metal processing in the block of kurgan cultures, researchers usually distinguish two large archaeological communities known as the Yamnaya (Pit-Grave) and Catacomb. The first is known primarily through study of materials from kurgan burials, as settlements are exceptionally rare. Within the territory represented by Catacomb tombs, settlements are better known, but, even so, the materials recovered from kurgan funerary complexes dominate the artifact assemblages recovered. The majority of researchers traditionally argued (and still do) that, in the steppe and forest-steppe zones of eastern Europe, the Catacomb community of the Middle Bronze Age (MBA) replaced the Yamnaya, which belonged then to the EBA. Recently, however, much more attention is being paid to the considerable amount of evidence for the synchronous existence of these complexes over a long period. The radiocarbon chronology strongly strengthens this position.

Comparative analysis shows the probability that, even if there were calendar distinctions between the different regions of both communities, the difference is not significant. This conclusion is based on a representative series of radiocarbon dates: 273 dates for the Yamnaya complexes and 191 for the Catacomb. A comparison of these dates yields a 68% probability that a long period of coexistence took place within the twenty-seventh to twenty-first centuries BCE (Fig. 8.6).

The formative period of the Yamnaya coincided with the end of the fourth century or within the first centuries of the third millennium BCE. Thus, it is curious that the earliest dates (the thirty-third to thirty-first centuries BCE) correspond with the sites in both the eastern and western peripheral regions of the distribution. This includes the Volga-Ural area (including sites of the Poltavka type) and also the northwestern Black

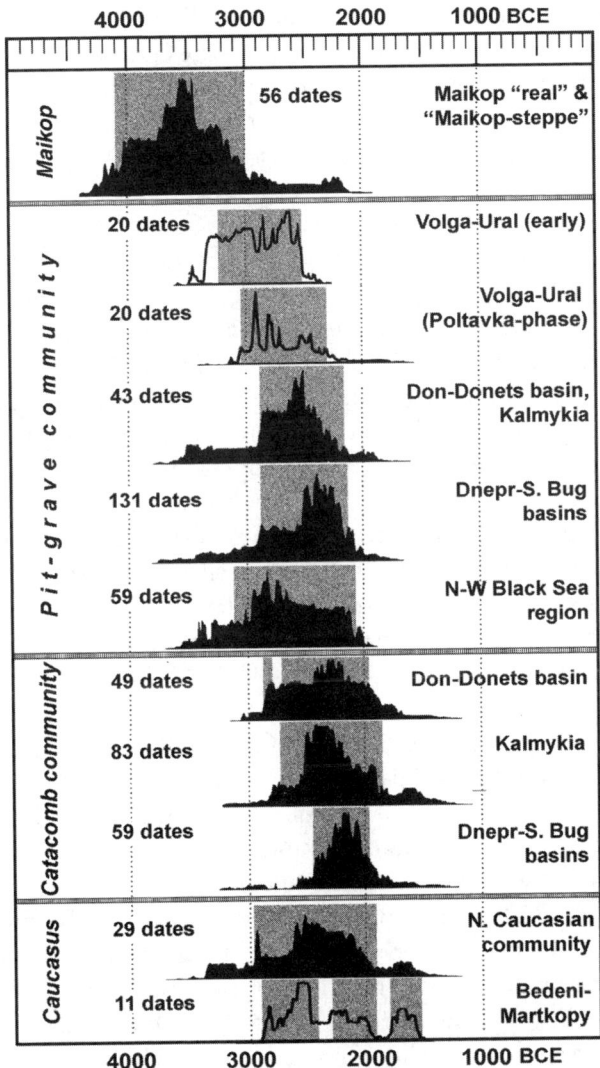

Figure 8.6. Sum probabilities of the radiocarbon dates of the stockbreeding archaeological cultures and communities of the Circumpontic metallurgical province (mainly the northern block of the CMP).

Sea coast (Figs. 8.6 and 8.7). At this early period (the twenty-eighth century BCE), several of the central regions of the Yamnaya community – from the basins of the Dnepr, Southern Bug, Don, and Donets, and also Kalmykia – are three to four centuries later by comparison to the funeral complexes mentioned previously.

In the three main areas of the Catacomb distribution, the picture is a little bit different. There, the complexes of the central geographic group localized in the basin of the Don and Severski Donets (Figs. 8.6 and 8.8) are the earliest. Materials from the peripheral regions – that is, from Kalmykia and especially from the basins of the Dnepr and Southern Bug – are from a later time.

On the whole, the Yamnaya culture sites date two to three centuries earlier than the earliest of the Catacomb sites. It is also necessary to shift the end of the last one for two to three centuries later, or, at 68% probability, those dates fall in the twentieth to nineteenth centuries BCE.

The new radiocarbon chronology dramatically changes our previous ideas about the calendar position of the main eastern European steppe communities. Moreover, the radiocarbon dates sharply correct the location of contact between the steppe peoples and the North Caucasian archaeological community, or the Transcaucasian successor of the Maikop community, or the kurgan culture of the Bedeni-Martkopi type (Dzhaparidze 1998). Diagrams of the distribution of the sums of their probabilities show a contemporaneous relationship among these complexes and the northern materials relating to the Yamnaya and Catacomb communities.

The Yamnaya and Catacomb Communities: The Problem of Chronological Correlation

In an extensive territory in southeastern Europe, including an area not less than 0.7 million square kilometers (compare Figs. 8.7 and 8.8), the cultures of two large archaeological communities (Yamnaya and Catacomb) co-existed for approximately 600 years. Beginning with V. A. Gorodtsov's classic works, we learned that the later position of the Catacomb cultures was based on the stratigraphic position of some catacomb graves. Subsequent research attested to a level of development of metal processing that was thought to secure the chronological sequence of cultures. The Catacomb community appeared to be more developed (Chernykh 1992: 83–91, 124–132). On the basis of similar conclusions, comparatively late dates were also assigned to sites of the Poltavka type, which were widespread in the Volga-Urals region (Chernykh 1992: 132–133).

The accumulation of radiocarbon dates and their systematic processing have led to complete contradictions of our former ideas not only about the absolute ranges of dates of materials analyzed here but also about their comparative chronological position. The consequence of

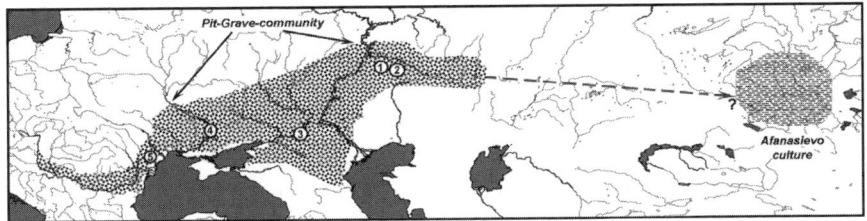

Figure 8.7. Map of the Pit-Grave (Yamnaya) archaeological community (second phase of the CMP) and its proposed relation with Altay region (Afanasievo culture). Space variants of the Pit-Grave community: (1) Volga-Ural (in essence Yamnaya culture); (2) Volga-Ural (Poltavka culture); (3) Kalmykia and Don-Donets basin; (4) Dnepr and Southern Bug basins; (5) northwest Black Sea area.

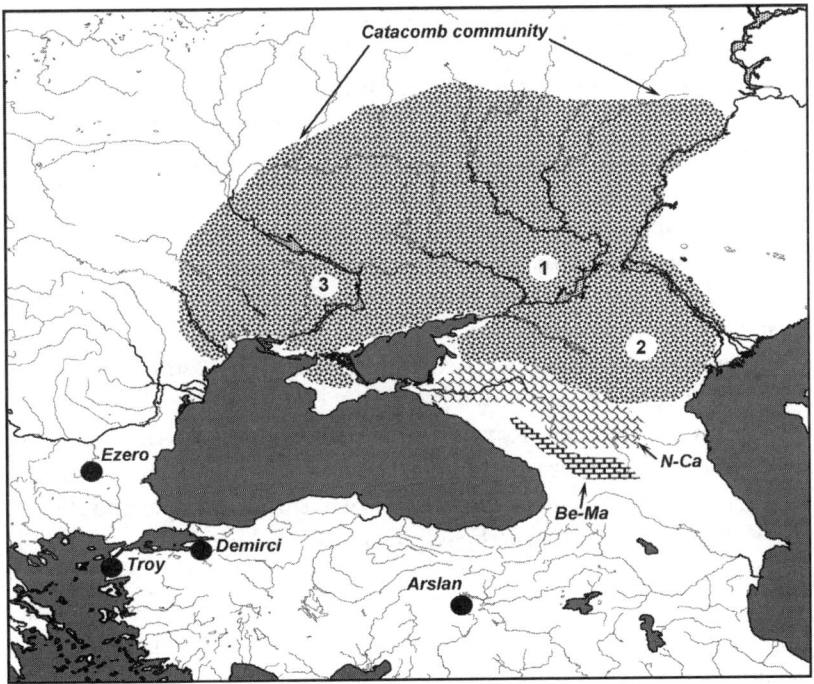

Figure 8.8. Map of the Catacomb archaeological community and other cultures (second phase of the CMP). Catacomb community: (1) Don-Donets basin; (2) Kalmykia; (3) Dnepr and Southern Bug basins. N-Ca = North-Caucasian archaeological community. Be-Ma = Bedeni-Martkopi kurgan culture.

this re-dating is that calendar ranges for a variety of cultures, in what seemed to be an unshakable archeological axiom, have changed. We thought it indisputable that more perfect forms and technology necessarily related to later times. According to our new data, however, this can no longer be accepted unconditionally. More technologically advanced ways of metal processing and technology can co-exist with less advanced ones. Confirmation of this phenomenon can be found in the Eurasian communities.[7]

In this section, we must make an essential distinction between the geographic distribution, influence, and inter-relation of the Yamnaya and Catacomb communities (Figs. 8.7 and 8.8). On the maps, it is clear that the known distribution of Yamnaya culture sites is greater. The huge latitudinal extent of this distribution from Pannonia (Ecsedi 1979) down to the southern Trans-Urals is not less than 3,000 kilometers (Fig. 8.7). The Catacomb community was settled in a much more compact territory and stretched from the Dniester down to the Middle and the Lower Volga region. The distribution of these sites from the west to the east did not exceed 1200–1400 square kilometers (Fig. 8.8).

Second, we need to pay attention to an important feature exclusive to the Yamnaya distribution – that is, that its spatial extent was amazingly far to the east and extended up to the Altai Mountains (Fig. 8.7). There, alongside the Afanasievo culture, the metallurgy of the Sayan-Altai mountain zone emerged and played a leading formative role in the metal production of the Late Bronze Age cultures of the steppe belt and in eastern Eurasian metallurgy as a whole.

The Yamnaya community is different from the Catacomb one also because it exploited its "own" copper ore sources. Archaeological research on the famous Kargaly ore field situated in the South Ural steppe periphery gave researchers graphic evidence of the mining and metallurgical production connected with the Yamnaya and Poltavka cultures localized in the northeast of the MP (Chernykh 2002: 128–139; 2005: 29–35; 2007: 57–70).

Unlike the extensive spread of the Yamnaya, the Catacomb culture group is more closely connected to the northern Black Sea area and especially to Caucasian metallurgy. The radiocarbon chronology suggests that the decline of the Catacomb community lasted for two to three centuries later than the Yamnaya. A very important conclusion follows from this; the "rolling away" of its eastern border marked the end of the Circumpontic province westward down to the Volga River (Fig. 8.8).

The Third Stage of Steppe Belt Formation: Eurasian and Eastern Asian Metallurgical Provinces

THE THIRD stage of steppe belt formation coincided with the transition into the Late Bronze Age (LBA). To a great extent, this period was crucial to the history of many Eurasian groups. The structure of the Eurasian world was formed and its basic features remained the same up to the turn of the first millennium BCE to CE. The territorial extent of metal-bearing cultures in Eurasia and its west, adjoining the Mediterranean Sea and including northern Africa, grew to a maximum of 40–43 square kilometers. Also, the domain of the archaeological cultures of the steppe belt reached its maximum extent of approximately 16–17 million square kilometers (Fig. 8.1).

One of the most important events of this period was the disintegration of the Circumpontic province and the development of new cultures. Basically, the CMP was a kind of primogenitor of the West-Eurasian model of mining metallurgical production. This period of the LBA was marked, however, by the beginnings of another model of this production, namely, the Eastern Eurasian one.

After the disintegration of the CMP, its northeast (eastern European) zone served as the base for the formation of a large Eurasian metallurgical province (EAMP). This block of related productive centers, except for eastern Europe, included the vast spaces of steppe and forest zones of Northwest Asia, and the majority of regions of Central Asia up to the deserts of the Kara Kum, the foothills of the Pamir-Tianshan, and even the region of Xinjiang (Chernykh 2007: 37–109). Its maximum territorial scope reached to between 7.5 and 8 million square kilometers. The chronological range of the EAMP was approximately 1,000 years – from the last centuries of the third up to the end of the second millennium BCE. In comparison to other provinces, the industrial centers of the Eurasian province probably preserved the basic morphological and technological standards of the CMP, although these standards had undergone some basic modifications during their development.

Within the EAMP, a huge system, which we call the East Asian metallurgical province, interfaced with the final formation of the steppe belt territorial contours. At this point, many spheres of the Eurasian province are studied in an incomparably more detailed way, whereas the East Asian province remains unknown to those who do not read Chinese.

Within the steppe domain and connected with borders of the EAMP, essential changes proceeded with the lifestyle of the cattle herders of the Bronze Age. Starting with the third and the second millennia BCE, a settled lifestyle began to supersede the nomadic and semi-nomadic one. Archaeozoological research shows that a mobile style of stockbreeding still remained (Antipina and Morales 2005: 41–42). Archaeologists have found traces of thousands of large and small settlements left by populations of the stockbreeders in these extensive spaces and have discovered that farming activities were not part of the lifeways of these communities (Lebedeva 2005). Cemeteries on the steppe belt are also not less numerous. Mounded kurgan burials, however, gradually gave way to flat cemeteries.

By the end of the second millennium BCE, stockbreeding activities again prevailed and the role of the nomadic and semi-nomadic way of life became stronger. The latest of these developments began to dominate in the Scythian-Sarmatian world and completely replaced the Late Bronze Age communities early in the first millennium BCE. The Scythian-Sarmatian world then renewed the LBA tradition of constructing large necropolises with massive "royal" kurgans that were furnished with precious inventories.

With the coming of the Late Bronze Age and the parallel formation of the Eurasian province came the discovery and the beginning of exploitation of huge amounts of copper and tin ores. This first took place in the Asian part of the province – stretching from the eastern Ural Mountains through Kazakhstan up to the Rudny or western Altai. At this time, the widespread production of tin bronzes also appeared. From this point on, the populations of the Eurasian province were able to meet completely the regional demands for metal production, and ties with the centers of the Caucasian metallurgy of the previous historical periods, during the days of domination of the CMP, stopped.

THE EURASIAN PROVINCE – EARLY PHASE: COUNTER WAVES OF CULTURES

Three phases can be distinguished in the Eurasian province. In the formative phase, the most essential phenomena include two countering and extremely swift waves of population movements that take place in northern Eurasia. Each movement was characterized by expressive but, at the same time, amazingly contrasting essential features.

The first wave moved from the west to the east and formed the large region of interaction known as the Abashevo-Sintashta, which included the well-known archaeological cultures of the Abashevo, Sintashta, and – it is reasonable to assume – the Petrovka culture, or the easternmost community. The distribution of these three cultures exceeded 1 million square kilometers, which encompassed a territory stretching from the

Upper Don basin and the forest Volga region down to the steppes and forest-steppes of western Siberia (Fig. 8.9).

Within this important development, a cultural continuity emerged. For example, it is hardly possible to draw a border between materials of the Abashevo sites, on the one hand, and the Sintashta, on the other. The same is true when trying to differentiate between attributes of the Sintashta and the Petrovka cultures. Metal items from these cultures are modifications of many categories of artifacts known in the disintegrated Circumpontic province.

The second wave from the east to the west was probably one of the most amazing phenomena in the ancient history of Eurasian communities. This has become widely known in the literature as the Seima-Turbino transcultural phenomenon, which takes its name from the cemeteries of Seima and Turbino (Chernykh and Kuz'minykh 1989; Chernykh 1992: 215–234). It can be argued that this phenomenon was rooted in a group of comparatively diverse (unfortunately not quite clear) cultures of the steppe, such as those from the mountain and even taiga zones of the extensive Sayan-Altai mountain area and other regions adjoining it mainly to the south and the west. The uniqueness of the antiquities from this area becomes obvious with the analysis of some key features.

The first is the absolute sudden and unexpected appearance of extreme-ly well-developed forms of metal weapons. Bronze Seima-Turbino

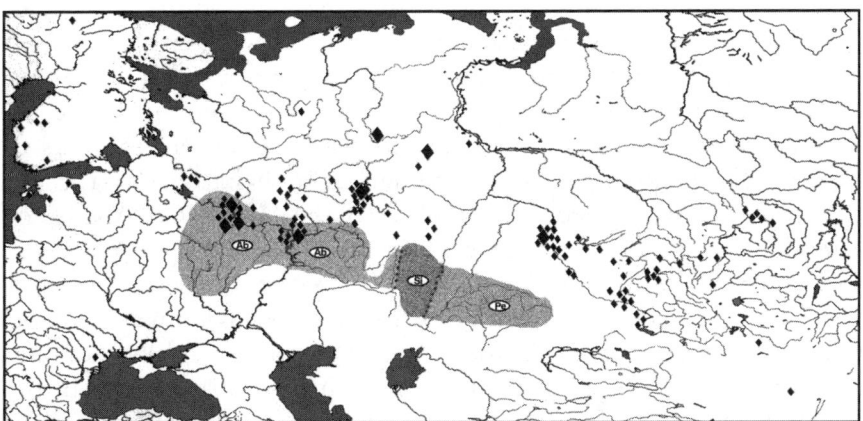

Figure 8.9. Map of the areas of the Seima-Turbino sites and metal (rhombic signs) and Abashevo-Sintashta archaeological community (the early formation phase of the Eurasian metallurgical province). Ab = Abashevo culture; Si = Sintashta culture; Pe = Petrovka culture.

weapons and the technology of their casting are absolutely different from those known in the west, within the CMP, and, in particular, from the Abashevo-Sintashta community to the east.

Second, the Seima-Turbino antiquities are interfaced with dissimilar types of sites, the Abashevo-Sintashta and the Petrovka settlements and necropolises. Settlements of this phenomenon are not known at all, and archaeologists have encountered primarily non-kurgan burials associated with these artifacts. In these necropolises, cenotaphs dominate, and graves are differentiated by sets of metal products and by stone inventories. The recovery of human remains in the Seima-Turbino cemeteries is rare, as is the association of ceramic utensils. A significant part of the metal finds known to us are connected to individual chance finds, even though they may be traces of destroyed tombs.

Third, it is not difficult to reconstruct the extremely swift movement of the military of the Seima-Turbino groups from the region of their concentration to the west. The richest set of finds are concentrated within the limits of the forest zone of eastern Europe, in the basins of the Upper Volga region and Kama (Fig. 8.9). Separate products of the Seima-Turbino type are known, however, up to the Baltic region and even to Moldova (i.e., the famous Borodino hoard). The distance between the furthest points of finds of metal products from the western – or even central – China up to the Finno-Scandinavian region exceeds 6,000 kilometers.

Both waves, western and eastern, suggest counter-courses for their movement and thus suggest parallel routes. The Abashevo-Sintashta sites were mainly spread in the forest-steppe region on the east side of the Urals and in the extreme south of the forest zone on the west side of the Urals. The Seima-Turbino metal shows us that its carriers either preferred the forest zone or were compelled to move into this safer environment. Nevertheless, between the two waves or movements one can clearly see evidence of contact. Some burials of obvious Seima-Turbino type can be distinguished at burial places of the Abashevo or Abashevo-Sintashta. As a rule, the later ones included Abashevo weapons and sometimes ceramics. It is also curious that burials of these communities follow Seima-Turbino mortuary ritual – that is, human remains were almost always absent in the graves. This fact establishes not only a relative but also an absolute chronology. It is remarkable that in the materials of the Abashevo-Sintashta necropolises traces of interaction, in reverse, cannot be found.

The calendar age of the Abashevo-Sintashta community is based on the analysis of 75 radiocarbon dates (Fig. 8.10). The distribution of

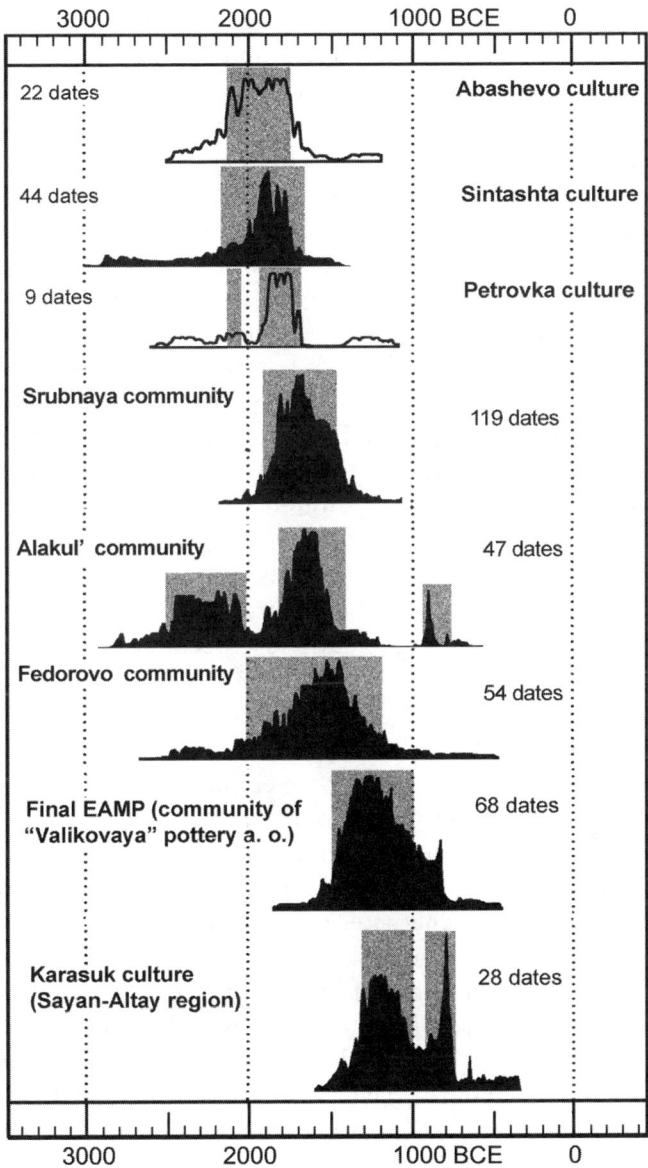

Figure 8.10. Sum probabilities of the radiocarbon dates of the stockbreeding archaeological cultures and communities of the Eurasian and East Asian (Karasuk culture) metallurgical provinces.

numbers for the three cultures is uneven: the Sintashta culture provides 44 dates, the Abashevo 22 dates, and the Petrovka 9. However, such distinctions have not affected the range of probability (68%). For example, the latter dates precisely coincide with each other – between the twenty-second and the eighteenth-seventeenth centuries BCE. Here it must be emphasized once again that the absence of appreciable difference between the chronological ranges of these three cultures that were removed from each other allows us to assume a quick distribution of the whole Abashevo-Sintashta community in an easterly direction.

Assigning radiocarbon dates for the sites of the Seima-Turbino type is more difficult. Today, only four dates are known to the author: one is from the burial of Satygha cemetery, found in the northern taiga and to the east of the low Ural ridges; and three others were recently found and investigated at the necropolis at Yur'ino (Ust'-Vetluga), which is situated in the basin of the Upper Volga. All four dates are within the limits established for the Abashevo-Sintashta chronological range.[8] Hence, we can with some confidence define the age of the Seima-Turbino phenomenon as similar to the Abashevo-Sintashta chronological range, or between the twenty-second and the eighteenth-seventeenth centuries BCE.

THE SECOND PHASE OF THE EURASIAN PROVINCE: STABILIZATION OF THE SYSTEM

The second phase of the EAMP development was characterized by a stabilization of the steppe belt cultures and communities. An appreciable fusion of the major features of the last cultures can be observed. In the enormous territories of the steppe belt provinces, uncountable and barely perceptible or characteristic settlement patterning is known. The same must be said about the infinite number of funeral sites, where there is little expression of social hierarchy among the deceased. This feature of the first phase of the Circumpontic province of the Early Bronze Age was completely erased. Because of the homogeneity of cultural remains, archaeologists have difficulty defining reliable territorial and chronological borders of this or that community. The framework of the previous ones frequently represents a rather blurred picture.

In the steppe belt, we most often pay attention to two enormous archaeological communities – the Srubnaya and Andronovo – and their remains have determined the character of the whole province (Fig. 8.11). The pattern of cultural continuity associated with the steppe belt in the second phase of the Eurasian province is striking. It is easy to see extensive zones of mixed character. For example, an extensive area began

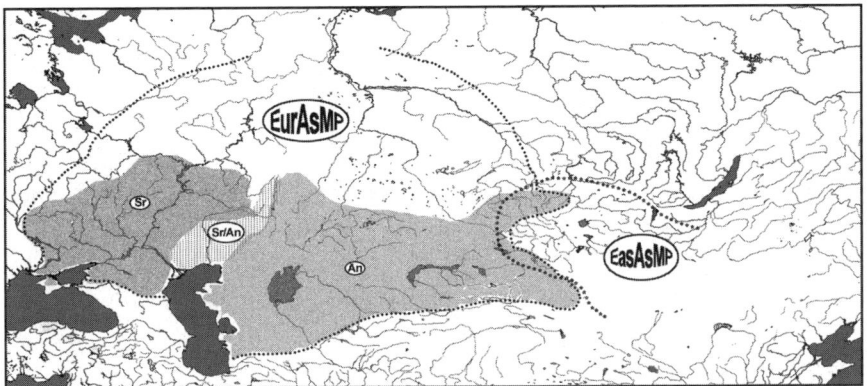

Figure 8.11. Map of the areas of major archaeological communities of the Eurasian metallurgical province (stabilization phase): Sr = the Srubnaya community; An = Andronovo community; Sr/An = mixed zone of Srubnaya and Andronovo communities.

to appear in the steppes and semi-deserts to the north of the Caspian Sea and stretched to the southern Ural forest-steppe (Fig. 8.11). On the whole, the degree of homogeneity of steppe cultures increased so greatly that at times archaeologists prefer to interpret this as a huge uniform Srubnaya-Andronovo community.

Western (or Srubnaya) sets of artifacts have been excavated mainly in eastern Europe in a zone of approximately 1.5 to 2 million square kilometers. The Eastern (or Andronovo) community is represented by two basic types of cultures: the Alakul' and the Fedorovo. The sites of the Eastern community are spread in a territory twice as large as that of the Srubnaya – or not less than 3 million square kilometers.

It is important to include in this development a number of smaller and localized cultures situated to the south in Central Asia near Kayrakkum, Tazabag'yab (Chernykh 1992: fig. 67). Archaeological remains in this region are very similar to those of the Alakul' culture. To the north, along the southern edge of the forest Eurasian zone, steppe-type communities were also situated, such as the Pozdniakovo and Prikazanskaya, which were similar to the Srubnaya; the Cherkaskul', which was similar to the Andronovo; and so on. Northern communities, as a rule, were marked by clear evidence of contact with steppe populations.

Regarding the question about the connection between Abashevo-Sintashta and the Seima-Turbino, discussed previously as counter-waves of interaction, it appears that the Abashevo-Sintashta became more dominant in the region through time. Evidence for this can be found in the overall decline of the Seima-Turbino pattern. Traces of its heritage remained only in the taiga zone of western Siberia in the settlements of the Samus'-Kizhirovo type (Chernykh and Kuz'minykh 1989: 144–162). Later, during the third EAMP phase, its influence became more noticeable and some of its characteristic features re-emerged. The initial western intrusion presented in these spaces by the Abashevo-Sintashta wave of movement obviously prevailed in the second phase.

The absolute chronology of this second phase is based on 220 calibrated radiocarbon dates. Most of them (119) are connected to materials recovered from Srubnaya contexts. The distribution of the sums of probabilities of this large sample is comprehensive (Fig. 8.10). The chronological range (68% probability) for the Srubnaya community is within 500 years from the twentieth to fifteenth century BCE. The distribution of the sums of probabilities for the basic cultures of the Andronovo community is, unfortunately, less reliable. It is especially difficult to interpret the corresponding frequency diagrams connected with materials of the Alakul' sites. In this case, a revision of the whole set of the materials included in the database is required to correct the problem. In comparison to the dates for the Srubnaya complexes, the frequency diagram of distribution of the sums of probabilities for the Fedorovo culture is much more ambiguous (Fig. 8.10). Most likely, these materials were influenced very strongly by the "pattern of cultural continuity" discussed previously for this time period.

THE THIRD PHASE OF THE EURASIAN METALLURGICAL PROVINCE: DISINTEGRATION OF THE SYSTEM

The dynamics of the development of the EAMP during the third phase in the second half of the second millennium BCE (Fig. 8.10) includes the transformation of cultures of both major communities. An extremely indistinct and amorphous community of cultures is identified with the so-called *valikovaya* pottery. These pottery assemblages always contain pots decorated under the rim, around the neck, or around the shoulders with a single applied cordon or band (valik). This detail is its main attribute. It can be found on utensils of various cultures spread across an amazingly wide space that stretches from the northern Balkans and Danube up to the Altai. The distribution of this ornamental pattern, most likely, should be connected to the western impulse, because in the

Carpatho-Balkan region this decorative element is consistently present on clay vessels dating from the Early Bronze Age.

There are fewer sites of this time in the steppe belt of the Eurasian province compared to the number in the previous phase, which includes both settlement and cemetery sites. The reason for this is most likely connected with the transition to a more mobile, nomadic, or semi-nomadic lifeway.

Forms of metallurgical and metal-processing production lose appreciably the initial features that are known from the previous period (Chernykh 1992: 235–252). The geographic framework of the Eurasian province began to diminish in distribution in both the West and the East. In the regions of the northern Balkans, Carpathians, and the Danube basin by this time were the productive centers of the European metallurgical province (Chernykh 1992: 252–263).[9] Western metal imports and the production of items that followed western models could be observed in the second half of the second millennium BCE far to the east and up to the Low and Middle Volga reaches (e.g., the Sosnovaya Maza hoard).

The eastern border of the Eurasian province, in the Altai region, is also difficult to define. On the eastern periphery of the Eurasian province, a considerable number of single-bladed knives alien to the province stereotypes was a distinctive feature of the industrial centers of the East Asian metallurgical province.

The explanation for the insignificant number of the radiocarbon dates known and registered by us, which total only 68 (Fig. 8.10), is the rather limited study of the final EAMP sites. Nevertheless, 68% of the probability of the sum of their probabilities confidently dates to the second half of the second millennium BCE. In all likelihood, this chronological range can be accepted because it compares to the dates of communities and cultures of the previous phase.

The Third Stage of Steppe Belt Formation: The East Asian Metallurgical Province

THE EAST Asian metallurgical province arises synchronously with the Eurasian one. Its major features and details are not studied thoroughly in Western languages. Here, I mainly touch on the northwestern East Asian metallurgical province zone located primarily within the limits of the Sayan-Altai mountain area and spaces surrounding it from the forested Mongolian plateaus to the stony spaces of the Gobi desert (Fig. 8.11).

The early phase of its beginnings was connected to the transcultural Seima-Turbino phenomenon mentioned previously. Its later stages can be characterized as a continuation of the Seima-Turbino traditions of metallurgy and metal processing. The most important materials of the stage are present in the sepulchral inventory of the widely known Karasuk culture (Chlenova 1972; Chernykh 1992: 264–271;[10] Legrand 2006). The recovery of metal objects is connected with graves that were disturbed.

The early Seima-Turbino and later Karasuk types of metallurgy are sufficiently inter-related. Attempts to reconstruct the dynamics of development of the metallurgy in the Sayan-Altai region are rare in languages available to me. The Seima-Turbino wave of population movement was definitely aimed in a western direction. Its chronological range, according to the known contacts of these populations with the Abashevo-Sintashta community, is limited to five centuries – from the twenty-second up to the eighteenth or seventeenth century BCE. The Karasuk complexes are not recorded in a series of radiocarbon dates but should fall within approximately a 500-year span, ranging from the last third of the second millennium BCE up to first third of the first millennium BCE (Fig. 8.10). The 300- to 400-year break between the Seima-Turbino and the Karasuk range dates is difficult to explain. Probably, only the appearance of new finds and a more detailed study of what has recently been recovered will stimulate progress on this issue.

There was a quick distribution of Karasuk forms toward the east, or the opposite of the Seima-Turbino movements to the west (Fig. 8.12). A significant number of imitations of the Karasuk metal forms are known in early dynastic China. In particular, these imitations are well presented even in the royal burial complexes of the cemetery at Anyang, which can be dated from the thirteenth to the eleventh century BCE, or the late Shang dynasty (Chang and Xu 2005: 150–176). It is probably from this time that an active opposition to the steppe world begins in ancient China. The Karasuk artifacts were undoubtedly left by nomadic cattlemen, but settlements of this culture are unknown in this region.

The morphology of Karasuk metal objects differs sharply from those of the ancient Chinese of the Shang and Western Zhou. Inhabitants of the Sayan-Altai included weapons, such as the well-known Karasuk curved, single-bladed knives with sculptural handles and, more rarely, daggers. The northern steppe (or taiga-steppe) forms are present at the Shang royal funeral complexes at Anyang.

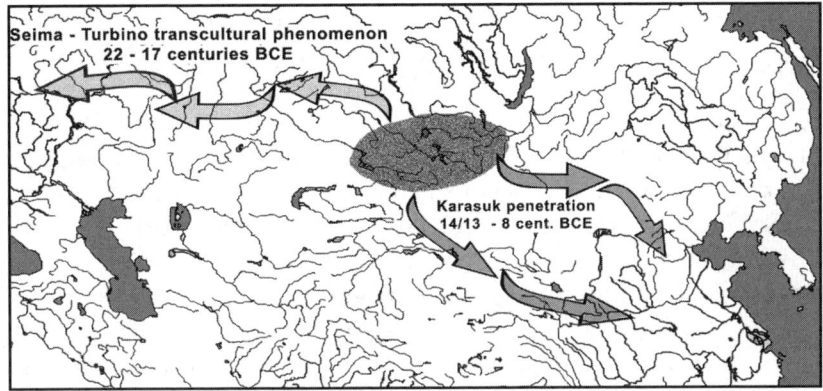

Figure 8.12. General ways of diffusion and penetration of Seima-Turbino and Karasuk metal forms.

Over the 3,500 kilometers (as the crow flies) from the known Karasuk sites to the Far East where Karasuk-type objects were found, the route was from the Sayan-Altai region through Xinjiang (Mei 2000, 2004), Mongolia (Erdenebaatar 2004), northern China (including Inner Mongolia; Linduff et al. 2000) up to the basin of Liao River and nearly up to Liaodong Gulf (Wagner 2006: 101–276). Another line of distribution of the steppe forms points a more southern and southeastern direction. Similar products are known in semi-arid and desert foothills of Altun-Shan and Shanxi-Shaanxi Plateau (Fig. 8.12). This area is adjacent to the sovereign territory of the Shang governors on the Yellow River (Chinese Archaeology 2003: 585–590).

One essential difference between the dynastic Chinese and steppe models is that most bronzes of the Shang and Western Zhou metallurgical production were made for sacral ritual purposes, whereras Karasuk metallurgy was used to make weapons (Chernykh 2007). The decoration of the curved knives with animal pommels or other characteristic ornamental patterns did not change the basic character of northern handicraft. The steppe founders and smiths produced many fewer plaques, pendants, and mysterious "bow-shaped objects" than weapons.

Final Remarks

BY THE second half of the second millennium BCE, the formation of a huge Eurasian steppe belt of cattle-breeding cultures was completed.

This process had a long trajectory of development and was character-ized by surprisingly powerful territorial "surges" as well as sudden declines. From this time, the "latitudal" structure of the basic types of the Eurasian cultures was formed. Until the turn to the first millennium CE, this structure shaped many major historical processes. The steppe belt served as a watershed between the cultural worlds of the southern and northern communities of the giant Eurasian continent.

Historical processes were often affected by the origin, existence, and destruction of communities and populations there. A "longitudinal" profile of the Eurasian world can also be formed, one that probably shows its effect on the formation of two metallurgical traditions – the West-Eurasian and the East-Eurasian. The great belt of steppe cultures often formed the bridge connecting these two dissimilar worlds.

An integrated system of statistically processed, calibrated radio-carbon dates has created a revision in the general picture of not only the absolute but also the relative chronology for various Eurasian cul-tures and communities. The many archeological postulates for relative chronological scales must be changed. It is clear, however, that discus-sion of these complicated problems cannot be solved within this single chapter. Rather, they will certainly demand focused new research in the future.

Notes

1. The limited length of this article dictates the presentation of the materials that are supported by summary diagrams of the distribution of the sums of proba-bilities of calibrated radiocarbon dates and general geographical maps. It is also necessary to concentrate only on key problems and archaeological communities, which may be covered within the space of this chapter.
2. However, this syndrome is inherent to no small degree, for example, in the cul-tures of the eastern European and western Siberian forest zones.
3. The debates reflecting disagreements of a similar sort can be found in a variety of books and articles. See, for example, Archaeology of Ukrainian SSR 1985: 204–205, 305–320; Chernykh 1991; Telegin 1991; Telegin et al. 2000; Kotova 2002: 5–11; and other works. The curious full report of materials from the northern Black Sea Coast necropolises of the fifth–fourth millennium BCE was published by J. J. Rassamakin (2004). On the basis of the analysis of one house and buri-als, the author allocated four groups of burial or funeral traditions. However, in my opinion, the four funeral traditions suggested by Rassamakin confirm this "syndrome of a cultural continuity" so characteristic of the block of the steppe communities.

4. It was necessary to select only the most essential dates for this chapter, but the complete set of radiocarbon dates for the whole CMP system is much larger.

5. Under the steppe Maikop, we mean kurgan complexes located mainly in the steppe zone to the north of the Kuban and Terek basins, between the Azov and Caspian seas. This is already out of the area of the "native" Maikop culture (Fig. 8.4). The inventory of the specified complexes contains characteristic objects, mainly ceramics, of Maikop types and styles.

6. We should notice that in the southern zone that it is nearly impossible to find anything equivalent to the set of metal products of the Maikop culture. The Nahal Mishmar hoard from Palestine (Fig. 8.4) is widely known, and its set of metal subjects and datings are atypical (Bar-Adon 1971). However, all of the 13 radiocarbon dates known to the author and connected with the various organics accompanying the hoard revealed an unexplained and rather impressive disorder of values: from 5000 up to 3500 years BCE – and this at 68% of probability.

7. This issue is touched upon here very briefly, and it cannot be expanded for discussion within the limits of this chapter. It certainly demands considerably more developed substantiation and systematic research.

8. Because of the scant number of the Seima-Turbino dates, the values of the last dates are not included in the diagrams (Fig. 8.10).

9. The western neighbor of the gigantic EAMP was the European metallurgical province. Distinctions between both systems are impressive, but we shall turn only to a pair of examples. Incalculable collections of metal, up to 80–90%, in the European province are concentrated in hoards. By contrast, the number of metal hoards in the EAMP is very limited. Besides, for example, all metal objects typical of the Seima-Turbino transcultural phenomenon are found distributed across some 1 million square kilometers – and are characterized by only 583 objects. Only one of the largest (most likely) hoards in the European metallurgical province is from Transylvania – Uioara de Sus – and contained more than 5,800 objects (socketed celts, sickles, arrow heads, ornaments, etc.) and weighed about 1,100 kilograms (Petrescu-Dîmboviţa 1977: 114–117).

10. In this book (Chernykh 1992), the eastern European is still called the Central Asian province. Now, however, I consider the term "eastern European" as more appropriate for that region.

References

Antipina, E., and A. Morales. 2005. Kovboi, Vostochnoevropeiskoi stepi v pozdnem bronzovom veke. OPUS, *Mezhdistsiplinarnye issledovaniya v arkheologii.* Vyp 4. Moscow: IA RAN, pp. 26–49.

Arkheologiya Ukrainskoi SSR. 1985. Tom 1. Kiev: Nauka.

Bar-Adon, P. 1971. The Cave of the Treasure. The Finds from the Caves in Nahal Mishmar. *Judean Desert Studies.* Jerusalem: Bialik Institute and Israel Exploration Society.

Chang, K., and P. Xu (eds.). 2005. *The Formation of Chinese Civilization: An Archaeological Perspective.* New Haven: Yale University Press.

Cernych, E. N. 1991. Frühestes Kupfer in der Steppen- und Waldsteppenkulturen Osteuropas. In J. Lichardus (ed.), *Die Kupferzeit als historische Epoche.* Bonn: Dr. Rudolf Habelt GMBH, pp. 581–592.

Chernykh, E. N. 1992. *Ancient Metallurgy in the USSR.* Cambridge: Cambridge University Press.

(ed.). 2002. *Kargaly.* Vol. 2. Moscow: Languages of Slavonic Culture.

(ed.). 2005. *Kargaly.* Vol. 4. Moscow: Languages of Slavonic Culture.

2007. *Kargaly.* Vol. 5. Moscow: Languages of Slavonic Culture.

Chernykh, E. N., L. I. Avilova, L. B. Orlovskaya. 2000. *Meallurgicheskie provintsii i radiouglerodnaya khronolgiya.* Moscow: IA RAN.

Chernykh, E. N., L. L. Avilova, L. B. Orlovskaia, and S. V. Kuz'minykh. 2002. Metallurgiya v Tsrkumpontiiskom areale: ot edinstva k raspadu. *Rossiiskaya Arkheologiya* 1: 5–23.

Chernykh, E. N., and S. V. Kuz'minykh. 1989. *Drvnya metallurgiya Severnoi Evrazii (Seiminsko-Turbinskii Fenomen).* Moscow: Nauka.

Chernykh, E. N., and L. B. Orlovskaia. 2007. Radiouglerodnaya Kronologiya Maikopskoi Arkheologicheskoi Obshnosti. In *Arkheologiya, Entografiya i Fol'kloristika Kavkaza. Materialy Mezhdurnarodnoi Nauchnoi Konferentsii.* Makhachkala: Izd. Dom "Epokha," pp. 10–28.

Chinese Archaeology. 2003. *Xia and Shang.* Archaeological Monograph Series, type A, no. 29. Beijing: China Social Sciences Press.

Chlenova, N. L. 1972. *Khronolgiya pamyatnikov karasukskoi epokhi.* Moscow: Nauka.

Comşa, E. 1991. Cucuteni und Nordpontische Verbindungen. In J. Lichardus (ed.), *Die Kupferzeit als historische Epoche.* Bonn: Dr. Rudolf Habelt GMBH, pp. 85–88.

Dzhaparidze, O. M. 1998. *K etnokul'turnoi istorii rguziinskikh plemen v III tysyacheletii do n.e. (Rannekurganskay kul'tura).* Tbilisi: Izdatel'stvo Tbiliskogo Universiteta.

Erdenebaatar, D. 2004. Burial Materials Related to the History of the Bronze Age in the Territory of Mongolia. In K. Linduff (ed.), *Metallurgy in Ancient Eastern Eurasia from the Urals to the Yellow River.* Lewiston: Edwin Mellen Press, pp. 173–188.

Formozov, A. 1965. *Kamennyi vek i eneolit Prikuban'ya.* Moscow: Nauka.

Jovanović, B. 1979. Stepska kultura u enolitskom periodu Jugoslavije. Praistorija Jugoslavenskih zemalja. *Eneolit.* Sarajevo: Akademija nauka i umetnosti Bosne i Hercegovine, pp. 381–396.

Kotova, N. S. 2002. *Neolitizatsiya Ukrainy.* Lugansk: Vidavnitstvo "Shlykh."

Lebedeva, E. 2005. Arkheobotanika i izuchenie zemledeliya epokhi bronzy Vostochnoi Evropy. OPUS, *Mezhdistsiplinarnye issledovaniya v arkheologii.* Vyp 4. Moscow: IA RAN, pp. 50–68.

Legrand, S. 2006. The Emergence of the Karasuk Culture. *Antiquity* 80(310): 843–879.

Linduff, K., R. Han, and S. Sun (eds.). 2000. *The Beginnings of Metallurgy in Ancient China.* Lewiston: Edwin Mellen Press.

Mei, J. 2000. *Copper and Bronze Metallurgy in Late Prehistoric Xinjiang: Its Cultural Context and Relationship with Neighboring Regions.* BAR International Series 865. Oxford: Archaeopress.

2004. Metallurgy in Bronze Age Xinjang and Its Cultural Context. In K. Linduff (ed.), *Metallurgy in Ancient Eastern Eurasia from the Urals to the Yellow River.* Lewiston: Edwin Mellen Press, pp. 173–188.

Munchaev, R. 1975. *Kavkaz na zare bronzovogo veka.* Moscow: Nauka.

Petrescu-Dîmbovitsa, M. 1977. *Depozitele de Bronzuri din România.* Bucharest: Editura Academiei RSR.

Rassamakin, Yu. 2004. Die nordpontische Steppe in der Kupferzeit. Gräber aus der Mitte des V. Jts. bis Ende des 4. Jts. Chr. Teil I – Text. Teil II – Katalog und Tafeln. *Archäologie in Eurasien.* Bd. 17. Mainz: Verlag Philipp Von Zabern, pp. 254. (I) and 278 (II).

Rezepkin, A. D. 2000. Das frühbronzezeitliche Gräberfeld von Klady und die Majkop-Kultur in Nordwestkaukasien. *Archäologie in Eurasien.* Bd. 10. Berlin: Verlag Marie Leidorf GmbH, 73, pp. 85.

Ryndina, N. V. 1998. *Drevneishee metalloobrabatyvaushee proizvodsvo Yugo-Vostochnoi Evropy.* Moscow: Editorial URSS.

Telegin, D. J. 1991. Gräberfeld des Mariupoler Typs und der Srednij Stog-Kyltur in der Ukraine (mit Fundortkatalog). In J. Lichardus (ed.), *Die Kupferzeit als historische Epoche.* Bonn: Dr. Rudolf Habelt GMBH, pp. 55–84.

Telegin, D., N. N. Kovaliukh I. D. Potekhina, and M. Lillie. 2000. Chronology of Mariupol Type Cemeteries and Subdivision of the Neolithic–Copper Age Cultures into Periods for Ukraine. In *Radiocarbon and Archaeology*, vol. 1. Saint Petersburg: Thesa, pp. 59–74.

Todorova, H. 1999. Die Anfänge der Metallurgie an der westlichen Schwarz-meerküste. *Der Anschnitt* 9: 237–246.

Wagner, M. 2006. Neolithikum und Frühe Bronzezeit in Nordchina vor 8000 bis 3500 Jahren. *Archäologie in Eurasien.* Bd. 21. Mainz: Verlag Philipp von Zabern, pp. 1–355.

CHAPTER 9

Late Prehistoric Mining, Metallurgy, and Social Organization in North Central Eurasia

BRYAN K. HANKS

I N RECENT years, one of the most vibrantly debated prehistoric developments in the Eurasian steppe region has been the emergence of the Sintashta culture, which is dated from 2100 to 1800 cal. BCE (G. B. Zdanovich 1988, 1989). The contributions in this volume by Anthony and Frachetti stress the importance of the Sintashta case study, and Anthony, in particular, argues for a strong connection between changing environmental conditions, an increase in the scale of warfare and chiefly competition, and the emergence of a broader economic pattern of exchange between Central Asia and the Sintashta settlements in the southern Urals zone.

This chapter investigates in more detail the relationship of Sintashta social organization to mining and metals production in that region. Many uncertainties still surround the scale and nature of bronze production connected with the emergence, development, and decline of the Sintashta archaeological pattern. These important issues are part of a much broader problem that exists within contemporary studies of the steppe region concerning the relationship between metal-producing communities, local micro-regions, and larger supra-local systems of trade and exchange. Traditional views of Eurasian metallurgy have often highlighted the development and diffusion of metal technology and objects within larger regional networks. In contrast, this chapter emphasizes the need to better understand the micro-regional impact of metallurgical production and its effect on the development of specific forms of social organization and complexity. From this perspective, it is argued that the Sintashta pattern still remains enigmatic in terms of the actual characteristics of its social complexity and the economic system connected with its development. It is also suggested that new conceptual models drawing on recent anthropological approaches to mining and metallurgy, coupled with the application of new multi-disciplinary field methods and data collection, must be employed before an improved understanding of the Sintashta emergence and decline can be achieved.

Modeling Early Metallurgy and Social Organization

ONE OF the primary themes of this volume is the evaluation of comparative models for the study of new patterns of social complexity in the Eurasian region. The role of early metals production and consumption is a crucial element in such studies. Traditional approaches to the study of metallurgy, stretching back to the broad syntheses of Childe (1936, 1944), were based on models of unilinear evolution and focused on the origins and diffusion of metallurgical industries as being connected with specific stages of social development. Through time, and as interest shifted toward the independent innovation and development of metallurgy in Europe, scholars began investigating in more detail the technological aspects of early mining and smelting, networks of trade, and socio-political control over metal commodities.

However, although many studies have contributed to our understanding of early metal technology, some scholars have argued that not enough attention has been focused on the social nature of the communities that were engaged in the extraction of ores and the production of metal commodities (Knapp et al. 1998; Gilman 1981; Shennan 1998; Topping and Lynott 2005). To be certain, metal production has been routinely associated with the emergence of hierarchy within and between early complex societies, but some scholars have suggested that too great an emphasis has been placed on modeling the elite control over such systems (Shennan 1993, 1998; Yener 2000). This reaction reflects an important transition that has occurred recently in the study of the social organization, settlement patterning, and social networking connected with early mining and metals production. With increasing frequency, scholars have emphasized the importance of examining the unique social and spatial characteristics connected with these activities from a broader anthropological perspective – one that includes not only the technological aspects of metallurgy but also the socio-economic, spatial, and ideological dimensions of the communities engaged in the production rather than just the consumption of metals (Godoy 1985; Knapp et al. 1998; Kassianidou and Knapp 2005; Levy 1993, 2003).

These issues have specific relevance for understanding the nature of metals production and consumption in prehistoric Eurasia, including the enigmatic Sintashta archaeological pattern. Although the nucleated, fortified nature of Sintashta settlements and the complex mortuary deposits are well represented archaeologically, an actual understanding of how these settlements functioned socio-economically, either as part of a larger regional system or as autonomous polities, has not been achieved. The remainder of this chapter examines specific issues connected with what is known about the extraction, production, and trade of ores in the southern Urals region and how these activities connect with regional settlement patterning and the social organization of Sintashta societies.

Modeling the Sintashta Archaeological Pattern

ARCHAEOLOGICAL INVESTIGATIONS of Sintashta sites have been ongoing since the 1970s, and twenty-two fortified settlements with Sintashta culture phases have been identified in an area 400 by 150 kilometers in the southeastern Ural Mountains of Russia (Fig. 9.1). The limited regional distribution of these sites and their seemingly high level of nucleated organization have led to this pattern being called the "country of towns" (G. B. Zdanovich and Zdanovich 2002).

Russian archaeologists have conducted basic aerial surveys of these settlements and carried out partial excavations at nine. Excavations at cemeteries associated with the settlements have yielded complex multi-phase burial constructions and some of the earliest evidence for light chariot technology in the Eurasian steppe region (Anthony and Vinogradov 1995; Anthony, chapter 4 in this volume). In contrast to the limited published data on settlement excavations, several monographs have been produced from cemetery investigations (e.g., Epimakhov 2002b, 2005; Gening et al. 1992; Vinogradov 1995, 2003). Furthermore, several recent publications in English have provided important summaries of the culture history, archaeological study, and various interpretations of the Sintashta development (Anthony 2007; Jones-Bley and Zdanovich 2002; Kohl 2007; Koryakova and Epimakhov 2007).

In recent years, as theories have been put forward to characterize the Sintashta culture, a wide variety of terminology has been introduced. As Koryakova recently noted (2002: 107), it has been perceived as: "1) a middle-scale society, based upon a proto-city structure and social

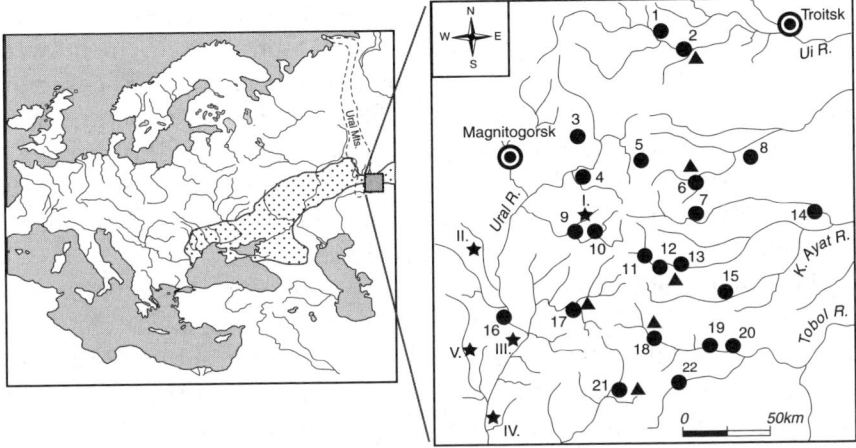

Figure 9.1. Map of southeastern Ural Mountains region indicating sites associated with Sintashta culture developments: Settlements: (1) Stepnoe; (2) Chernorech'ye; (3) Shikurtay; (4) Bakhta; (5) Parizh; (6) Ust'ye; (7) Rodniki; (8) Chekatai; (9) Kuisak; (10) Sarym-Sakly; (11) Konoplyanka; (12) Zhurumbai; (13) Ol'gino; (14) Isiney; (15) Kamysty; (16) Kizilskoe; (17) Arkaim; (18) Sintashta; (19) Sintashta 2; (20) Andreyevka; (21) Alandskoye; (22) Bersuat. Triangles represent investigated cemeteries. Ancient Mines: (I) Vorovskaya Yama; (II) Bakr-Uzyak; (III) Sokolki; (IV) Ishkino; (V) Ivanovka Dergamysh.

organization (Zdanovich 1997), 2) a simple chiefdom (Beryozkin 1995; Epimakhov 2002a, 2002b), 3) a complex chiefdom (Koryakova 1996), 4) a proto-urban or proto-state society (G. B. Zdanovich 1995; G. B. Zdanovich and Zdanovich 2002), and 5) a variant of a theocratic society (G. B. Zdanovich 1995)."

All of these models suggest that Sintashta societies were based on some form of complex economic organization and political leadership that was structured according to institutionalized hierarchy. Such a view seems in part to be substantiated by the complex nature of Sintashta cemeteries. For example, the presence of prestige items such as stone mace heads, interment of some individuals with chariots, lavish animal sacrifice, and the inclusion of weaponry seem to indicate variation in wealth and funerary treatment. However, other scholars (Koryakova and Epimakhov 2007; D. G. Zdanovich 2002) have argued that the presence of social stratification is much more difficult to determine from the variability seen within Sintashta mortuary ritual.

Rough demographic calculations of the settlements suggest that at best only about one-third of the people inhabiting Sintashta settlements were buried within kurgan complexes (Epimakhov 2002a, 2002b). Such factors have challenged clear interpretations of Sintashta social structure based solely on the mortuary evidence. Furthermore, the variability present in mortuary ritual, coupled with the uniform, "egalitarian" nature of the households within the fortified settlements, does not strongly support a "chiefdom-like" model for Sintashta societies.

The use of such analogous terminology, however, is symptomatic of the larger problem that exists within Sintashta studies, that is, the desire to find a model that "fits" most closely with the archaeological evidence rather than striving to gain a better understanding of the many social and economic variables that have often been at the core of comparative studies on social complexity. Such variables include the nature of craft specialization, settlement patterning, storage and redistribution, proximity to environmental resources and trade routes, and the relationship between subsistence and productive economies, just to name a few. I would argue that more intensive research in any one of these areas would lead to a better understanding of how and why Sintashta societies emerged, developed, and declined in the ways that they did. I would also argue that *no single model* might account for the Sintashta pattern as it is observed. Just as Anthony has suggested in this volume, the Sintashta pattern is a multi-variable phenomenon wherein climate, social conflict, metallurgy, and regional and inter-regional interaction played significant roles. More work must be done in all of these areas in the future to improve our understanding of this development.

As an example, the role of climate change is an important one. However, even though a widespread climate change has been documented for much of Eurasia after 2500 BCE, numerous specialists have suggested that more detailed micro-regional climate studies must be undertaken in order to understand how broader climate trends relate more specifically to micro-regional climate change and/or stability (Blyakharchuk et al. 2004; Peck et al. 2002; Tarasov et al. 1997). At present, such detailed studies are not available for the Sintashta area, and therefore local variation in environment and ecology must be more firmly established. Such environmental studies must also overlap with local catchment studies that establish proximity to natural mineral resources and ancient forest cover for fuel and settlement construction. All of these factors were undoubtedly an important part of the differential resource distribution that influenced Sintashta settlement patterning and longevity of occupation in certain zones.

The prevalence of warfare among Sintashta communities also seems a very well-established trait for this culture. The ubiquitous nature of settlement fortifications, many with episodic evidence of fire, the inclusion in graves of weaponry fashioned from bronze and stone, and the appearance of spoke-wheeled chariot technology all seem to support a high level of endemic regional warfare. With such archaeological evidence, it is therefore very surprising that *little to no* trauma has been observed during the analysis of human remains recovered from Sintashta mortuary sites. For example, a recent detailed bio-archaeological analysis of 118 human remains from the site of Kamennyi Ambar 5 (Judd, Kovacik, and Rajev 2008) revealed no evidence of trauma connected with inter-personal violence or injury. The analysis of 31 human remains from Kurgan 25 at the Bol'shekaraganskii cemetery, which is associated with the Arkaim settlement, also has not indicated any pattern of inter-personal violence (Lindstrom 2002). This lack of correspondence between direct and indirect evidence of warfare is a vexing problem, and even though it is certain that the potential for violence did exist, the frequency and scale of such confrontations is not well understood. In fact, it is not known if Sintashta communities fortified their sites against other fortified Sintashta settlements or against other cultural groups occupying the region. The fact that these questions cannot be answered is directly related to our lack of understanding for how Sintashta settlements functioned within the region and whether stress existed over access to and control of grazing lands, ore and timber resources, and trade.

Many of these questions seem tied in some way to Sintashta metal production. It is often assumed that because of the ubiquitous recovery of evidence for metallurgy (slag, metal droplets, furnaces, etc.) from settlements that Sintashta communities undertook a high scale of bronze production. As Anthony argues in this volume, this may have been one of the main motivating factors for inter-regional trade between the southern Urals and Central Asia. Unfortunately, our present understanding of Sintashta metallurgy is also not well substantiated because of several questions surrounding the nature of mining and production of metals in the region. For example, what was the actual scale of bronze production at Sintashta sites? Were all fortified settlements undertaking the primary processing of copper ores, or were those settlements closest to the ore resources functioning in this way? If there were hierarchies of access and control over such resources, were bulk ores and metal ingots being traded between settlements in the Sintashta region for later processing, secondary smelting, and refinement? How long did timber and ore resources last in any one given area? Did the exhaustion of these resources force certain Sintashta communities to move to new catchment zones, thus creating conflict over local resources?

All of these questions connect with the need to form more-explicit strategies for the investigation of Sintashta mining and metal production and related socio-economic patterning, especially because these factors appear to have a substantial relationship to both the archaeology of the settlement sites and the interpretation of Sintashta social complexity.

Sintashta Metallurgy

THE URAL Mountains are one of the key geological resources within Eurasia, as they contain an abundance of mineral wealth, including rich deposits of oxidized sulfide ores with copper carbonates. Such mineralogical deposits were crucial for the exploitation of copper and iron ore deposits in later prehistory (Chernykh et al. 2000; Chernykh 1992; Grigor'yev 1999).

One of the most intriguing aspects of Sintashta settlements is that nearly all the households excavated within the fortified zones yield evidence of one or more wells or cisterns and what are believed to be cupola-shaped furnaces connected with smelting activities (Gening et al. 1992; G. B. Zdanovich and Zdanovich 2002; G. B. Zdanovich 1997). Grigor'yev has argued that the use of the wells in combination with the furnaces provided increased draft (cold air induction) for the smelting process (Grigor'yev and Rusanov 1995). Other objects connected with copper smelting, such as pestles, slag, and metal droplets, also have been routinely recovered from households. The widespread nature of these finds suggests that metallurgy was not an activity undertaken by different individuals or groups within the fortified zones but was instead practiced in all households (Epimakhov 2002a, 143). While there is evidence for metallurgy within the household contexts, there has been great debate over the actual scale of production. Some scholars have suggested a high level of specialized community production (Vinogradov 2004; D. G. Zdanovich 2002), and others have argued for a lower level of household production for local utilitarian items and weaponry (Grigor'yev 2000; Chernykh 2004b).

Another significant issue surrounding Sintashta metal production is the lack of evidence for mining or quarrying near the settlements. This has led to speculation over whether Sintashta societies exploited local ore resources for their immediate needs or traded bulk, or partially reduced, copper ore or processed metals (ingots) for later on-site smelting and refining by other groups. This type of trade with widely dispersed small-scale metal production has been discussed in detail recently for the Late Bronze Age in the southwestern Urals, and it is clear that such strategies existed among Bronze Age communities in the steppe region (Peterson et al. 2006, 325; Peterson, chapter 11 in this volume; Diaz del Rio et al. 2006).

Geologists and archaeologists have undertaken several systematic geological surveys in the Sintashta region, and the results of these studies have been published (Zaikov et al. 1995, 2000, 2002). These surveys have identified several ancient mines in the southeastern Urals that are believed to have been important sources of arsenical copper for Sintashta communities (Fig. 9.1). However, all of these sites are situated on the western and northern periphery of the Sintashta settlement distribution, and therefore the cost-benefit of accessing ore resources at these sites is an important question. One would expect that proximity to ore resources would have been a major consideration, especially given the nucleated and fortified characteristics of Sintashta settlements that suggest conflict between groups.

The examination of many of these mines has shown extensive Late Bronze Age activities, which has largely destroyed any evidence of earlier Middle Bronze Age exploitation by Sintashta or other cultural groups. The site of Vorovskaya Yama is a good example of this. Located on the southwestern edge of the Sintashta settlement distribution, the site is situated 6 kilometers from Kuisak, 10 kilometers from Sarym-Sakly, and 40 kilometers from Arkaim (Fig. 9.1). It would therefore have been a potentially important site of ore extraction for the communities occupying these settlements.

The amount of ore extracted from this open-pit feature is calculated to have been 6,400 tons with a possible 2% copper concentration. This estimate is based on the geological survey and archaeological trench excavations of the large surface pit feature at the site, which measures 30–40 meters in diameter and 5 meters in depth (Fig. 9.2). Trench excavations at the site revealed a few pottery sherds, approximately 20 domestic animal bones (cattle, horse, and sheep), and stone tools (Zaikov et al. 2000). The pottery sherds recovered are representative of Late Bronze Age cultural types, specifically Alakul' and Kul'sko-Srubnaya, and are believed to be the traces of a temporary work camp at the site. No Sintashta ceramics were recovered in the trench excavations. Therefore, while Vorovskaya Yama clearly represents an important mining site for the Bronze Age,

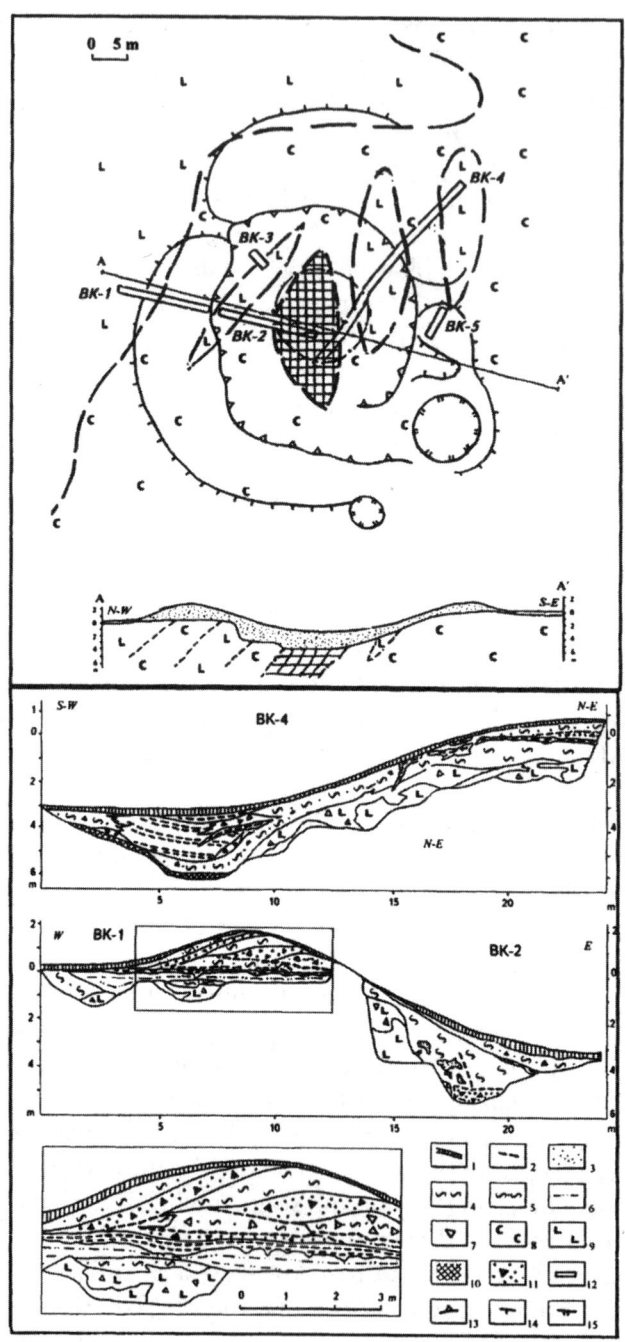

distinguishing how much of it was connected with either Middle Bronze Age or Late Bronze Age exploitation has not been determined. This problem exists for many of the other mining sites where evidence of Late Bronze Age mining has overshadowed earlier exploitation.

The current lack of evidence for intensive mining by Sintashta groups also has a bearing on estimating the scale of production for bronze metals. As noted, scholars are divided over the quantity of metal actually being produced by these societies. This problem is in part accentuated because of the lack of publication of excavated Sintashta settlements. Detailed monographs on the excavations of Sintashta, Arkaim, Ust'ye, and other settlements with Sintashta cultural features have not been produced, and this restricts the possibility for a comparison of the archaeological and archaeometallurgical evidence.

In contrast, recent work at the Late Bronze Age complex of Kargaly by Chernykh and colleagues has been published in several volumes (Chernykh 2002a, 2002b, 2004a). This research has revealed the existence of a very specialized metal-producing site in the southwestern Urals. Chernykh has compared the archaeological evidence for mining and metal production at the settlement of Gorny (part of the Kargaly mining complex) with available published data from the Sintashta settlement of Arkaim (Table 9.1). As these data suggest, there is a marked difference in the amount of materials being excavated from these two sites in terms of metallurgical debris, faunal remains, and ceramic vessels. This is rather startling, as the cultural area excavated at Arkaim is 10 times that of Gorny, yet Gorny has produced a much higher density of material culture.

Recent excavations by Vinogradov at the settlement site of Ust'ye have revealed a substantial density of materials connected with what is a preliminary phase of Sintashta settlement and later Petrovka habitation features (Fig. 9.3). A total of 1,200 objects linked with metallurgy (copper ore, slag, droplets, ingots, etc.) have been recovered from an excavated

---

Figure 9.2. Plan and profile views of the Vorovskaya Yama mine: (1) modern topsoil level; (2) layers of buried soils; (3) drifts; (4) clay soils; (5) loam soils; (6) sandy loam; (7) grass and rock debris; (8) serpentinites; (9) basalts; (10) rodingites with fluccans of copper ores; (11) rock debris from rodingites; (12) trenches; (13) contours of the open pit; (14) contours of the spoil; (15) contours of house depressions (after Zaikov et al. 2002: 429).

Table 9.1. Comparison of archaeological materials from the middle
Bronze Age settlement of Arkaim and the Late Bronze Age of Gorny

Comparative items	Arkaim	Gorny
Total site area in hectares	1.7–1.75	3.5–4.0
Excavated area (square meters)	8,055	892 (main site)
Ceramic sherds: reconstructed vessels	9,000:304	~11,000: 755
Copper samples: objects	?: 15?	3,131: ~400
Casting molds	?	172
Stone hammers	~40	1,184
Slag (pieces)	?	4,416
Animal bone fragments: artifacts and half-finished items	1,1834: ?	>2,250,000: 18,000

Source: After Chernykh 2004a: 235.

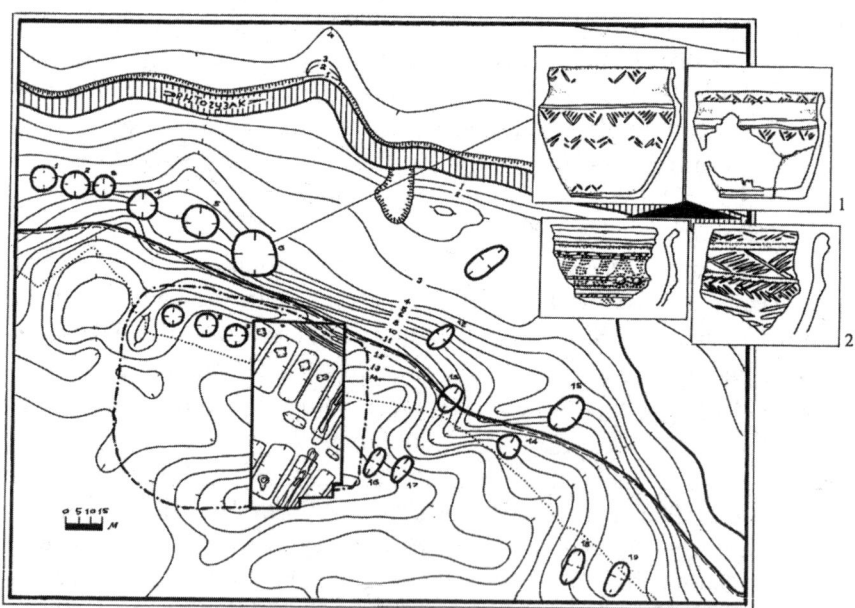

Figure 9.3. Plan of Ust'ye settlement indicating earlier Sintashta settlement and later Petrovka house depressions – with associated pottery
types (after Vinogradov 2004: 266).

area of 3,000 square meters (Vinogradov 2004: 267). The forthcoming publication of the Ust'ye excavations will undoubtedly have an impact on our current understanding of Sintashta metal production, particularly the nature of the relationship between the Sintashta and Petrovka phases represented at the site (N. B. Vinogradov personal communication).

As noted, interpretations of the actual scale of production by Sintashta communities are severely hampered by the lack of available published evidence. Site catchment analyses also have not been systematically undertaken in many cases, and there is a strong possibility that the exploitation of surface outcroppings near the settlements have gone unrecorded. This overlaps importantly with Grigor'yev's observation that Sintashta metallurgists appear to have smelted ores principally from ultra basic ore-bearing rocks and serpentine, which contain lower concentrations of copper (2000: 143). This, as Grigor'yev suggests, is a "curious paradox of Sintashta metallurgy," especially because other deposits that are rich in copper exist within the southern Urals region (ibid.). Such strategies seem out of place if one considers the "industrial" scale model of bronze production by Sintashta communities.

The typological and spectrographic analysis of bronze objects produced by Sintashta populations has formed an important component of study for the southern Urals Bronze Age, and scholars have argued for a clear connection between Sintashta groups and the Abashevo (Balanbashevo), Potapovo, and Seima-Turbino cultural groups in the Urals region and the western Siberian forest zone (Degtyareva 2006; Zaikov et al. 2002). Similarity in the types of weapons and tools recovered from these regions has led to active debate over regional production characteristics and the scale and nature of inter-regional trade and the diffusion of technology and ideas. The spectrographic analysis of objects, however, cannot always account for prehistoric practices of smelting and recycling, as indicated in comments on the study of early metallurgy in the Eurasian steppe region (Kohl 2007, 166; Cernykh 2003). Such problems do not, of course, discount the value of undertaking such analyses, as large-scale analyses by Chernykh and colleagues have provided a substantial amount of data for understanding the emergence, development, and transmission of metallurgical technology within the Eurasian region (Chernykh and Kuz'minykh 1989; Chernykh 1992). Nevertheless, the analysis of objects must go hand in hand with more-detailed studies of slags and furnaces within Sintashta house features and more-comprehensive surveys of local catchment zones surrounding Sintashta settlements.

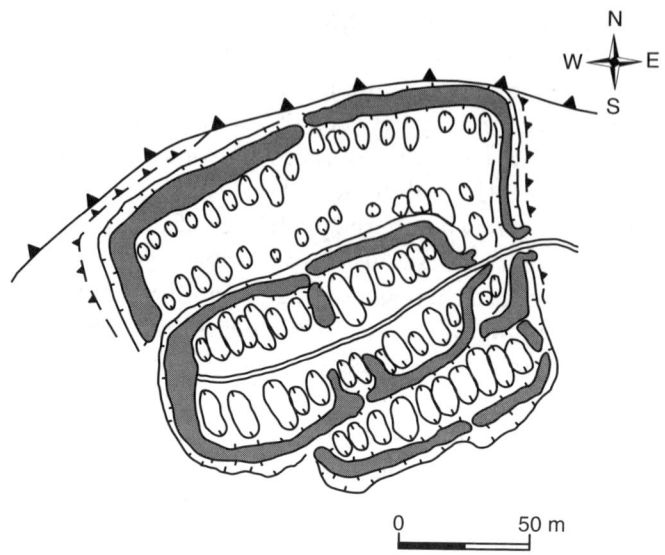

Figure 9.4. Plan of Andreevskoe settlement illustrating circular house depressions with fortification bank (*shaded*) and ditch complexes (adapted from G. B. Zdanovich and Batanina 2002).

While scholars have actively supported the theory that Sintashta societies acted as autonomous polities, there are many details about the variation in settlement patterning and the individual sites themselves that deserve more attention. For example, Sintashta settlements are not distributed evenly throughout the southeastern Urals region. Some sites are grouped within 8–15 kilometers of each other along the same river courses (Fig. 9.1). Moreover, many of the sites show clear evidence of being enlarged to accommodate additional house structures, and fortifications were extended to protect these new additions (Fig. 9.4). Could the initial phase of some of these settlements represent the "colonization" of new catchment zones for ore and timber resources by family groups or specialists from settlements? Or could joint efforts in the exploitation and processing of copper ores explain the grouped nature of settlements in certain areas? On the basis of available evidence, such questions cannot yet be answered for the Sintashta region.

It may be productive to draw inspiration from comparative studies on mining societies in other regions of the world in order to develop a more-specific strategy for field research in the Sintashta region. Such studies have sought to develop more-nuanced understandings of ancient

COPPER WORKING PROCESS

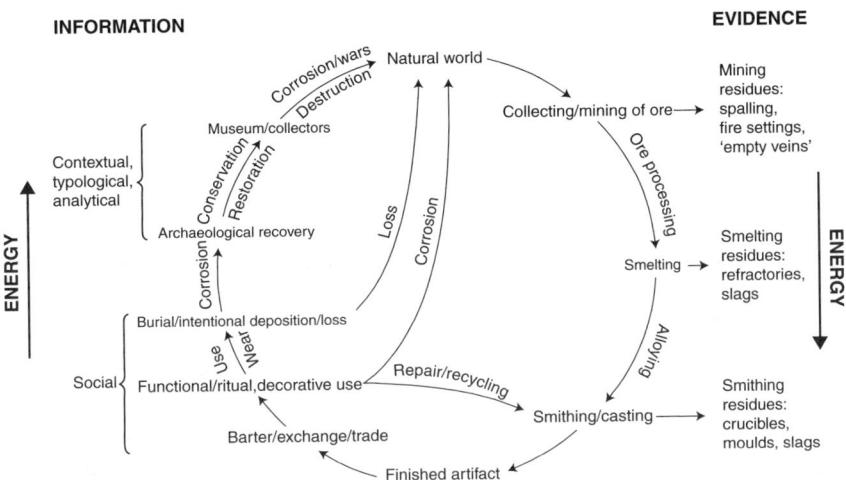

Figure 9.5. Cycle of copper production and working (redrawn from Ottaway 2001: 88).

metallurgy as part of complex social processes including "community" organization, craft specialization, and supra-local networks of interaction (Knapp et al. 1998; Shennan 1998).

New Approaches to Sintashta Archaeology

THERE IS a vast body of literature on the technological nature of metal production, but far less has been published on metal production as a social phenomenon. As Ottaway (2001: 90) has outlined, the whole process of production, starting with the prospecting of metalliferous copper ore to the final finished metal object, includes "prospecting, mining, beneficiation, smelting, or roasting *and* smelting, refining, alloying, casting and smithing." All of these activities must be seen as part of a much larger cycle of copper production with specific sequences that have crucial spatial, technological, and social contexts that must be considered (Fig. 9.5). The various stages of the copper production cycle may also lead to a segmented network wherein different individuals, groups, or communities play a role in how metals and metal objects are produced, recycled, and re-produced.

The chapter by Peterson in this volume, based on his detailed metallurgical analyses and regional survey, includes a flow chart for the various stages in the process of metal production and recycling for the Late Bronze Age in the Samara Valley region. Such charts can be very useful for modeling the spatial, technological, and social aspects of metals production. However, the creation of such a flow chart for Sintashta metallurgy is at present impossible because of the lack of data needed to identify most of the stages outlined here. For example, very little concrete evidence for prospecting, mining, and beneficiation has been detected thus far, and archaeological data to differentiate roasting and smelting processes from other stages of refinement have not been clearly identified. Although most of the evidence recovered in Sintashta household contexts would appear to relate to refining, smelting, and smithing, the recovery of tuyeres, crucibles, and form molds is rare. This puzzling lack of evidence challenges the homogeneous model for substantial metal production among Sintashta settlements. Rather, a more-complex pattern is likely to have existed with much greater variation between the settlement communities.

This heterogeneous view fits well with Frachetti's discussion in this volume concerning "non-uniform complexity" for Bronze Age pastoralist societies. As he suggests, the nature of variability in economic and sociopolitical strategies produced cycles of growth and decline among populations. In this way, it may be helpful to consider the Sintashta pattern as part of longer-term scalar change within the southeastern Urals region. Recent calibrated radiocarbon dating has substantiated the chronological relationship between the Sintashta phase and later Petrovka, Alakul', and Srubnaya cultures, which are often subsumed under the broader category of "Andronovo Horizon" (Hanks et al. 2007).

The fact that many Sintashta settlements, such as Ust'ye, have cultural levels connected with Late Bronze Age phases (Petrovka, Alakul', etc.) suggests that long-term occupation of previous Sintashta settlement zones existed during the latter half of the second millennium BCE. However, the pattern of settlement was different with a marked decline in emphasis on fortification and circular planning. Also, where metallurgical analysis has provided data for comparison, there appears to be a difference in the composition of metals being produced, as objects from the Late Bronze Age include pure copper, tin bronzes, and arsenical bronzes (Koryakova and Epimakhov 2007: 83). This change in technology would appear to reflect the obtainment of new ores and the development and diffusion of new techniques for smelting and alloying (Grigor'yev 2000: 143). During the Late Bronze Age, the nature of interaction and trade changed remarkably (Frachetti 2002, 2004;

Frachetti, Chapter 3 in this volume). Many sites within the southern Urals, western Siberia, and Kazakhstan show an aggregation of differing pottery types, non-local forms of metal objects, and other material culture changes. The diffusion of new forms of technology and metals was clearly happening at a much greater scale (Degtyareva et al. 2001). This, coupled with a decrease in fortification at settlements, reflects very different types of relationships between communities of the Late Bronze Age from what is observable for the Middle Bronze Age Sintashta culture.

Future studies of the Sintashta pattern should focus on these dynamic regional changes with a greater emphasis placed on the recovery of more-detailed economic, technological, and environmental data within local catchment zones. These studies can produce invaluable information on diachronic shifts in local socio-economic patterns and strategies. More-detailed environmental studies must also be undertaken, as communities involved in the extraction of base metals and production of metals require sustainable resources in the form of water, fuel, subsistence, and ore. Such exploitation should leave more traces within local landscapes than what has been observed thus far for the Sintashta region.

With these important issues in mind, more-detailed studies in the following areas may yield more-productive empirical data for understanding the Sintashta development and for modeling its social complexity:

1. Rather than conceptualizing Sintashta fortified settlements as representative of singular polities within a broader cultural zone, they should be thought of as communities with potentially broader scalar connections. The implementation of a community approach provides a useful mid-level scale of analysis that operates between the household unit of analysis and larger regional settlement patterning (Canuto and Yeager 2000; Godoy 1985; Knapp et al. 1998: 3). This is precisely the scale that is underrepresented in previous scholarship on the Sintashta development. Furthermore, the variation seen between sites should be more carefully examined as too great an emphasis has been placed on reinforcing homogeneous aspects of Sintashta settlements.

2. The implementation of more-intensive site-catchment studies at several Sintashta settlements may provide a fuller picture of such variation. Given the problems noted within this article for

the identification of mining and quarrying, more-systematic pedestrian surveys may help in detecting the smaller-scale exploitation of surface ore deposits. The probability of such exploitation patterns seems high with the suggestion by Grigor'yev that Sintashta communities may have exploited low-grade copper-bearing ores rather than the rich deposits that are known for the southern Urals region.

3. The application of multi-element soil chemistry and soil geomicromorphology for the study of cultural floor levels in Sintashta houses, in addition to post-Sintashta cultural phases at settlements, may provide new data on metallurgical activities and other craft specialization. In particular, geomicromorphology may provide evidence of different time scales of activities connected with metals production, such as seasonality.

4. More-detailed bio-archaeological studies, particularly those combining bone chemistry, may also be applied to investigate possible social differentiation within communities and between communities. Metals processing produces high levels of heavy metal pollutants, and the study of arsenic and lead levels in human bones may provide another line of evidence in the analysis of mortuary data for reconstructing Sintashta societies and for detecting variation in community health (Lindh et al. 1980; Oakberg 2000).

Conclusion

THE STUDY of prehistoric mining, production, and exchange has formed an important part of modeling early social organization and the rise of complexity within many regions of the Old World. In recent years, much more attention has been focused on the central Eurasian region and the important dynamic that existed between the steppe environmental zone and neighboring territories. One of the most unique developments in this region was the emergence of the Sintashta archaeological pattern with its strongly fortified settlements and complex mortuary patterns, including early chariot technology. While warfare and climate played a significant role in this development, as discussed by Anthony in this volume, the nature of organization of the Sintashta communities and their relationship to their local environmental contexts can be understood only through more-detailed archaeological analyses. While Sintashta settlements have been viewed as representative of early complex societies, no clear evidence has been established previously for settlement pattern hierarchies or clear social inequality in domestic house organization. Conventional mortuary studies also have failed to provide unambiguous evidence for a clear model of social stratification. Therefore, future research along the lines suggested in this chapter may provide more-comprehensive empirical data for understanding how Sintashta communities functioned and what factors lead to the emergence, development, and decline of this important archaeological pattern in north central Eurasia.

References

Anthony, D. 2007. *The Horse, the Wheel, and Language: How Bronze-Age Riders from the Eurasian Steppes Shaped the Modern World*. Princeton: Princeton University Press.
Anthony, D. W., and N. B. Vinogradov. 1995. Birth of the Chariot. *Archaeology* 48(2): 36–41.
Beryozkin, Yu. E. 1995. Arkaim kak tseremonialny tsentr: vzglyad americanista. In V. S. Bochkaryov (ed.), *Konvergentsiya i Vostochnoi Evropy*. Saratov: Saratovskiy Universitet, pp. 29–39.
Blyakharchuk, T. A., H. E. Wright, P. S. Borodavko, W. O. van der Knaap, and B. Ammann. 2004. Late Glacial and Holocene Vegetation Changes on the Ulagan High-Mountain Plateau, Altai Mountains, Siberia. *Palaeogeography, Palaeoclimatology, Palaeoecology* 209: 259–279.
Canuto, M. A., and J. Yeager (eds.), 2000. *The Archaeology of Communities: A New World Perspective*. London: Routledge.
Cernych, L. 2003. Spektralanalyse und Metallverarbeitung in den früh – und mittel – bronzezeitlichen Kulturen der ukrainischen Steppe als Forschungsproblem. *Eurasia Antiqua* 9: 27–62.
Chernykh, E. N. 1992. *Ancient Metallurgy in the USSR: The Early Metal Age*. Cambridge: Cambridge University Press.
(ed.), 2002a. *Kargaly*. Vol. 1: *Geological and Geographical Characteristics: History of Discoveries. Exploitation and Investigations of Archaeological Sites*. Moscow: Nauka.
(ed.), 2002b. *Kargaly*. Vol. 2: *Gorny: The Late Bronze Age Settlement, Topography, Lithology, Stratigraphy, Household, Manufacturing, Sacral Structures, Relative and Absolute Chronology*. Moscow: Nauka.
(ed.), 2004a. *Kargaly*. Vol. 3: *Gorny Site: Archaeological Materials, Mining and Metallurgy Technology, and Archaeobiological Studies*. Moscow: Nauka.
2004b. Kargaly: The Largest and Most Ancient Metallurgical Complex on the Border of Europe and Asia. In K. Linduff (ed.), *Metallurgy in Ancient Eastern Eurasia from the Urals to the Yellow River*. Lewiston: Edwin Mellen Press, pp. 223–237.
Chernykh, E. N., and S. V. Kuz'minykh. 1989. *Drevnyaya metallurgiya Severnoi Evrazii (seiminsko-turbinskii fenomen)*. Moscow: Nauka.

Chernykh, E. N., L. N. Avilova, and L. B. Orlovskaya. 2000. *Metallurgicheskiye provintzii i radiouglerodnaya khronologiya.* Moscow: Institut Arkheologii RAN.

Childe, V. G. 1936. The Axes from Maikop and Caucasian Metallurgy. *Liverpool Annals of Archaeology and Anthropology* 23: 113–119.

———. 1944. Archaeological Ages as Technological Stages. *Journal of the Royal Anthropological Institute of Great Britain and Ireland* 74: 7–24.

Degtyareva, A. D. 2006. Metallicheskiye orudiya truda Sintashtinskoi Kul'tury. *Vestnik Arkheologii, Antropologii i Etnografii.* 7: 49–74.

Degtyareva, A. D., S. V. Kuz'minykh, and L. B. Orlovskaya. 2001. Metallo-proizvodstvo petrovskikh plemen (po materialam poseleniya Kulevchi III). *Voprosy arkheologiyi, antropologiyi i etnografiyi* 3: 23–54.

Diaz del Rio, P., P. Lopez Garcia, J. Antonio Lopez Saez, M. Isabel Martinez Navarrete, A. L. Rodriquez Alcalde, S. Rovira-Llorens, J. M. Vicent Garcia, and I. de Zavala Morencos. 2006. Understanding the Productive Economy during the Bronze Age through Archaeometallurgical and Palaeo-Envrionmental Research at Kargaly (Southern Urals, Orenburg, Russia). In D. Peterson, L. Popova, and A. Smith (eds.), *Beyond the Steppe and the Sown: Proceedings of the 2002 University of Chicago Conference on Eurasian Archaeology,* pp. 343–357.

Epimakhov, A. 2002a. Complex Societies and the Possibilities to Diagnose Them on the Basis of Archaeological Data: Sintashta Type Sites of the Middle Bronze Age of the Trans-Urals. In K. Jones-Bley and G. B. Zdanovich (eds.), *Complex Societies of Central Eurasia from the Third to the First Millennia BC,* vol. 1. Washington, DC: Institute for the Study of Man, 139–147.

———. 2002b. *Yuzhnoe Zaural'ye v period srednei bronzi.* Chelyabinsk: Izd-vo Yuzhno-Ural'skovo gosudarstvennogo universiteta.

———. 2005. *Ranniye Kompleksniye Obshyestva Severa Tzentral'noi Evrazii (po materialam mogil'nika Kamennyi Ambar-5), Kninga 1.* Chelyabinsk: Nauk.

Frachetti, M. 2002. Bronze Age Exploitation and Political Dynamics of the Eastern Eurasian Steppe Zone. In K. Boyle, C. Renfrew, and M. Levine (eds.), *Ancient Interaction: East and West in Eurasia.* Cambridge: McDonald Institute, pp. 161–170.

———. 2004. Bronze Age Pastoral Landscapes of Eurasia and the Nature of Social Interaction in the Mountain Steppe Zone of Eastern Kazakhstan. Ph.D. dissertation, University of Pennsylvania.

Gening, V. F., G. B. Zdanovich, V. V. Gening. 1992. *Sintasthta. Archeologicheskii pamyatnik Ariiskikh plemen Uralo-Kazakhstanckikh stepei.* T. 1. Chelyabinsk: Uzhno-Ural'skoe knizhnoe Izd-vo.

Gilman, A. 1981. The Development of Social Stratification in Bronze Age Europe. *Current Anthropology* 22: 1–23.

Godoy, R. 1985. Mining: Anthropological Perspectives. *Annual Review of Anthropology* 14: 199–217.

Grigor'yev, S. A. 1999. *Drevniye indoevropeitzi. Opit istoricheskoi rekonstruktzii.* Chelyabinsk: Institut istorii i arkheologii UrO RAN.

———. 2000. Investigation of Bronze Age Metallurgical Slag. In J. Davis-Kimball, E. Murphy, L. Koryakova, and L. Yablonsky (eds.), *Kurgans, Ritual Sites, and Settlements – Eurasian Bronze and Iron Age.* BAR International Series 890. Oxford: Archaeopress, pp. 141–149.

Grigor'yev, S. A., and I. A. Rusanov. 1995. Experimentalnaya rekonstruksiya drevnego metallurgischeskogo proizvodstva. In G. Zdanovich (ed.), *Arkaim: Issledovaniya, Poiski and Otkritiya*. Chelyabinsk: Kamennyi Poyas, pp. 147–158.

Hanks, B., A. Epimakhov, and C. Renfrew. 2007. Towards a Refined Chronology for the Bronze Age of the Southern Urals, Russia. *Antiquity* 81: 333–367.

Jones-Bley, K., and D. G. Zdanovich (eds.). 2002. *Complex Societies of Central Eurasia from the 3rd to the 1st Millennium BC*. Vols. 1 and 2. Washington, DC: Institute for the Study of Man.

Judd, M., M.-E. Kovacik, and D. Rajev, 2008. Community Health at Kamennyi Ambar 5. Unpublished report on the bioarchaeology analysis of 118 individuals from the Middle Bronze Age Kamennyi Ambar 5 cemetery. University of Pittsburgh.

Kassianidou, V., and A. B. Knapp. 2005. Archaeometallurgy in the Mediterranean: The Social Context of Mining, Technology and Trade. In E. Blake and A. Bernard Knapp. (eds.), *The Archaeology of Mediterranean Prehistory*. Oxford: Blackwell, pp. 213–251.

Knapp, A. B., V. C. Pigott, and E. W. Herbert. 1998. *Social Approaches to an Industrial Past: The Archaeology and Anthropology of Mining*. London: Routledge.

Kohl, P. 2007. *The Making of Bronze Age Eurasia*. Cambridge: Cambridge University Press.

Koryakova, L. 1996. Social Trends in Temperate Eurasia during the Second and First Millennia BC. *Journal of European Archaeology* 4: 243–280.

2002. The Social Landscape of Central Eurasia in the Bronze and Iron Ages: Tendencies, Factors, and Limits of Transformation. In K. Jones-Bley and G. B. Zdanovich (eds.), *Complex Societies of Central Eurasia from the 3rd to the 1st Millennium BC*, vol. 1. Washington, DC: Institute for the Study of Man, pp. 97–117.

Koryakova, L. N., and A. V. Epimakhov. 2007. *The Urals and Western Siberia in the Bronze and Iron Ages*. Cambridge: Cambridge University Press.

Levy, T. E. 1993. Production, Space and Social Change in Protohistoric Palestine. In A. Holl and T. E. Levy (eds.), *Spatial Boundaries and Social Dynamics: Case Studies from Food-Producing Societies*. Ann Arbor: International Monographs in Prehistory, pp. 63–81.

Lindh, U., D. Brune, G. Nordberg, and P.-O. Wester, 1980. Levels of Antimony, Arsenic, Cadmium, Copper, Lead, Mercury, Selenium, Silver, Tin and Zinc in Bone Tissue of Industrially Exposed Workers. *Science of the Total Environment* 16: 109–116.

Lindstrom, R. 2002. Anthropological Characteristics of the Population of the Bol'shekaraganskii Cemetery, Kurgan 25. In D. Zdanovich (ed.), *Arkaim: Nekropol' (po materialam kurgana 25 Bol'shekaraganskogo mogil'nika)*, Kniga 1. Chelyabinsk: Institut istorii i arkheologii Ural'skogo Otdeleniya RAN, pp. 159–166.

Linduff, K. (ed.), 2004. *Metallurgy in Ancient Eastern Eurasia from the Urals to the Yellow River*. Lewiston: Edwin Mellon Press.

Oakberg, K. 2000. A Method for Skeletal Arsenic Analysis, Applied to the Chalcolithic Copper Selting Site of Shiqmim, Israel. *Journal of Archaeological Science* 27: 895–901.

Ottaway, B. 2001. Innovation, Production and Specialization in the Early Prehistoric Copper Metallurgy. *European Journal of Archaeology* 4: 87–112.

Peck, J. A., P. Khosbayar, S. J. Fowell, R. B. Pearce, S. Ariunbileg, B. C. S. Hansen, and N. Soninkhishig. 2002. Mid to Late Holocene Climate Change in North Central Mongolia as Recorded in the Sediments of Lake Telmen. *Palaeogeography, Palaeoclimatology, Palaeoecology* 183: 135–153.

Peterson, D., P. Kuznetsov, and O. Mochalov, 2006. The Samara Bronze Age Metals Project: Investigating Changing Technologies and Transformations of Value in the Western Eurasian Steppes. In D. Peterson, L. Popova, and A. Smith (eds.), *Beyond the Steppe and the Sown: Proceedings of the 2002 University of Chicago Conference on Eurasian Archaeology*. Leiden: Brill, pp. 322–342.

Shennan, S. 1993. Commodities, Transactions and Growth in the Central European Early Bronze Age. *Journal of European Archaeology* 1(2): 59–72.

1998. Producing Copper in the Eastern Alps during the Second Millennium BC. In A. B. Knapp, V. C. Pigott, and E. W. Herbert (eds.), *Social Approaches to an Industrial Past: The Archaeology and Anthropology of Mining*. London: Routledge, pp. 191–204.

Tarasov, P. E., F. Jolly, and J. O. Kaplan. 1997. A Continuous Late Glacial and Holocene Record of Vegetation Changes in Kazakhstan. *Palaeogeography, Palaeoclimatology, Palaeoecology* 136: 281–292.

Topping, P., and M. Lynott (eds.). 2005. *The Cultural Landscapes of Prehistoric Mines*. Oxford: Oxbow Books.

Vinogradov, N. B. 1995. Chronologia, soderzhaniye i kul'turnaya prinadlezhnost' pamyatnikov Sintashtinskogo tipa Bronzovogo veka v Yuzhnom Zaural'ye. *Vestnik Chelyabinskogo gosudarstvennogo pedagogicheskogo instituta. Istoricheskiye nauki*. No. 1. Chelyabinsk: Nauka, pp. 16–26.

2003. *Mogil'nik bronzovogo veka Krivoe Ozero v Yuzhnom Zaural'ye*. Chelyabinsk: Yuzhno-Ural'skoe knizhnoe izd-vo.

2004. Sintashtinskie i Petrovskie Drevnosti Uzhnogo Yrala. Problema Sootnosheniya i Interpretatsii. In A. N. Gei (ed.), *Pamyatniki Arkheologii I Drevnego Iskusstva Evrazii*. Moscow: RAN, pp. 261–284.

Yener, K. A. 2000. *The Domestication of Metals: The Rise of Complex Metal Industries in Anatolia*. Boston: Brill.

Zaikov, V. V., G. B. Zdanovich, and A. M. Uminov. 1995. Mednyi rudnik Bronzogo veka "Vorovskaya Yama." In Rossiya i Vostok: Problemy Vzaimodeistviya, Chast V, Kniga 1: *Kul'tury Eneolita-Bronzy Stepnoi Evrazii*. Chelyabinsk: 3-ya ezdunarodnaya Nauchnaya Konferentsiya, pp. 157–162.

Zaikov, V. V., G. V. Zdanovich, and A. M. Uminov, 2000. Vorovskaya Yama – Novyi Rudnik Bronzovogo Veka na Uzhnom Urale. In G. V. Zdanovich and S. Ya. Zdanovich (eds.), *Arkheologicheskii Istochnik i Modelirovaniye Drevnikh Tekhnologii*. Chelyabinsk: Institut Istorii I Arkheologii Ural'skogo Otdeleniya RAN, pp. 112–129.

Zaikov, V. V., A. M. Uminov, A. Ph. Bushmakin, E. V. Zaikova, A. D. Tairov, and G. B. Zdanovich. 2002. Ancient Copper Mines and Products from Base and Noble Metals in the Southern Urals. In K. Jones-Bley and G. B. Zdanovich (eds.), *Complex Societies of Central Eurasia from the Third to the First Millennia BC*, vol. 2. Washington, DC: Institute for the Study of Man, pp. 417–442.

Zdanovich, D. G. 1997. *Sintashtinskoye obshchestvo: sotzialnye osnovy (kvazigorodskoi) kultury Uzhnogo Zauralya epokhi srednei bronzy.* Chelyabinsk: Chelyabinskiy Gosudarstvenny Universitet.

(ed.), 2002. *Arkaim: Nekropol'*, Kn. 1. Chelyabinsk: Uzhno-Ural'skoe knishnoe izd-vo.

1988. *Bronzovii vek Uralo-Kazakhstanskikh stepei (osnovi periodizatzii).* Sverdlovsk: Izd-vo Ural'skogo universiteta.

1989. Phenomen protogorodskoi tsivilizatsii bronzovogo veka Uralo-kazakhstanskikh stepei. Kulturnaya I sotsialnaya obuslovlennost. In V. M. Masson (ed.), *Vzaimodeistvie kochevykh kultur I drevnikh tsivilizatsiy.* Alma-Ata: Nauka, 179–189.

1995. Arkaim: Arii na Urale ili nesostoyavshayasya tsivilizatsiya. In G. B. Zdanovich (ed.), *Arkaim: Islledovaniya. Poiski. Otkrytiya.* Chelyabinsk: Kamenny Poya, pp. 21–42.

1997. Arkaim – kur'turnii kompleks epokhi srednei bronzi Yuzhnogo Zaural'ya. Rossiiskaya Arkheologiya, no. 2: 47–62.

Zdanovich, G. B., and I. M. Batanina. 2002. Planography of the Fortified Centers of the Middle Bronze Age in the Southern Trans-Urals according to Aerial Photography Data. In K. Jones-Bley and G. B. Zdanovich (eds.), *Complex Societies of Central Eurasia from the 3rd to the 1st Millennium BC*, vol. 1. Washington, DC: Institute for the Study of Man, pp. 121–147.

Zdanovich, G. B., and D. G. Zdanovich. 2002. The "Country of Towns" of Southern Trans-Urals and Some Aspects of Steppe Assimilation in the Bronze Age. In A. C. Renfrew and M. Levine (eds.), *Ancient Interactions: East and West in Eurasia.* Cambridge: McDonald Institute, pp. 249–263.

CHAPTER 10

The Bronze-Using Cultures in the Northern Frontier of Ancient China and the Metallurgies of Ancient Dian Area in Yunnan Province

RUBIN HAN AND XIAOCEN LI

DURING THE period of the Shang through the Han dynasty (c. 1250 BCE – 220 CE), mobile pastoralists who lived in the *beifang*, or the Northern Frontier of ancient dynastic China, contributed to and strongly influenced the emerging Chinese civilization. Their effect was felt not only in the northern territories and Central Plain of China but also in the southwest. The examination of the impact of these mobile peoples on local bronze cultures has ignited the interest and attention of many scholars; however, little attention has been given to their contribution to the distinctive culture, or cultures, of what is now southwestern China.

The ancient text called the *Shiji* (Records of the Grand Historian) suggested that during the period of the Warring States and the Han Dynasty (fifth century BCE–second century CE), there was a state called Dian near Lake Dianchi in present-day Yunnan Province in southwestern China. There have been many different views about the formation and origination of the metal culture of the Dian State. One recent proposal is that the Dian was made up of a mixture of peoples from the Baipu (from the Chu State in south-central China), of Di and Qiang (pastoral peoples from the north), and a local population (*Kunming yangfutou* 2005). The purpose of this chapter is to investigate metallurgical technologies used in the state of Dian and to explore the implications of that in relation to interaction with peoples from the *beifang*, or the Northern Frontier.

In the past 50 years, thousands of metal objects were unearthed from dozens of Bronze Age cemeteries excavated in the ancient Dian region (Fig. 10.1). The study of these important materials should lead to a deeper understanding of the bronze-using cultures in this area. With

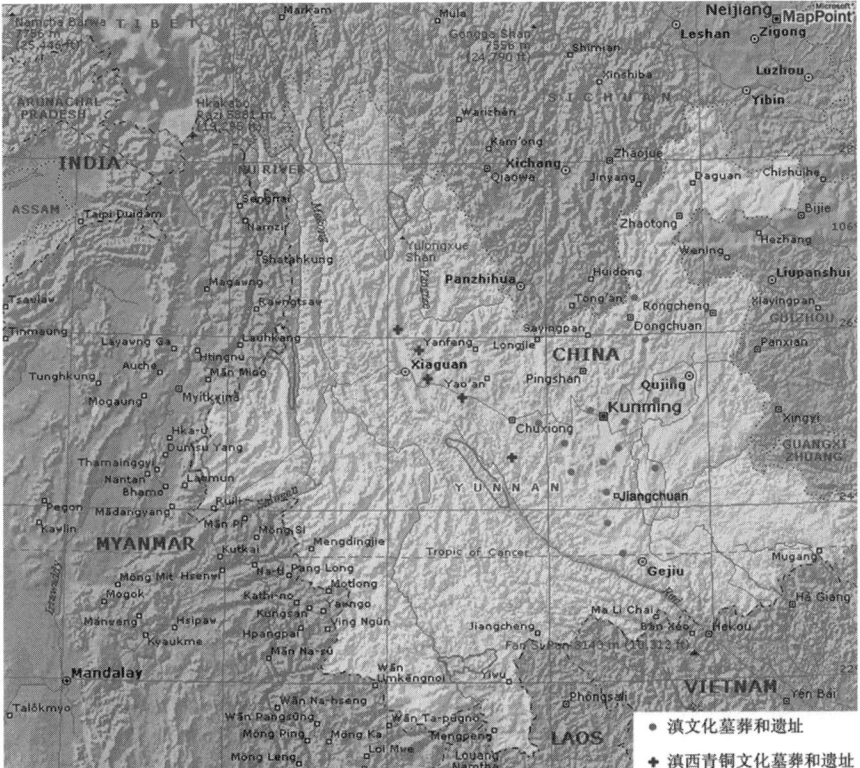

Figure 10.1. The distribution of cemeteries of the Bronze Age in ancient Dian region, Yunnan. • = cemeteries and sites of the Dian culture; + = cemeteries and sites of the bronze cultures in the western Dian region.

Lake Dianchi as the center, the bronze-using cultures of the ancient Dian region were distributed in the present-day cities and counties of Kunming, Jinning, Jiangchuan, Chenggong, Qujing, Anning, and so on. Remains from this area are found in cemeteries and are dated from the early stage of the Warring States period (fifth century BCE) to the early period of the Eastern Han Dynasty (first century CE). This evidence documents activity that lasted more than five centuries – that is, longer than that recorded in the *Shiji*.

It is generally believed that the Qiang were the earliest people to enter Yunnan and that they moved into the northwest of the province as the first residents of the area. One branch of the Di and Qiang groups, the Bo people, came into Yunnan during or after the middle of the Warring States period (c. fourth century BCE); they were mobile pastoralists, like many peoples in the Northern Frontier. Yang Fan has suggested that they came into the region along what was called the "ethnic corridor," or the Tibetan plateau (Yang and Mei 2003). Because they possessed the most advanced culture in the region at that time, their metallurgical abilities were at relatively high technological levels.

In the 1990s, many bronze objects were unearthed at Yangfutou, Shizhaishan, and Lijiashan cemeteries; they included several categories of objects, some of which had locally characteristic shapes and decorative patterns. In order to examine the bronze-using cultures in the ancient Dian region and to attempt to explain the appearance of metal artifacts analogous to ones known in the Northern Frontier, we selected 135 bronze objects unearthed from eight sites around Lake Dianchi and researched their metallurgical components, structure, and manufacturing technology. The first results showed that 61.5% of the bronze objects tested were made of a copper-tin alloy and 81.5% were cast. In general, the trident-shaped iron swords with bronze hilts and the objects with tinned surfaces were very likely related in type and manufacture to comparable objects used by the cultures of the Northern Frontier steppe. Those from the south also exhibited strong local features. These tests suggested that the brilliant bronze culture in the Lake Dianchi area was the result of the intersection of several bronze production technologies and probably cultures.

The Application and Diffusion of Tin-Plating Technology

Ordos Bronze Ornaments

The ancient peoples in the steppe regions of the Northern Frontier were distinct from their neighbors living in the territories of the Shang and Zhou dynasties, especially in their design and uses of metal objects. Bronze ornaments, tools, weapons, and other utilitarian objects unearthed in the Ordos region under the Great Bend of the Yellow River displayed zoomorphic designs and patterns. The Ordos region is adjacent to the Eurasian steppe, and the artifacts found there shared typological features and visual ornamentation with those used by Eurasian groups. These analogies have interested scholars in China and abroad and much research on the style and iconography of these so-called Ordos-style bronzes has already been done (E. Wu 1983, 1985, 1994). In particular, some Ordos-style bronzes were made with a tinned surface not found outside of this region. Moreover, it is generally thought that the silver sheen on the surfaces was the result of silver-plating or inverse segregation during the casting of bronze with relatively high tin content. This technology has been found to be unique to the region.

Emma Bunker of the Denver Art Museum in the United States has noted that the shape and designs of a tin-plated, tiger-shaped bronze plaque published in the Maoqinggou archaeological excavation report (Neimenggu wenwu gongzuodui 1986) is very similar to another one in a private collection, but the report did not provide information about surface treatment of that plaque (Han and Bunker 1993). To carry out advanced research on the Ordos-style bronzes with tin-enriched surfaces, she went to the archaeological sites at Qingyang (Gansu Province), Tongxin and Guyuan (Ningxia Hui Autonomous Region), and Hohhot (Inner Mongolia) in 1989 and 1990 to conduct visual surveys of the materials there with the help of the local archaeological departments. During the surveys, she found bronze zoomorphic plaques, double-bird shaped plaques, buckles, button-shaped ornaments, and sword hilts that had a silver sheen and tin-enriched surfaces (Fig. 10.2).

Preliminary investigations showed that a majority of the bronze objects unearthed from the archaeological sites in Qingyang, Guyuan, and Liangcheng (Inner Mongolia) had tin-enriched surfaces. For example, among the 114 double-bird shaped plaques unearthed from eight tombs in Maoqinggou cemetery observed by Han and Bunker, 99 objects (86.8%) had silvery, tin-enriched surfaces. Most of them from Qingyang and Guyuan had only one side enriched, and only part of the plaques in the same shape and design were tin enriched. The rectangular bronze plaques with tin-enriched surfaces that were unearthed from Maoqinggou bore designs on one side and were undecorated on the reverse. Both sides were tin enriched. These bronzes were made between the sixth century and the end of the fourth century BCE; similar objects were not found among the bronzes of the later Warring States period and the Qin and Han dynasties (third century BCE–first century CE). With the invention and popularization of the gilding technique in the fourth century BCE, gilt bronzes as well as gold and silver ornaments were produced and increased in numbers, while the bronzes with tin-enriched surfaces declined in numbers and eventually vanished. To study the manufacturing techniques of the bronzes with tin-enriched surfaces unearthed in China, Chinese and American researchers represented by Rubin Han and Emma Bunker conducted a cooperative study of this special subject beginning in 1990. Eleven Ordos-style bronze objects were selected for in-depth analyses of their metallographic features. The study concluded that all of them were products of the tinning technique.

The tinning would seem to be a way of enriching the sheen of the surface that added an opulent and distinctive appearance to certain objects in the otherwise ordinary inventory. Although it is clear from study of

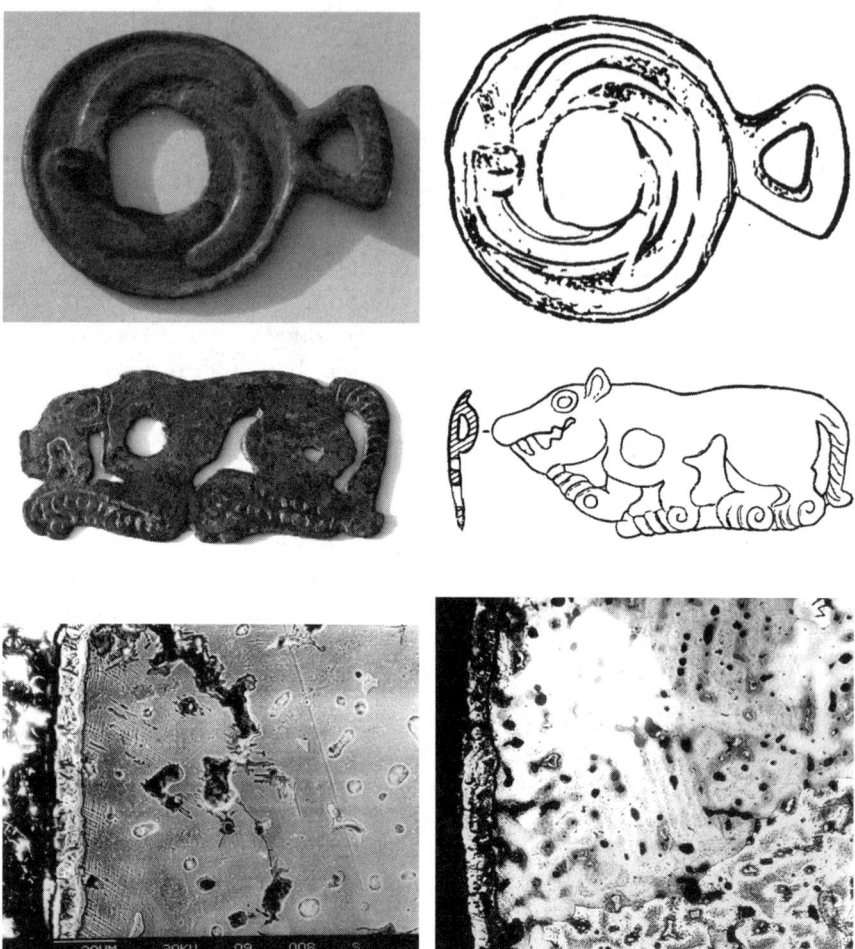

Figure 10.2. The bronze plaque ornaments unearthed from Ordos area and their structures.

the Maoqinggou cemetery that social inequality according to gender and age in the community can be documented (X. Wu 2004), just what the distinctive technological preparation of tinning might signify in burial has not yet been systematically studied.

The Tiger-Stripe Pattern on the Surfaces of Ba-Shu–Style Weapons

THE MOST recent research on the Ba-Shu–style bronzes unearthed in the Chengdu and Sanxia areas (in present-day Sichuan) shows that weapons, mainly *ge*-daggers, *yue*-axes, and willow-leaf-shaped swords, made up the majority of bronzes from the ancient Ba-Shu regions (Yao Zhihui 2005). Most of these weapons were decorated with designs covered with a "tiger-stripe" pattern peculiar to the Ba-Shu style. Yao Zhihui and Sun Shuyun observed and recorded 92 samples during their investigations of the Ba-Shu–style bronze weapons (including swords, spearheads, and *ge*-daggers) that are collected and housed in more than 10 institutions (Yao et al. 2005). Forty-eight of these samples, most of which were made in the middle and later stages of the Warring States period, were decorated with the tiger-stripe pattern in black and silver colors. Each pattern had two versions: regular and irregular, circular and half-circular, floral shapes, and so on (Fig. 10.3).

Because most of the weapons with tiger-stripe patterns were intact, we conducted only non-destructive component analysis and structure analysis on 12 of them. The main components of the black and silver patterns consisted of copper-tin alloy and SnO_2, and no obvious differences have been found in the tested samples. Zeng Zhongmao and He Tankun (Yao 2005) provided three fragments of bronze swords with tiger-stripe patterns unearthed in the Chengdu area and dated from the Warring States period so that we could test the internal components and structure of the alloy. After the in-depth and close analysis by Yao Zhihui and Sun Shuyun, we know that the parts with patterns were tinned and that the rust corrosion of the parts without patterns was more severe than that of the parts with patterns. Moreover, Yao Zhihui, Dong Yawei, and Sun Shuyun designed and conducted many tests that indicated that the tiger-stripe patterns on the surfaces of the bronzes were made by using the tinning technique (Yao et al. 2005).

The Tin-Plated Bronzes of the Ancient Dian State in Yunnan

THE BRONZES, such as cowry-shell containers, armor, kneeling figurines, hoes, and so on, unearthed from the Shizhaishan cemetery in Jinning, Yunnan, all had a silver sheen on the surface. Some scholars have thought it was the result of inverse segregation of bronzes with a high tin component, whereas others thought it was the result of tin plating. Neither opinion has been previously tested scientifically. Scholars in the fields of archaeology and history of science and technology have paid close attention to the cause of the silver luster. During a recent survey of the bronzes unearthed in the area of ancient Dian State, Xiaocen Li and Rubin Han found that the bronzes with silver sheen

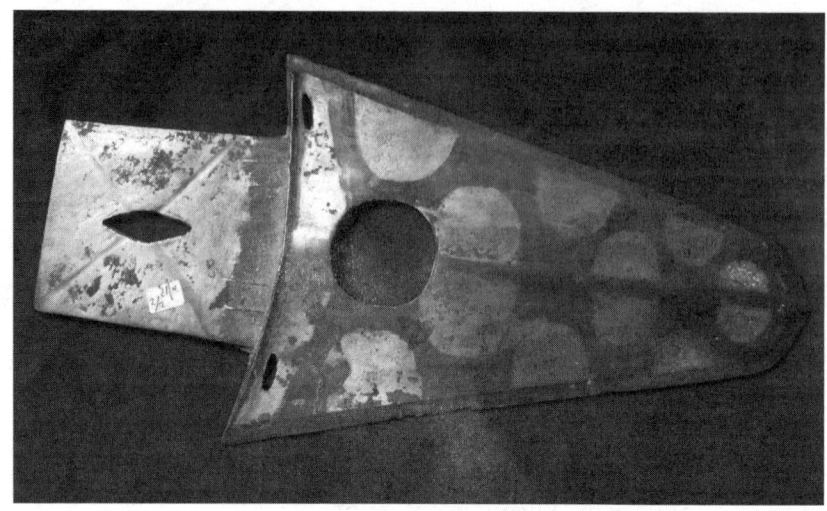

Figure 10.3. Bronze weapons decorated with tiger-stripe patterns unearthed from Ba and Shu areas in present-day Sichuan Province.

were found in the tombs at Shizhaishan (Jinning County), Lijiashan (Jiangchuan), and Batatai (Qujing) cemeteries (Figs. 10.4 and 10.5) (Li 2004). With the help of the colleagues from the Yunnan Provincial Museum and Yunnan Institute of Cultural Relics and Archaeological

Figure 10.4. Cowry container (*top*) and tin-plated bronze drum (*bottom*) unearthed from Lijiashan, Yunnan.

Figure 10.5. Tin-plated kneeling figurine unearthed from Shizhaishan, Yunnan.

Research, we collected samples from 18 of the tombs and conducted metallographic tests (see Table 10.1). The tests confirmed that all 18 samples were tin plated.

Among the 18 samples, 3 button-shaped ornaments, 1 drum, and 1 cowry container were tinned on a single side, and the other 13 were tinned on both sides. All of them also exhibited metallographic features of tinning. In the five samples from the Shizhaishan cemetery, one spear (*Shicai* 6 [5]) was gathered from the surface; however, all four of the others were unearthed from M71, a large-sized tomb. The hoe, spear, button-shaped ornament, and horse fittings were all tinned, perhaps marking the status and position of the deceased. The tinned hoe and spear, which were not practical tools or weapons and show no wear,

Table 10.1. Sample-collection of the tinned objects in Dianchi Area, Yunnan

Object	Shizhaishan, Jinning	Lijiashan, Jiangchuan	Batatai, Qujing	Total
Scabbard	1			1
Button-shaped ornament	1 (single-side tinned	2 (single-side tinned)		3 (single-side tinned)
Hoe	2			2
Spear	1	2		
Sword blade		4	1	5
Dagger-ax		1		1
Drum		1 (single-side tinned)		1 (single-side tinned)
Cowry container		1 (single-side tinned)		1 (single-side tinned)
Round-shaped object		1		1
TOTAL	5	12	1	18 (5 are single-side tinned)

Note: All of the objects without special mention are tinned on both sides.
Source: Li Xiaocen 2004.

were probably used only in ceremonies and special rites (Fig. 10.6). The 10 samples from the Lijiashan cemetery were all from M68, a large-sized tomb including a *ge*-dagger, spear, sword, and button-shaped ornaments that bore features similar to those from Shizhaishan (M71). Rare gilt objects, a gold scabbard, and ornaments made of gold-silver and silver-copper alloys were unearthed from the tombs M71 at Shizhaishan and M68 and M61 at Lijiashan and also likely suggest the high status and position of the occupants in tombs with abundant grave goods.

Trident-Shaped Iron Swords with Bronze Hilts

TRIDENT-SHAPED IRON swords with bronze hilts have been found in Gansu and Ningxia, where pastoral groups lived during the pre-Qin period and the Han Dynasty. Luo Feng has done a typological study on nine iron swords with bronze hilts unearthed in these areas and dated

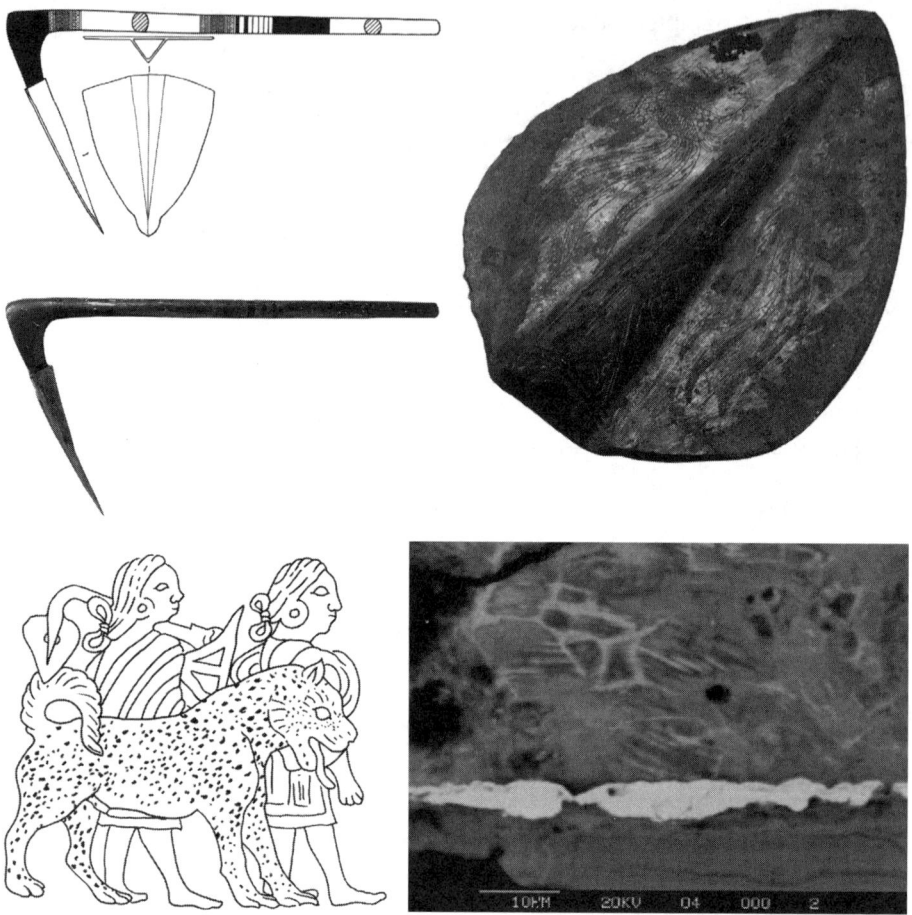

Figure 10.6. Bronze hoe unearthed from Shizhaishan, Yunnan, and its structure.

them to the sixth–fifth century BCE (Luo 1993). The results of metal-lographic tests on four of them showed that they were made of solid-smelted iron enhanced with cementation (Fig. 10.7).

Similar swords have also been found in the Ba-Shu regions (Fig. 10.8). In the Yangfutou cemetery of the ancient Dian State near Lake Dianchi, many trident-shaped iron swords with bronze hilts have been unearthed (Fig. 10.9). All of them were made in or after the fourth century BCE, but unfortunately studies have not been conducted on them. They do suggest, however, the southern movement of some group of pastoral peoples from the Northern Frontier.

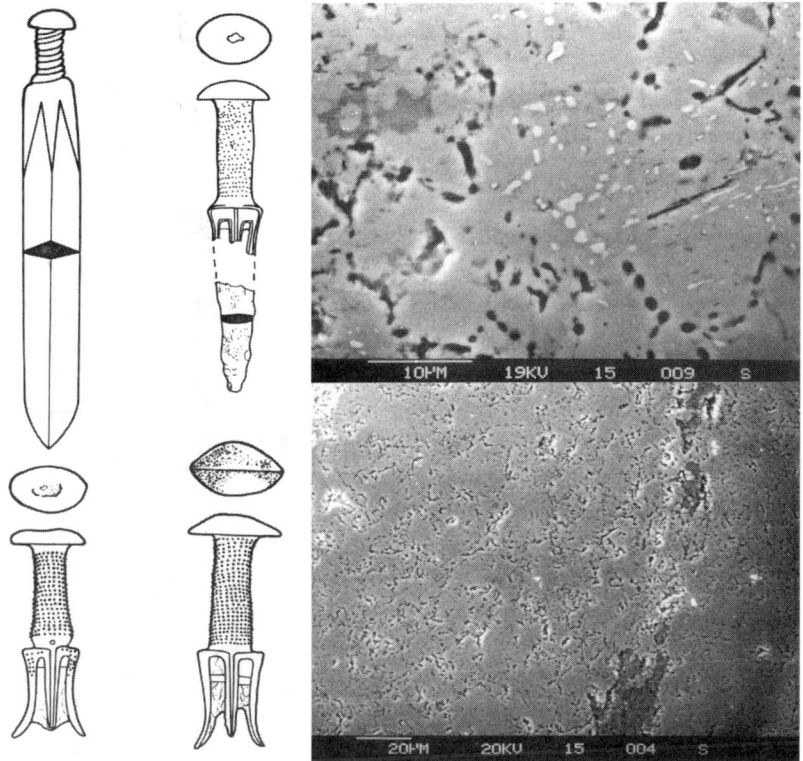

Figure 10.7. Trident-shaped iron sword with bronze hilt unearthed from Longshan area of Gansu and Ningxia and its structure (the iron part).

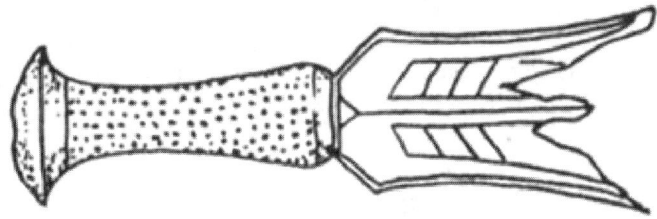

Figure 10.8. Trident-shaped iron sword with bronze hilt unearthed from Ba and Shu areas in present-day Sichuan.

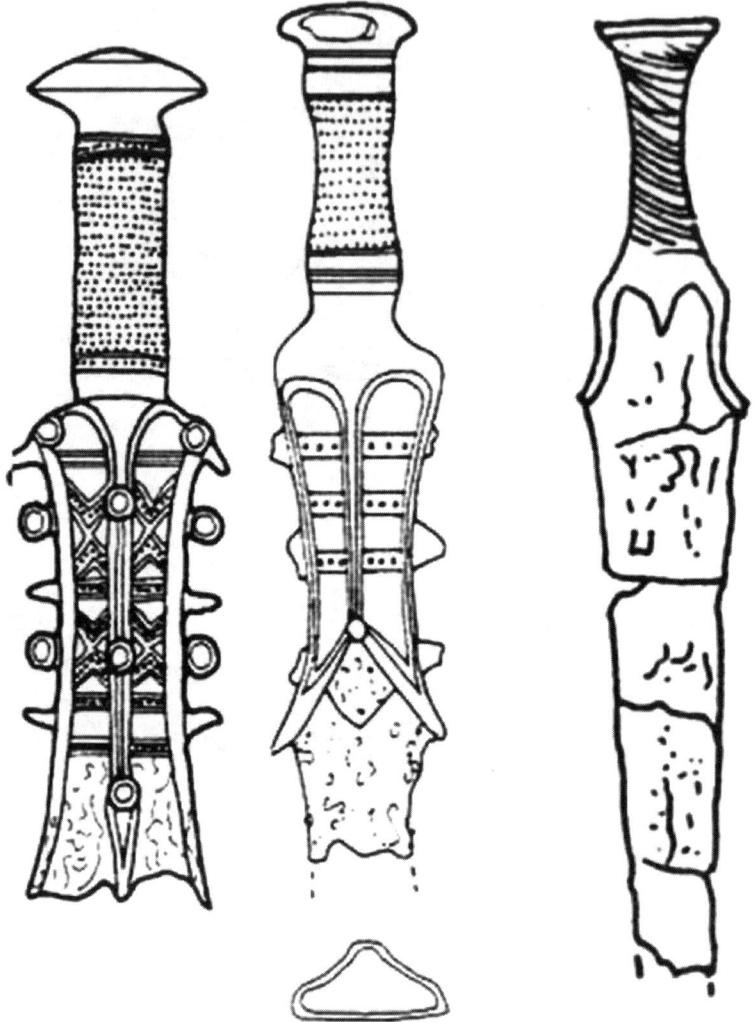

Figure 10.9. Trident-shaped iron sword with bronze hilt unearthed from Yangfutou cemetery in Dianchi region, Yunnan.

Discussion

WHAT WAS the purpose of tinning on ancient bronzes? Was it merely for decoration? For rust prevention? For display of status of the occupants of the tombs and/or the owners of these bronzes? Or for all of these reasons? The tinned bronzes were indeed able to maintain the shine on bronzes at low cost and maintenance, to keep the bronzes from rusting quickly, and to extend the life of the objects. The excavated tinned bronzes have provided proof of these advantages. The appearance of tinned bronzes (and other products) unquestionably documents experimentation with the appearance of an "outsider" technology. The tinning technique can be seen as one important invention of the bronze cultures in the steppe area of the Northern Frontier.

At the Institute for Historical Metallurgy, we took cross-section samples of tinned bronze objects, excavated from three different areas, and tested them by optical and scanning electronic microscopy (SEM-EDX) to characterize their microstructure and composition. All the samples had common metallographic features. An interface exists between the tinned layer and bulk metal, and tests showed that the surface tin concentration is three to four times higher than that from its bulk. Either a slip band or re-crystallized grains constitute the surface treatment. The tin gradient varies from high to low in the layer. Thickness of the tinned layer is between 3 and 40 μm, and included phases ε、η、and δ. The methods of creating the tinned layer involved a hot tin bath for both tinned surfaces and the application of liquid tin on hot bronze for a single tinned surface. In addition, the categories of tinned bronzes in different regions varied; however, all of them were created using the tinning technique. The study of buckles, plaques, and other ornaments produced in the Ordos; weapons decorated with tiger-stripe patterns in Ba-Shu regions; and bronze drums, cowry containers, and kneeling figurines holding umbrellas, hoes, *ge*-daggers, swords, and button-shaped ornaments from the region of the Dian has shown that the transmission of the tin-plating technique engaged the aesthetic taste of people in different areas and that the migration of peoples, their ideas, and/or objects from the Northern Frontier passed south to the ancient Dian State through the Ba-Shu regions of Chengdu and Sanxia in Sichuan. In the sixth–fourth century BCE, only the pastoral peoples living in what is now present-day northwest China and southwestern Inner Mongolia marked tomb occupants by using tinned bronzes, presumably only for burial, whereas no tinned bronzes have been found in tombs excavated in present-day northeast China and northeastern Inner Mongolia. This suggests that it was a regional technology that spread south but not east.

No tinned grave goods were found at the Yangfutou cemetery in the Lake Dianchi region, Yunnan, but trident-shaped iron swords with bronze hilts and claw picks, popular with the Di and Qiang groups in northwest China, were found there. After the emergence of gilding technology, tinning disappeared as a surface treatment on bronzes in the Ordos. In the Ba-Shu and Dian regions, tinned objects co-existed with gilded ones during the later Warring States period through the middle and later stages of the Western Han Dynasty (third–first century BCE). Thus, it is quite possible that the use and diffusion of trident-shaped iron swords with bronze hilts and the bronze objects treated with the tinning technique imply a change in the migration routes of the mobile pastoralists from the Northern Frontier (Gansu and Ningxia) to the southwest.

The exquisite and unique tiger-shaped *ge*-dagger unearthed in an early Western Zhou (eleventh century BCE) tomb at Baicaopo Village, Lingtai County, Gansu Province (Cultural Relics Group of Gansu Provincial Museum 1972) has a silver sheen on its surface and is the earliest tin-plated bronze object discovered to date in present-day Chinese territory (Gansu Provincial Museum, Fig. 10.10). Ma Qinglin

Figure 10.10. Tin-plated bronze yue-axe unearthed from Baicaopo, Lingtai County, Gansu Province (Western Zhou, eleventh century BCE).

and others systematically analyzed the tin-plated bronzes unearthed from the pre-Qin tombs in Li Xian and Tianshui counties in Gansu Province, such as shovels, knives, rings, arrowheads, band-shaped belt hooks, trappings, and bells and published their results (Ma et al. 1999). Do these bronzes suggest a relationship between this area and the bronze cultures in the Northern Frontier? And if so, what was the relationship? This question is yet to be examined, so the date and regional origins of the tin-plating technique and its transmission to the Dianchi area, as well as an understanding of the type of contact or exchange with the pastoralists of the Northern Frontier, need to be researched in the future.

Yunnan has rich mineral resources, and written records of mining can be found in the *Dili Zhi* (Records of Geography) of the *Hanshu* (Western Han Chronicle) and the *Nanzhong Zhi* (Records of Nanzhong Area – the area to the south of Ba and Shu, or present-day Yunnan) of the *Hua Yang Guo Zhi* (Records of the States to the South of Mount Hua). Incomplete statistics show that at present there are at least 160 copper mines within the province of Yunnan. Archaeological discoveries in many locations have also yielded casting molds made of stone, clay, pottery, and metal (Wang 1998; the distribution of molds discovered is shown in Fig. 10.11). The rich mines and the presence of ancient molds demonstrates the possibility and conditions for bronze manufacturing in the ancient Dian State.

Research on the contributions of the bronze cultures of the pastoral peoples from the Northern Frontier steppe areas to those in southwest China is just beginning. Many issues are still unstudied, including ones that address the application of horse fittings, the domestication of horses, the changes in the function of bronze drums, the manufacture of cowry containers, and the uses of cowries. We hope these issues will draw the attention of more scholars in the future.

Conclusion

THE DEVELOPMENT of metal manufacturing in the ancient Dian State in the middle and later stages of Warring States period (fourth century BCE) was due to the arrival of Di and Qiang peoples, who were pastoralists from the steppe area of the North Frontier. The cultural relics of this period unearthed in Yunnan show the diversity of the Dian culture and display a mixture of features from peoples living in the Northern

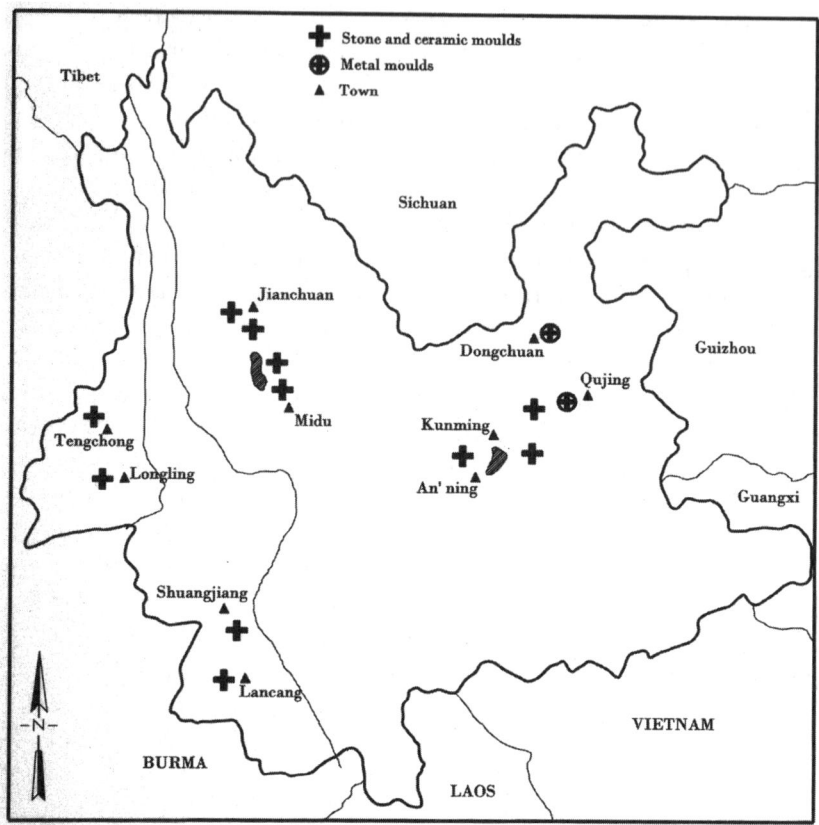

Figure 10.11. The locations in Yunnan where the casting molds were discovered.

Frontier, the Chu State in present-day central China, and locally in Yunnan.

Acknowledgments

THE AUTHORS hereby express our deep gratitude to the leaders and archaeological colleagues in the Yunnan Provincial Institute of Cultural Relics and Archaeology, Yunnan Provincial Museum, and Kunming Municipal Museum who have offered us strong support and cooperation. We would also like to thank Qiu Zihui who sponsored this research and our colleagues and graduate students from our institute. In addition, many thanks also to Wang Dongning at Lehigh University, who gave us important support, and to Katheryn Linduff and her students, who helped to translate this paper into English. We could not have published this work without their efforts.

References

Cultural Relics Group of Gansu Provincial Museum 甘肃省博物馆文物组. 1972. Lingtai Baicaopo Xizhou mu 灵台白草坡西周墓 (The Western Zhou Tombs in Baicaopo Village, Lingtai County). *Wenwu* 文物 (Cultural Relics), no. 12: 2–8.

Han, Rubin 韩汝玢, and Emma C. Bunker 埃玛. 邦克. 1993. Biaomian fuxi de E'erduosi qingtongqi shipin de yanjiu 表面富锡的鄂尔多斯青铜饰品的研究 (The Study of Ancient Ordos Bronze with Tin-Enriched Surface in China). *Wenwu* 文物 (Cultural Relics), no. 9: 80–96.

Li, Xiaocen 李晓岑. 2004. *Gu Dian Diqu Chutu Jinshuqi de Jishu Yanjiu* 古滇地区出土金属器的技术研究 (A Technical Study on the Metallic Artifacts Excavated in Ancient Dian Region). 北京科技大学博士论文 (Doctorial Dissertation of History of Science and Technology, University of Science and Technology, Beijing). (Advisor: Rubin Han), pp. 76–89.

Luo, Feng 罗丰. 1993. Yi Longshan wei zhongxin Ganning diqu Chunqiu Zhanguo shiqi beifang qingtong wenhua de faxian yu yanjiu 以陇山为中心甘宁地区春秋战国时期北 方青铜文化的发现与研究 (The Discoveries and Research on the Northern Frontier Bronze Cultures around Longshan in Gansu and Ningxia). *Neimenggu Wenwu Kaogu* 内蒙古文物考古 (Inner Mongolia Cultural Relics and Archaeology), no. 1–2: 29–48.

Ma, Qinglin, et al. 马清林等. 1999. Chunqiu Shiqi Du Xi Qingtongqi Du Ceng Jiegou he Nai Fushi Jili Yanjiu 春秋时期镀锡青铜器镀层结构和耐腐蚀机理研究 (Evaluation of Tin Coatings on Bronzes of Ancient China). *Journal of Lanzhou University* (Natural Science Edition) 35(4): 67–72.

Neimenggu Wenwu Gongzuodui 内蒙古文物工作队. 1986. Maoqinggou mudi 毛庆沟墓地 (The Cemetery of Maoqinggou). In *E'erduosi Qingtongqi* 鄂尔多斯青铜器. Wenwu chubanshe 文物出版社, pp. 227–315.

Wang, Dadao 王大道. 1998. Yunnan chutu qingtong shidai zhufan ji qi zhuzao jishu chulun 云南 出土青铜时代铸范及其铸造技术初论 (A Tentative Study of Casting Molds and Casting Technology of the Bronze Age in Yunnan). In *Yunnan Kaogu Wenji* 云南 考古文集 (Collection of Papers on Yunnan Archaeology). Kunming 昆明: Yunnan Minzu Chubanshe 云南民族出版社, pp. 114–127.

Wu, En 乌恩. 1983. Zhongguo beifang qingtong toudiao daishi 中国北方青铜透雕带饰 (Bronze Belt Ornaments with Cutout Designs in Northern China). *Kaogu Xuebao* 考古学报, no. 1 第一期, 25–37页.

1985. Yin zhi Shang chu de beifang qingtongqi 殷至周初的北方青铜器 (Bronze Vessels found in Northern China during the Late Shang Period). *Kaogu Xuebao* 考古学报, no. 2 第二期, 135–156页.

1994. Luelun guaiyi dongwu wenyang ji xiangguan wenti 略论怪异动物纹样及相关问题 (Brief Discussion on Patterns of Fantastic Animals and Related

Questions). In *Gugong Buwuyuan Yuankan*, vol. 3 故宫博物院院刊, 第三期 (Journal of the Palace Museum, Beijing), 27–30页.

Wu, Xiaolong. 2004. Female and Male Status Displayed at the Maoxinggou Cemetery. In Katheryn M. Linduff and Yan Sun (eds.), *Gender and Chinese Archaeology*. Walnut Creek: AltaMira, pp. 203–235.

Yang, Fan, and Liqiong Li. 2003. Dian wenhua gaishu yu yanjiu (Outline and Study of Dian Culture). In Chinese National Museum and Culture Department of Yunnan Province (ed.), *Yunnan wenwu zhi guang: Dianwanggui wenwu jingpinzhan* (Dawn of the Civilizatin in Yunnan: Gems of the Dian Kingdom). Beijing: Chinese Social Sciences Press, pp. 30–40.

Yao, Zhihui 姚智辉. 2005. 晚期巴蜀文化青铜器技术研究及兵器表面斑纹工艺探讨. (A Technical Study of Bronzes from the Ba-Shu Region in the Fifth–Second Centuries B.C.E and a Discussion on the Surfacing Craft of Bronze Weapons). 北京科技大学博士论文. (Doctorial Dissertation of History of Science and Technology, University of Science and Technology, Beijing). (Advisor: Shuyun Sun), pp. 54–69页.

Yao, Zhihui 姚智辉, Shuyun Sun 孙淑云, and Lin Xiao 肖磷. 2005. Bashu binqqi biaomian hubanwen xingcheng jishu de yanjiu 战国巴蜀兵器表面斑纹技术研究 (A Technical Study of Silver or Black Mottle in the Surface of Ba-Shu Bronze Weapons). In *Zhongri diwujie jixieshi yu jixie sheji guoji xueshu yantaohui jiaolou* 中日第五届机 械史与机械设计国际学术研讨会议论文集 (Proceedings of the Fifth China-Japan International Conference on History of Mechanical Technology and Mechanical Design), Chiba, Japan, Nov. 27–30, pp. 102–109.

Yang, Fan 杨帆, and Liqiong Mei 梅丽琼. 2003. Dian wenhua gaishu yu yanjiu 滇文化概述与研究 (Outline and Study of Dian Culture). In *Yunnan wenwu zhi guang: Dianwangguo wenwu jingpinzhan* 云南文物之光: 滇王国文物精品展, (Dawn of the Civilization in Yunnan, Gems of the Dian Kingdom). Beijing: Chinese Social Sciences Press, pp. 30–40.

2005. *Kunming yangfutou mudi quansan* 昆明羊甫头墓地卷三 (Kunming Yangfutou Cemetery, vol. 3), pp. 864–867. Beijing 北京: Science Press 科学出版社. Zhongguo Qingtongqi Quanqi Bianji Weiyuanhui 中国青铜器全集编辑委员会.

1997. *Zhongguo Qingtongqi Quanqi* (6) 中国青铜器全集 (6) (Collection of the Chinese Bronzes). Xizhou (2) 西周 2 (Western Zhou). Beijing 北京: Wenwu chubanshe 文物出版社.

CHAPTER 11

Production and Social Complexity
Bronze Age Metalworking in the Middle Volga

David L. Peterson

R ESEARCHERS CONSIDER metal production to have been second only
to pastoralism as a form of productive economic activity in the
Middle Volga region throughout the Bronze Age (Vasil'ev et al. 2000:
8). This presents an interesting paradox. There is virtually universal
acceptance of the high social significance of early metal making in the
steppes, as shown particularly by the integral role of metal artifacts in
kurgan burial rites. However, metal appears to have been produced on
a relatively small scale in many areas, including the Middle Volga, for
much if not all of the Bronze Age. Although many archaeologists have
assumed that metal was important in the past because of its "self-evi-
dent usefulness or inherent attractiveness" (Sherratt 1997: 103), its value
was foremost a product of human knowledge, labor, and interactions,
in which even very small-scale production could have been of considerable local importance. The association of metalwork with kurgan burial
rites indicates that metal was not perhaps necessary for survival but may
have been a social necessity, at least for a certain level of social identity
that could be negotiated or maintained in interactions surrounding the
production and acquisition of metalwork. What is less clear is how participation in metal making, as opposed to maximization of output or
control of product, contributed to the emergence of social complexity in
the Middle Volga or elsewhere.

The metal-making process involved a temporally extended sequence
of activities that were often carried out by different hands in different
places (Fig. 11.1). The efforts of a number of producers (miners, smelters, metalworkers, those who raised the food they ate, etc.) contributed

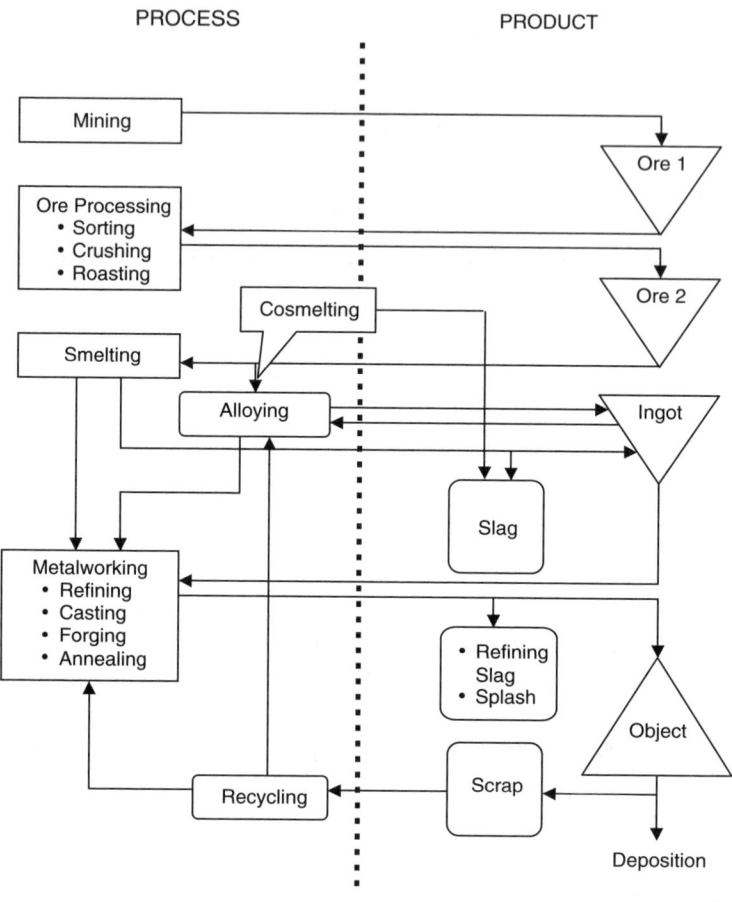

Figure 11.1. The overall metal-making process for ancient copper and bronze, including activities and products associated with recycling.

to the production of copper and bronze metalwork in Eurasia during the Bronze Age. Such geographic divisions of labor are what distinguish "metallurgical focuses" from "metalworking focuses" in E. N. Chernykh's (1992) metallurgical province system. If we judge from the small scale of metal production known from the Middle Volga before the specialized production known at Mikhailo-Ovsyanka during the Late Bronze Age (LBA) II period (Matveeva et al. 2004; Fig. 11.2), the efforts of producers, and the interactions that brought them together in the fabrication of finished objects, were more critical to the creation of metalwork than

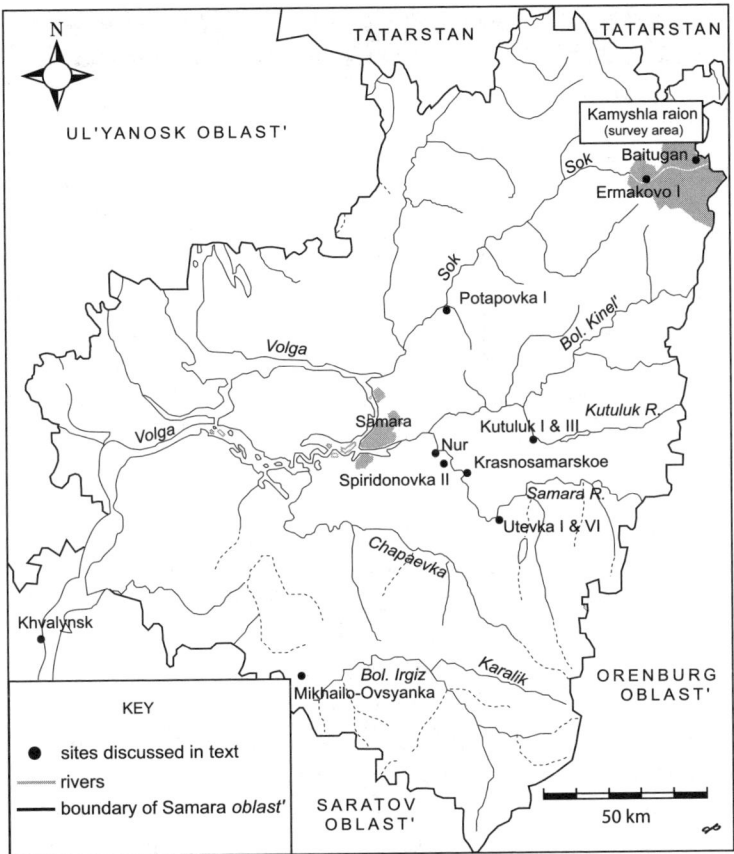

Figure 11.2. Map of Samara Oblast' showing sites and localities discussed in the text (drawing by David Peterson).

limits of supply alone. This does not mean that metalwork should be approached as simply the outcome of the time and energy that went into making it. In most cases, the knowledge, skills, and experience needed for metalworking could be transmitted and acquired only through the participatory learning that occurs in apprenticeship (Keller and Keller 1996). Therefore, the labor power associated with metal making could have been easily controlled through social restrictions on the transmission of skills and knowledge and on who could be a producer. Social distance and distinctions could also be created in the metal-making process itself, if divisions of labor and specializations arose in which some parties were forced into more onerous or hazardous occupations, like mining,

Period	Years BCE	Horizon Style
LBA II	1500-1300	Late Srubnaya
LBA I	1800-1500	Early Srubnaya
MBA II	2200-1800	Potapovka
MBA I	2800-2200	Poltavka
EBA	3300-2800	Yamnaya

Figure 11.3. Chronology of the periods in Middle Volga prehistory and archaeological style horizons that are discussed in the article.

while others were better rewarded for their efforts (cf. Shennan 1999a). Under these circumstances, it is likely that expansion in the scale of production and exchange of metal by one group would have been viewed as a threat to the status quo by many of their existing partners.

In this chapter, I examine some important aspects of the relationship between the production of metalwork and the emergence of social complexity in the Eurasian steppe, utilizing results of my recent investigation of Bronze Age metal making in the Samara Oblast', in the Middle Volga region of present-day Russia (Fig. 11.2). I examined metalwork from contexts relating to several Early to Late Bronze Age horizon styles in Samara (Fig. 11.3). Here I pay special attention to objects from the Potapovka I and Utevka VI kurgan cemeteries (Fig. 11.2). These are the principal sites of the Potapovka horizon, which has close affinities to sites of the contemporary Sintashta-Arkaim horizon further east. The latter are associated with the early complex societies of the southern Urals region that are the topic of several contributions to this volume.

The examination of metalwork from Samara, including pieces containing arsenical bronze that originated from the southern Urals, provides important details on the inter-regional relations that surrounded metalwork in the Volga-Ural in the late third to early second millennium BCE. Despite the intensification in metal production by at least some communities in the Urals at that time, the Middle Bronze Age (MBA) II inhabitants of the Samara area did not simply acquire the metalwork they placed in kurgan burials by exchange with more "metal-rich" communities to the east. There is good reason to believe that local artisans in the Middle Volga region of the Volga-Ural steppe and forest-steppe practiced recycling and maintained their own metal pools even within different areas of the Middle Volga. It was not simply the local ownership of metalwork but also the performance of metalworking in the creation of finished pieces that was key to providing the copper and bronze objects placed in kurgan burials.

It is possible that at least some of the increased output from the Urals was destined for exchange with new partners in Central Asia. What is just as significant, or maybe more so, is the evidence that copper began to be cast in consolidated ingots at this time (Zaikov and Zdanovich 2000: 83, figs. 4.6, 85), an indication that the commoditization of metal itself was underway. In this way, an old form of mystification may have been traded for a new one, as authority vested in socially restricted practices may have begun to be superseded by production for exchange, and the closer association of the value of metal products with the material itself. Processes like this, which entail the emergence of new regimes of value, are not merely economic but are inevitably political as well (Appadurai 1986). Such developments may have been viewed by other partners in Samara and neighboring regions as a threat to older systems of status and authority that incorporated metal making. This might help to explain why the shift that occurred in the production and consumption of metal in Eurasia at this time appears to have contributed to growth in militancy and conflict in the steppes by the late third millennium BCE. This includes the implications of the mass grave at Pepkino (Chernykh 1992: 201; Koryakova and Epimakhov 2007: 62–63, 66), as argued by Anthony in this volume, as well as the emphasis on weaponry that is a hallmark of metalwork from Potapovka I and Utevka VI (Agapov and Kuz'minykh 1994; Kuznetsov and Semenova 2000; Vasil'ev et al. 1992, 1994).

Metal and Authority

AN EXAMINATION of the cross-cultural significance of metal making suggests the kind of significant relationship between metal and political ritual authority that may have also existed in the Bronze Age, and what may have been at stake in the shift in production and consumption that was underway in the MBA II period in the Volga-Ural. Myths and religious texts indicate the strong associations that people in the past drew between metal and divinity, temporal authority, and moral imperatives, in which the significance of metal and metalwork extended well beyond their sheer economic importance. For example, in the Scythian creation story reported by Herodotus, the Scythian *ethnea* came into being as the first king, Kolaxais, seized four golden objects that had been fashioned by the gods as they came down from the sky (*Histories* 4.5–7). Thereafter, the lives of the Scythian kings are said to have depended on proper treatment of the sacred gold, which they were obligated to display and renew annually with public sacrifice and feasting lest they meet with an untimely death in the year that followed (Lincoln 1991: 192).

This legend encompasses a number of associations that ethnographers and historians commonly link with metal making. One is the spirit that is frequently associated with an object's donor(s) (Mauss 1990). Another attributes metal with a supernatural origin or traffic with outside forces, either divine or sinister, including the positive or negative status traditionally given to metal producers. Although these elements often lack clear material correlates, Budd and Taylor (1995) and Haaland (2004) also blame unproductive modern biases for the frequent hesitance to consider these aspects in archaeological accounts of ancient metalwork. Ethnographic and historical accounts of the ritual, symbolic, and transformative significance of metal, metalworking, and its products have a role to play in improving the present understanding of the relationship between metal production and early social complexity in the Eurasian steppe, particularly at the level of explanation and interpretation.

A common feature in such accounts is the animistic association of metal production with gestation, in which metal producers are often attributed with the power to unnaturally accelerate this process in earthly matter (Eliade 1978; Childs 1991: 40). One of the most important figures in the history and anthropology of Eurasia is the shaman, whose drum and the forge are sometimes metaphorically interchangeable. The trance sought through drumming might also be achieved in the hammering associated with metalworking. Shamans are commonly associated with mastery of fire, including direct or symbolic participation in working iron (a material credited with the potential to control hostile spirits), including the creation of a "body of iron" that made the shaman resistant to hostile spiritual forces (Eliade 2004). This is a particularly literal example of the anthropomorphic analogies that frequently crop up in relation to smelting and metalworking, in which metal making may be viewed at the same time as the perfection of the self. This further indicates the attraction of controlling the circulation of even a small level of output, and the potential that participation in production often held for legitimizing authority.

This may be true of earlier periods as well. As Budd and Taylor (1995: 139) have stated, the shamanic charisma of the metalworker may have coincided with public authority in Bronze Age Europe, as "the ability to put on a show of colorful, transmogrifying pyrotechnics may have commanded considerable respect." Copper-based figurines from Galich and elsewhere in the pre-Urals region (Chernykh 1970: fig. 63.17–18; 1992: pl. 20) may be evidence of similar associations between metal making and political authority in Bronze Age Eurasia. These anthropomorphic statuettes have what appear to be axe blades and tanged knives sprouting from the head and shoulders, suggesting the incarnation of forces associated with metal in human form. Alternatively, they may depict a divinity or culture hero as the giver of bronze tools and weapons, which are represented as emerging from the very body of the figure itself.

There is no way to definitively prove that these understandings were at play in the Bronze Age steppes, but they are suggested by the tantalizing glimpses that archaeological remains provide of the ceremonial importance of metal and metalworking tools, such as founder's burials in which casting molds and other tools were placed in the grave (e.g., the collective grave at Pepkino, Chernykh 1992: 201; Koryakova and Epimakhov 2007: 62–63, figs. 1.2 and 2.10). In this way, the apparent relationship between metalwork and social identity may have extended to metal *working* and social identity as well (Peterson 2007: chap. 3). In kurgan burial ceremonies, not only metal objects but also molds and other pieces that objectified the power associated with the fabrication of metalwork may have served as an index of the social prominence of the deceased and their survivors. The social significance of metal may have been situated not only in metalwork as a "fetishized commodity" (McLellan 2000: 435–443) but, like shamanism, also in the knowledge and practice of the activity itself.

Metalworking

THE RESULTS of the archaeometallurgical analysis of Bronze Age metalwork from Samara that are summarized in this chapter allow me to provide some sense of the complex political economy that surrounded metal making in the Middle Volga and its relationship to broader developments in the Volga-Ural (Peterson 2007). In drawing associations between metal and early social complexity, researchers frequently emphasize large-scale mining, the generation of surpluses, long-distance exchange, and the accumulation of wealth in burials. However, just as the activities that were part of Eurasian pastoralism were not uniform across space and time, and a certain level of complexity was not a blanket condition at any given (pre)historical moment (see respective contributions by Popova and Frachetti in chapters 16 and 3 in this volume), the practices involved in metal production were also subject to change and a wide degree of variation.

Identifying the metal-making activities that were carried out in a site or region during a particular period is important to understanding the relationship between metal and social complexity in that instance. Without such an understanding, there is a risk of evaluating every case in terms of a standard of large-scale, industrial production. While there is evidence in some cases for production on a substantial scale (e.g., Chernykh 2002), in others it may be less representative of how early metal production was carried out than of our own historical experience and anachronistic ideas of how production should have operated (Budd and Taylor 1995).

In general, metalworking is examined far less often than other aspects of metal production, even though it provides an essential link between mining and smelting, on the one hand, and the consumption of finished metal objects, on the other, in the overall metal-making process. Metalworking was a key activity in the earliest exploitation of gold and native copper (Muhly 1980; Renfrew 1986) and remained the dominant form of metal production in some regions during not only the Late Eneolithic but also the Bronze Age (Taylor 1999). There is a variety of evidence such as metal scrap and small, unconsolidated ingots, as well as data from metallography and EPMA-WDS analysis, that indicates that metalworking was practiced locally within Samara during the Bronze Age and involved the maintenance of local metal pools and frequent recycling (Peterson 2007: chap. 7).[1] Although discussions of ancient Eurasian metalwork have frequently assumed a direct correlation between the element composition of the metal in objects and their sources, practices such as alloying and recycling can alter the element profiles of copper and bronze to the point that the origin of the materials found in individual pieces is hopelessly obscured. Recycling also has significant socio-economic implications, as it changes the production cycle from one that is initiated with mining to another in which metalworkers may operate more independently from miners and smelters (Fig. 11.1). A detailed understanding of the metalworking practices that occurred after the "primary" production of metal from ore is therefore crucial to addressing basic questions concerning the origin of materials and objects, as well as the socio-cultural significance of metalwork to the Bronze Age inhabitants of the steppes. This significance included the impact of broad shifts in production and consumption, and the divisions of labor and society that accompanied emerging social complexity.

Bronze Age Metalworking in the Middle Volga

MOST OF the Middle Volga region is encompassed by the Samara Oblast', covering an area of 53,600 square kilometers at the Volga "bow" (Russ. *luk*) and its confluence with the Samara River, which begins as an offshoot of Ural River to the east (Fig. 11.2). The Samara Valley links the Middle Volga to extensive ore deposits in the Urals, which provided copper for metalwork in the western steppes by the third millennium BCE, as, for example, in the early utilization of Kargaly (Chernykh 2002). The Volga also connects the Samara area, by way of the Caspian Sea and its western shoreline, to sources of copper and arsenic bronze in the Caucasus. Arsenic bronze appeared in steppe metal assemblages by the Middle Bronze Age. The foundation of the city of Samara dates to 1586, when the Samarsky fortress was built at a strategic point in relations with steppe nomads to the southeast (Pavlov 1996: 11, 15). Perhaps it is its strategic position at the crossroads of Europe and Asia, and the southern steppes and boreal forests, that made the Volga-Ural so important in the earlier development of expansive prehistoric networks, especially those which appeared in the Late Bronze Age in association with the Srubnaya horizon style (Bochkarev 1991). Previously, 70–80% of some 300 Bronze Age metal artifacts that had been recovered in Samara had never been analyzed. Altogether, these factors have made Samara a promising setting for a long-term social history of early metalwork in Eurasia.

The present research has had two main components (Peterson et al. 2006; Peterson 2007: chaps. 6 and 7). The first is an archaeological survey performed in the Kamyshla *raion* of northeastern Samara, with the goal of investigating the contribution of local copper production to the consumption of copper-based metalwork during the Bronze Age.[2] The second is the archaeometallurgical analysis of the materials, techniques, and properties associated with samples of Bronze Age metalwork examined from Samara.[3] This analysis centered on element composition, utilizing electron probe microanalysis with wavelength-dispersed spectrometry (EPMA-WDS, or simply WDS); Vickers hardness testing, for comparison of the relative hardness of different materials as they were worked in different objects; and metallography, for the micro-structural examination of the metalworking techniques that were used by the smiths who made the objects (Goldstein et al. 2003; Northover 1998, 2002; Scott 1991). Techniques are an important area of research in the anthropology of technology for which there may be little or no site-based evidence, but they can still be addressed in metal artifacts through a program of analysis that includes metallography. The combination of these methods allowed for a more detailed characterization of the fabrication of ancient copper-based metalwork than would have been possible by any individual mode of analysis alone (Northover 1985, 1989).

The objectives of the WDS analysis were to determine the material selected for each object and class of object, to ascertain how these preferences changed with time, and to attempt to identify sources of materials by patterns in the impurities present in the metal. These patterns did not show the 1:1 correspondence between objects and individual sources of ore. Instead, they provided important evidence of the maintenance of metal pools by different groups of smiths operating in the Middle Volga in the late third to early second millennium BCE.[4] In addition, relationships were sought between impurity patterns and periods, alloys, and object class. Metallography was used to examine other relationships between techniques of metalworking and material, object class, and context. It has often been assumed that bronze was utilized for its potential for greater hardness than unalloyed copper, but the properties achieved in individual instances depended upon how the object was worked. Malleability and toughness instead appear to have been among the most important properties that metalworkers exploited in early bronze in the Middle Volga (Peterson 2007: chap. 7).

Samples from 86 objects were examined from Samara.[5] Samples were taken from two different areas on two of the objects, bringing the total number of analyzed samples to 88. The forms included a variety of tools, weapons, fasteners, and ornaments (Fig. 11.4). The major metal groups identified by WDS were copper (45 samples), arsenic bronze (26 samples), and tin bronze (13 samples). One small ring was made of 99 wt% silver (Fig. 11.4: 11), and "spiral" pendants were covered over with a very thin foil of electrum (gold and silver), only 0.1 millimeter thick (e.g., Fig. 11.4: 16). For each sample, at least 5, and in many cases, 10 analyses were performed for 16 different elements: Fe, Co, Ni, Cu, Zn, As, Sb, Sn, Ag, Bi, Pb, Au, S, Al, Si, and Mn. The normalized results for each element were averaged for each object and form the basis of the present discussion of element composition.

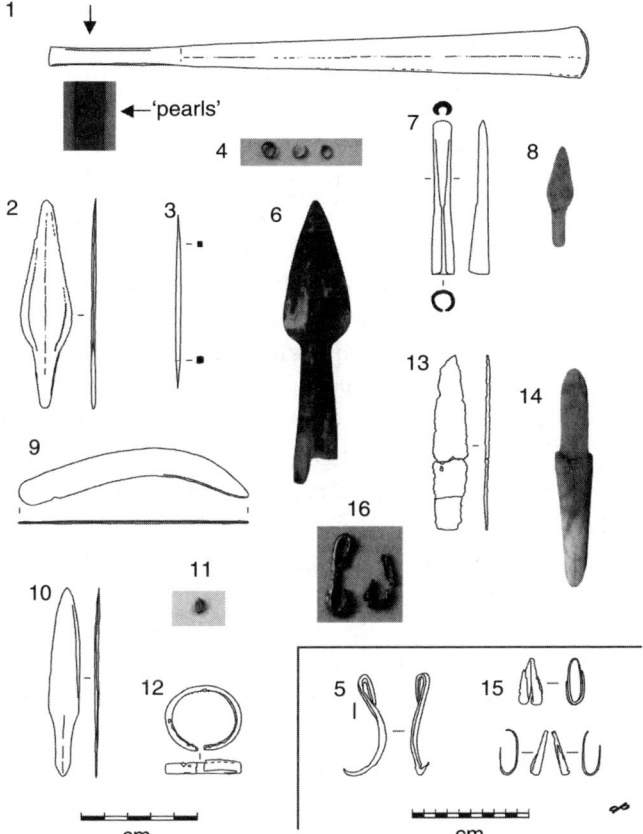

Figure 11.4. Examples of the metalwork examined in the study. *Yamnaya-Poltavka horizon:* (1) Kutuluk I, kurgan 4, grave 1; (2–3) Kutuluk III, k. 1, g. 2; (4) Nur, k. 2, g. 4; *Potapovka horizon:* (5) Utevka VI, k. 6, g. 2; (6) Utevka VI, k. 6, g. 4; (7) Utevka VI, k. 6, g. 6; (8) Potpovka I, k. 3, g. 5; (9) Utevka VI, k. 6, g. 6; (10) Utevka VI, k. 6, g. 1; (11) Potapovka I, k. 3, g. 5; (12) Utevka VI, k. 6, g. 11; (13–14) Grachevka II, k. 8, g. 8; *Srubnaya horizon:* (15) Spiridonovka II, k. 6, g. 35; (16) Nizhnyaya Orlyanka, k. 3, g. 8 (1–3, 14: copper; 4–10: arsenical bronze; 11: silver; 12–13, 15: tin bronze; 16: electrum foil over copper/bronze) (drawings by David Peterson).

Out of the 86 objects examined, 5 were from the Early and Middle Bronze Age Yamnaya and Poltavka or Yamnaya-Poltavka horizons (Figs. 11.3 and 11.4: 1–4). During the Yamnaya and Poltavka periods (EBA and MBA I), metalworkers in the Middle Volga produced copper implements common to the Circumpontic region in the third millennium BCE (Chernykh 1992: 83–91, 132–133). These included tanged knives and daggers, tetrahedral awls and chisels, adzes, and occasionally shaft-hole axes (Fig. 11.4: 2–3). The social significance of these forms is indicated by their limited presence in burials and general absence from other contexts, which belie their deceptively utilitarian nature. An early example of an unusual object that appears to have had high social significance is a large copper staff from a kurgan burial at Kutuluk I (Figs. 11.2 and 11.4: 1), radiocarbon dated to circa 2930 BCE (Kuznetsov 1991). It was found cradled scepter-like in the left arm of the deceased, apparently as a symbol of status and/or authority (Kuznetzov 1991: 138). The staff has stylistic affinities with some shaft-hole axes in the form of a few worn-down, pearl-like beads in the grip area (Chernykh 1992: fig. 28.24). These are similar to the "pearls" on some Yamnaya ceramics in the Volga-Ural, skeuomorphs of the rivets in contemporary sheet-cauldrons in the northern Caucasus (Mochalov 2008), and daggers from the northern Black Sea region (Anthony 1996).

Following the conventional identification of an alloy by the presence of admixtures at levels of 1% or more by weight (wt%), the WDS values show the presence of 26 arsenic bronzes among the samples, with arsenic content as high as 14 wt%, but mostly centered on 1.5–2.5 wt% (Fig. 11.5). Three date to the MBA I (Poltavka) period and are from the Nur cemetery, kurgan 2, grave 4 (Figs. 11.2 and 11.4: 4). Nineteen are from Potapovka horizon burials (MBA II), in the form of tools, weapons, and ornaments, including the chisel and socketed spearhead from burials in kurgan 6 at Utevka VI (Figs. 11.2 and 11.4: 5–7, 9–10). In relation to the concentration of arsenic values at the lowest end of distribution, the numbers that cluster around 1.5–2.5 wt% at first seem as though they could possibly be part of the higher peak for values in a bimodal distribution, in which the higher peak would represent bronze, and the lower would correspond to the presence of arsenic as an impurity in the copper. However, the log-normal distribution curve for arsenic concentrations (Fig. 11.5) instead suggests that concentrations of arsenic in the overall metal pool were in the process of diminishing to practically nothing, as occurs when copper and bronze with higher and lower levels of arsenic are mixed together numerous times over an extended period (P. Northover personal communication). The way in which values above 1 wt% blend into the group below indicates the probability of frequent recycling.[6]

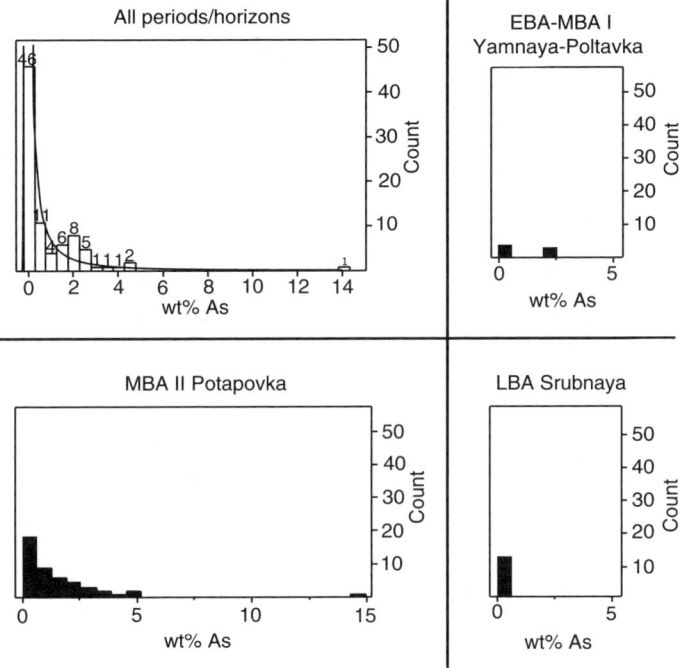

Figure 11.5. Arsenic concentrations in the Bronze Age metalwork from Samara based on WDS results. The log-normal curve for all is superimposed at top left.

There are fewer tin bronzes among the samples, 17 in all. One of the main reasons is that the majority of the objects examined (N = 62) belong to the MBA II Potapovka horizon, and tin bronze is more strongly associated with the Late Bronze Age. Most of the tin bronzes are Srubnaya ornaments (e.g., Fig. 11.4: 15). The exceptions are a knife from Grachevka II, kurgan 8, grave 8 (Fig. 11.4: 13), and five brace-lets from other Potapovka horizon graves (e.g., Fig. 11.4: 12). Unlike the WDS results for arsenic, in those for tin there is a sharp cutoff between the levels that indicate alloying and those associated with the presence of tin as an impurity. The tin bronzes in this case start around 3 wt%, while the highest tin value below that is only .08 wt% (Fig. 11.6a). The wide spread of alloy-level values for tin may be a further indication of recycling practices, but in which tin bronze was recycled less frequently than arsenic bronze, and in which the two alloys were kept separate.

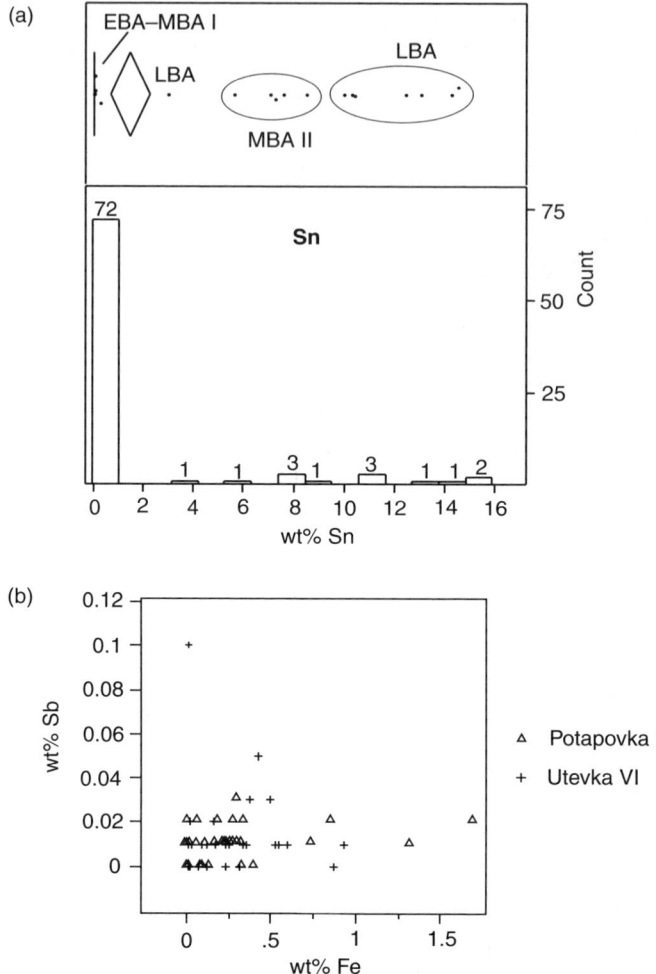

Figure 11.6. (a) Tin concentrations in the Bronze Age metalwork from Samara based on WDS results. As noted, the highest values are for MBA II and LBA, the lowest for EBA–MBA I. (b) Potapovka horizon: scatterplot of values for concentrations of antimony (Sb) v. iron (Fe), from WDS results for metalwork from Potapovka I and Utevka VI.

This prevented the formation of tin-arsenic bronze as each was apparently mixed with unalloyed copper but not with one another.[7]

A multivariate analysis of the 15 elements measured besides copper showed the strongest correlation between nickel and arsenic (.55, or a little more than 50%). Within the scatterplot of values for nickel and arsenic are groupings related to both period and cultural affiliation (Peterson 2007: fig. 7.81). These suggest that the metalworkers who forged the pieces analyzed from the Potapovka horizon drew on a greater number of sources than those who came before and after them.[8] In light of the increase in metal production that began in the southern Urals around the turn of the second millennium BCE, it is quite likely that this pattern is in part the result of the circulation of copper from a greater number of sources in the MBA II period than previously in the Middle Volga.

When values for iron are plotted against those for antimony in the WDS results for samples from Potapovka I and Utevka VI, it appears that the smiths who made these two groups of objects may have drawn from different, overlapping metal pools (Fig. 11.6b).[9] The Potapovka I group tends toward higher iron and lower antimony, whereas that for Utevka VI tends toward higher antimony and lower iron. One way to explain these differences is to attribute the objects in each cemetery to different groups of metalworkers, who maintained their own networks for the acquisition of materials and engaged in slightly different practices in fabricating the pieces. These metalworkers may have lived and worked within separate pastoralist communities centered on the Sok and Samara river systems on which the cemeteries are respectively located (Fig. 11.2).

The metallographic analysis yielded equally intriguing results. The samples were examined at 75x to 750x magnification. The micrographs (Peterson 2007) are summarized here in terms of work pattern in Figure 11.7. In the 81 samples from Samara for which work pattern (WP) was identified, four basic patterns were encountered: WP 1, cast and only lightly cold-worked objects (rare – in only 1 sample or ~1% of the total); WP 2, objects that were cast and then annealed, restoring the malleability of the cold-worked metal, followed by additional cold work (42 samples or ~52%); WP 3, objects that went through all the preceding steps but were also heavily cold worked to the point of heavy cold-work reduction, creating a "flattened" or "feathered" appearance in the crystal grains (34 samples or ~41%); and WP 4, objects with a micro-structure of relatively small crystals overall (Fig. 11.7). The last was probably a result of many bouts of alternating cold work and annealing (infrequent – 4 samples or ~5%).

The one sample that was made utilizing WP 1 is the knife from kurgan 8, grave 8 in the Grachevka II cemetery, which is assigned to MBA II or the transition from MBA II to LBA I (Figs. 11.4: 13 and 11.7: 1).

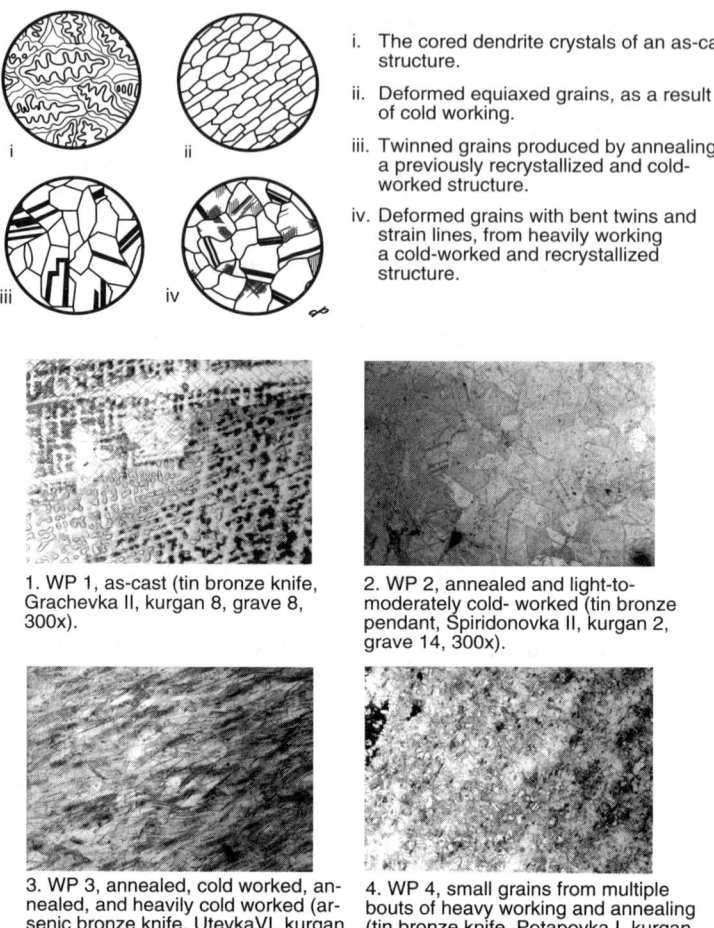

i. The cored dendrite crystals of an as-cast structure.

ii. Deformed equiaxed grains, as a result of cold working.

iii. Twinned grains produced by annealing a previously recrystallized and cold-worked structure.

iv. Deformed grains with bent twins and strain lines, from heavily working a cold-worked and recrystallized structure.

1. WP 1, as-cast (tin bronze knife, Grachevka II, kurgan 8, grave 8, 300x).

2. WP 2, annealed and light-to-moderately cold- worked (tin bronze pendant, Spiridonovka II, kurgan 2, grave 14, 300x).

3. WP 3, annealed, cold worked, annealed, and heavily cold worked (arsenic bronze knife, UtevkaVI, kurgan 6, grave 4, 300x).

4. WP 4, small grains from multiple bouts of heavy working and annealing (tin bronze knife, Potapovka I, kurgan 1, grave 4, 300x).

Figure 11.7. Work patterns (WPs) identified in the Bronze Age metalwork from Samara. Drawings are representations of some of the important microstructures associated with the work patterns (drawings by David Peterson after Scott 1991: fig. 12).

The object is a medium tin bronze (Sn 5.7 wt%) that is heavily corroded. The form is unusual, apparently a one-sided blade with a straight or slightly curved back, and is found in the Sintashta-Arkaim horizon (Gening et al. 1992: fig. 153.20) but is more typical of the East Urals in the Late Bronze Age (Chernykh 1992: fig. 82.14, 16). Although the sample from this knife was the only one showing the dendritic structure of unworked (or very lightly worked) metal, which had also not been annealed, there were 21 (~26% of samples) with the residual coring (or "ghosts") associated with partial retention of the "as-cast" structure.[10] These samples may provide clues to differences in the control of annealing temperatures with copper and bronze. Twelve samples, or more than half the total with residual coring, are of unalloyed copper, while, of the rest, 6 are arsenic bronze and 2 are tin bronze (Peterson 2007: table 7.5). This presents an interesting pattern, because out of a combined total of 45 objects of arsenic and tin bronze, only 8 were found to have residual coring (~18%), as compared to a total of 12 out of 44 for unalloyed copper (~27%). It suggests that the metalworkers who made the objects may have taken greater care in annealing bronze than copper, even though the former requires a temperature of 2 to 2.5 times greater than that needed to re-crystallize the latter.[11]

The number of samples that date earlier than MBA II is small; therefore, observations for them do not carry the same weight as those for the Potapovka horizon materials. Of the Yamnaya and Poltavka metalwork examined metallographically, two are of unalloyed copper fabricated by WP 2 (Fig. 11.4: 1–2). The other three are the arsenic bronze rings from the Nur cemetery (Fig. 11.4.4), relatively rare examples of Poltavka ornaments and early low-arsenic bronzes (~2 wt% As). Two of the rings from Nur were worked with WP 2, the third with WP 3. The Potapovka horizon objects include a large majority of the metalwork from the Potapovka I and Utevka VI cemeteries, the two most heavily investigated sites of the Potapovka horizon (Kuznetsov and Semenova 2000; Vasil'ev et al. 1992, 1994). Out of a combined total of 60 samples examined from both cemeteries, the frequency of WP 2 and WP 3 is quite close, with 27 samples showing WP 2 (46%), as compared to 29 with WP 3 (49%). Only 4 represent WP 4 (5%).

It is more informative to compare the results for samples from the Potapovka I and Utevka VI cemeteries separately from the others (Fig. 11.8). Samples from nearly as many objects were examined from Potapovka I (N = 29) as from Utevka VI (N = 30), but there is a notable contrast in the frequency of WP 2 and WP 3 in the samples from these sites that almost amounts to a reversal in pattern for the two sets. As Figure 11.8 shows, the greatest number of samples for Potapovka I exhibit WP 2 (N = 18), whereas there are almost the same number for WP 3 from Utevka VI (N = 19). In contrast, there are only half as many samples exhibiting WP 3 as there are WP 2 from Potapovka I (N = 9). The situation for Utevka VI is reversed, with only 10 samples that show

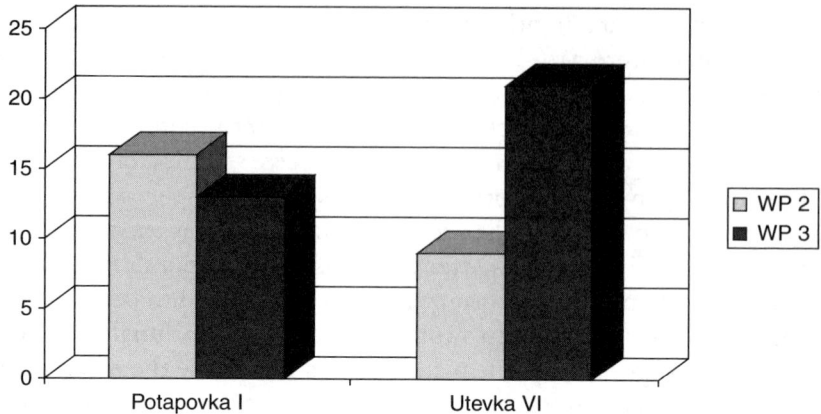

Figure 11.8. Potapovka horizon: comparison of frequency of work patterns (WP) in the metalwork from the Potapovka I and Utevka VI cemeteries.

WP 2, or approximately half the number for WP 3. These differences in degree of final cold working, and the reversal in pattern, correspond to the contrasting pattern discussed earlier in the EPMA-WDS values for iron and antimony in the samples from these sites (Fig. 11.6b). What initially seemed to be important differences in networks for the acquisition of materials also appear to relate to more fundamental distinctions in metalworking practices. This adds support to my argument that the metalwork from these cemeteries was made by different groups of workers. There appears to have been a greater tendency toward moderately cold-worked copper objects in the assemblage from Potapovka I and for heavily cold-worked arsenic bronzes at Utevka VI. For the assemblages from both sites, the majority of samples was of implements (tools, weapons) rather than ornaments (rings, bracelets, pendants, and sheet ornaments), but the group from Utevka VI contains the greatest number of copper and arsenic bronze implements exhibiting WP 3 as opposed to WP 2. The forms in these two groups are very similar, and the differences in metalworking were not related to functional differences in the objects from the two sites. Arsenic bronze is more prevalent in the metalwork from Utevka VI, which would have been more malleable and therefore more easily worked than copper. The nearest well-known source of arsenic bronze utilized in the Bronze Age is Tashkazgan, in the southern Urals region of Kazakhstan (Chernykh 1970, 1992). Any distinctions that the owners made between objects fabricated with arsenic bronze and those with copper may have further related to the networks that were needed to acquire the former, which was less readily available than the latter.[12]

The remaining samples date to the MBA II (Grachevka II) and LBA I periods (Spiridonovka II). Of the former, the tin bronze knife with WP 1 was already described. The other four samples examined from Grachevka are also from knives. Three are copper and one is arsenic bronze. One copper knife was fabricated with WP 3, while the other three samples exhibit WP 2. The group of samples from Spiridonovka II is composed of ornaments, with the exception of one copper needle. The majority was worked by WP 2.

Discussion and Conclusions

WHAT IS the relationship of metalwork and associated practices in the Middle Volga to broader social developments, not only in the area of Samara but in the greater social landscape of the Volga-Ural? The social standing of some individuals in the Middle Volga at the end of the Middle Bronze Age, and their elevated position in relation to a wider society, has been calculated in terms of the efforts made in constructing kurgan burials, grave goods, and the animals consumed in funeral feasts and sacrifices (Vasil'ev et al. 1992, 1994). Much the same might be said of the findings for the Sintashta excavations, which appeared in print just as the research on Potapovka I and Utevka VI was completed and being prepared for publication (Gening et al. 1992). Mortuary rites, including feasting, sacrifice, and burial assemblages, are all potentially important indices of social identity and complexity. However, "social complexity" and "complex society" are labels usually reserved for situations of urbanization or early state formation. Scholars of the Bronze Age in the steppes have been largely left to consider the relatively modest remains of so-called middle-range societies of tribes and chiefdoms (Shennan 1999b), which lack the institutional framework of early cities and states. With the general rejection of neo-evolutionism, they also have lost their former relevance as a universal stage in state formation (Carneiro 1981; Kirch 1984; McGuire 1983; Bawden 1989; Earle 1991, 1997; Shennan 1993; Yoffee 1993; Wright 1994; Smith 2003: 95–96). This is perhaps no more true than in the prehistoric steppes, where there were cyclical alternations between more and less complex social forms, and in which the nature of local complexity was variable and changing (Koryakova 1996, 2002; see Frachetti, Chapter 3 in this volume). In fact, some have gone so far as to suggest that Eurasian steppe pastoralists are better approached in terms of social "stagnation" than evolution (Gellner 1994: x). This is problematic in judging developments in the steppes according to the standards established for other regions.

With the Sintashta-Arkaim phenomenon, archaeologists working in the Eurasian steppe have encountered a more nuanced form of complexity beginning in the MBA II period. The site evidence for Sintashta-Arkaim includes not only burials (Gening et al. 1992; D. G. Zdanovich 2002; Epimakhov 2005) but also settlements, which, if not proto-urban, still exhibit a new degree of centralization relative to other Bronze Age steppe horizons and the institutionalization of planned, permanent settlement for relatively large groups (G. B. Zdanovich and Zdanovich 2002). Although their appearance has been linked to pressures associated with climatic deterioration, they appear to have been accompanied by social problems with social foundations: new attitudes of hostility relative to the earlier Bronze Age, and more pronounced cultural differences and an increasing sense of social distance between neighboring groups. These changes in the social landscape of the prehistoric steppes in the MBA II period (Koryakova 2002) were accompanied by changes in the scale of the production of metal. The establishment of Sintashta-Arkaim settlements, which heralded the beginning of a new era of permanent habitations and sedentism in the Volga-Ural, accompanied the appearance of intensified metal production in the southern Urals. This was not a blanket condition of communities in the region. At many sites, the evidence indicates that output remained relatively modest (see Hanks, Chapter 9 in this volume). In this volume (Chapter 4), Anthony details the chronology of potentially inter-related developments that led up to the heightened militarism of the MBA II period and walled, fortress-like construction at some Sintashta-Arkaim settlements. Another feature may have been the initiation of the production of copper for exchange with the contemporary Bactria-Margiana oasis settlements in Central Asia.

In the absence of early state formation or unambiguous indications of urbanization, aside from burial sociology, evidence for inter-regional interactions seems to be one of the most promising avenues for augmenting current knowledge of emerging complexity in the Volga-Ural (including contacts to the south and north beyond the steppe or forest-steppe zones). The evidence presented here indicates that, during the MBA II in the Middle Volga, not only the placement of metal in burials but also the practice of metalworking itself were socio-culturally significant activities important to establishing authority at a local level. It has long been recognized that copper from the southern Urals region contributed to consumption in surrounding regions, especially to the west, including the area of present-day Samara (or Kuibyshev, as it was known in the Soviet era) (Chernykh 1970, 1992; Agapov and Kuz'minykh, 1994). This might have placed leaders in the Middle Volga in a very tenuous position. Given the lack of the indices of complexity associated with the Sintashta-Arkaim settlements, one might assume that the MBA II inhabitants of the Middle Volga were junior partners in relation to their counterparts in the southern Urals. Although many copper sources are not easily distinguished from each other by trace elements (Chernykh 1970: 37–38; 1992: 19), Tashkazgan has been established as one of the primary sources of early arsenic bronze imported to the Samara area (Agapov and Kuz'minykh 1994: 171–172) and elsewhere in the Eurasian steppe (Chernykh 1970). According to the results of Agapov and Kuz'minykh (1994: 171–172), however, the apparent shift in networks for the acquisition of metal to the Volga-Kama that occurred by the LBA I period (especially bronze) was already underway by MBA II. Even for arsenic bronze, the degree of dependence of Middle Volga pastoralists on Tashkazgan or other eastern sources is far from certain.

While my main concern here has been with developments in the Middle Volga, it is no less important to scrutinize the evidence for metal-making practices in other regions before their relationship to broader social developments can be fully appreciated. I have been very general in my treatment of the MBA II social landscape in the southern Urals, if only to point out the latitude for heterarchical relations that existed in dealings with communities in the Middle Volga, even if metal production at a substantial scale had become a significant source of wealth and power to the east (Vinogradov 2003; D. G. Zdanovich 2002). Judging from the evidence from many Sintashta sites such as Ol'gino and Stepnoye, this is far from certain. In many instances, smaller-scale activities may have been as important as they were in the Samara area (see Hanks, Chapter 9 in this volume).

Despite the incorporation of metal that originated from other regions, recycling and the maintenance of local metal pools by metalworkers in the Middle Volga created a certain degree of autonomy in their access to materials, as opposed to blanket dependence on distant sources. What could not be replenished easily through local production was bronze, which became diluted in local metal pools through recycling. While this was acquired through networks centered on Tashkazgan, it was also available from sources in the Volga-Kama as well. If partners in the southern Urals sought to impose conditions on the access to bronze and copper that were disagreeable to their counterparts in the Middle Volga, the latter could turn to other partners to the north. Therefore, the increase in social complexity in the southern Urals region at the end of the Middle Bronze Age cannot be attributed to merely an "organic" dependence of metalworking regions on mining centers. It instead involved the purposeful re-orientation of society in the southern Urals and the development of a new regime of value that may have included the commoditization of metal, production for exchange with new partners in Central Asia, and a stronger association of the value of metalwork with the material itself, rather than the skills and knowledge of metal-workers.[13] The evidence for the lengths taken to maintain local metal pools in the Middle Volga could be a sign of resistance to such changes in communities at the eastern end of the Volga-Ural. The renewed emphasis on metalworking associated with the contemporary Seima-Turbino phenomenon (Chernykh 1992; Chernykh and Kuz'minykh 1987) may further represent an invigoration of traditional modes of interaction and envaluation of metal, which incorporated public ceremony, sacrifice, and exchange of elaborate and finely crafted pieces of metalwork, rather than a more "disembedded" trade in materials. The latter, coupled with an increased scale of production, was only one possible means of establishing authority, social prominence, and/or dominance through control of the production and consumption of metal. The detailed investigation of metal-making practices in the Middle Volga shows the importance of metalworking to achieving such ends locally in the MBA II, even under circumstances of small-scale production. Similar research may eventually reveal that the same was true of many communities in the southern Urals and other regions.

Acknowledgments

I WISH to thank the organizers of the Eurasian Steppe Symposium at the University of Pittsburgh, faculty and students alike, especially Bryan Hanks and Katheryn Linduff, for their invitation to participate, and their thoughtful comments and efforts in relation to this contribution. The final version benefited greatly from my interactions with the symposium's participants, particularly those whose contributions I have cited, and from the comments of two anonymous reviewers. I am also grateful to the many individuals and institutions that provided assistance and support for the research presented here, as part of my doctoral dissertation in anthropology at the University of Chicago. Funding was generously provided through a grant from the Wenner-Gren Foundation for Anthropological Research. This material is based on work supported by the National Science Foundation under grant no. 0431940. In addition to co-directing the survey portion of the research with me, my good colleagues Pavel F. Kuznetsov and Oleg D. Mochalov (Institute for the History and Archaeology of the Volga, and Samara State Pedagogical University) made metal artifacts from Samara available for me to sample. I carried out the EPMA-WDS analysis and hardness testing at the Department of Materials, University of Oxford, with the principal assistance of Peter Northover and Chris Salter. Any errors are, of course, my own.

Notes

1. The ingots are unconsolidated in the sense that they appear unworked and "fresh" from the furnace or crucible (Vasil'ev et al. 1994: fig. 30.11), unlike the bun- or lozenge-shaped ingots noted earlier from Arkaim and elsewhere.
2. The results of the survey are discussed only briefly in this chapter, but are already available in Russian and English (Kuznetsov et al. 2005; Peterson et al. 2006; also Peterson 2007: chap. 6).
3. Comparative samples of Early Bronze Age metalwork from Velikent, Daghestan (northeastern Caucasus) were also included (Peterson 2007: chaps. 5, 7, and 8). Regrettably, these had to be omitted here because of limits of space. I am grateful to the directors of the Daghestan-American Velikent Expedition, Philip Kohl (Wellesley College), and Rabadan G. Magomedov and the late Magomed G. Gadzhiev (Institute of Archaeology and Ethnography, Makhachkala, Daghestan), for their generous efforts in providing these samples.
4. Despite new applications, researchers are now generally skeptical of trace element analysis of sources, because of the alterations that concentrations of elements undergo in the reduction of metal from ore, variations in element concentrations within the same ore deposits, and the alterations that result from alloying and recycling, which often mixed together metal that originated from different sources. In the recent study of a second-millennium BCE copper ingot and pin from Kargaly, Chernykh et al. (2006) show that relatively high concentrations of sulfur, which have been traditionally associated with primary sulfide ore deposits, can still occur in metal derived from the secondary weathered zones located closest to the surface, which were often favored by ancient miners and metallurgists. Sulfur levels no longer appear to be a secure means of distinguishing between metal produced from these respective types of ore. Nonetheless, the patterns found in element profiles still require an explanation. I have attempted to provide one here in relation to practices that surrounded metalworking, especially recycling.

In recent years, the ambiguities associated with trace element analysis have led researchers to favor stable lead isotope ratio analysis (LIA) as an alternative method of identifying sources of ancient copper, whether unalloyed or in bronze. Trace element patterns now have something new to contribute, as neutron activation analysis is currently being used to resolve ambiguities in the results of LIA in Armenia and elsewhere (e.g., Meliksetian et al. 2007).

5. Also analyzed were 11 comparative samples of Early Bronze Age metalwork from Velikent, Dagestan (see note 3). Samples from an additional seven bracelets from Samara had to be excluded after initial analyses found them to be too corroded to provide reliable results.

6. Although limits of space prevent showing it here, like arsenic, the distribution in iron concentrations in the samples as a whole also appears to relate to recycling (Peterson 2007: fig. 7.79).

7. More than 30 wt% tin was detected in one example. This high level of tin is probably owed to the heavy corrosion of the object. It therefore has been omitted from Fig. 6a and the discussion.

8. Because of limitations of space, this figure could not be reproduced here.

9. Spectral-chemical analysis of the majority of objects examined from the Potapovka I cemetery (but not Utevka VI) was performed previously by Agapov and Kuz'minykh (1994). Their results are largely in agreement with those of the WDS. However, the present research was the first in which metallography and hardness testing had been conducted on metalwork from Potapovka I and the first time that any archaeometallurgical analysis was performed for the other objects, including those from Utevka VI and Grachevka II.

10. This is the result of incomplete crystallization caused by annealing at a temperature too low for full re-crystallization, and/or for a duration too short to complete the process (P. Northover personal communication).

11. Annealing temperatures used with copper and bronze in antiquity typically ranged from 500 to 800° C (Scott 1991: 7). While bronze has a lower melting point than copper, alloy levels of arsenic or tin lower the conductivity of copper, raising the temperature at which re-crystallization may occur within one hour of heating. For copper, that temperature is 200° C, while it is 375° C for Cu + 4% As, and 450–500° C for Cu + 10% Sn. Samples with residual coring may have been annealed at a high-enough temperature for a duration sufficient for re-crystallization but not sufficient to erase all traces of the as-cast structure.

12. The Tashkazgan ores are relatively high in arsenic, so that this arsenic bronze would have been the product of smelting rather than the addition of arsenic to the copper (Chernykh 1970; 1992: 18).

13. This is a rather different notion of regime of value than the original (Appadurai 1986), in that I am insisting that a greater role was played by producers, while Appadurai (1986: 57) was more strictly concerned with circulation and consumption. One could argue that a profitable way of approaching emerging social complexity in Eurasia is in terms of changes in regimes of value, including the involvement of producers, and new understandings of the importance of people, groups, and activities, but that is a topic for another time.

References

Agapov, S. A., and S. Kuz'minykh. 1994. Metall Potapovskogo mogil'nika v sisteme Evraziiskoi Metallurgicheskoi Provintsii. In I. B. Vasil'ev, P. F. Kuznetsov, and A. P. Semenova, *Potapovskii kurgannyi mogil'nik indoiranskikh plemen na Volge*. Samara: Samara University, pp. 167–173.

Anthony, D. W. 1996. V. G. Childe's World System and the Daggers of the Early Bronze Age. In B. Wailes and K. Ryan (eds.), *Craft Specialization and Social Evolution: In Memory of V. Gordon Childe*. Philadelphia: University of Pennsylvania Museum, pp. 47–66.

Appadurai, A. 1986. Introduction: Commodities and the Politics of Value. In A. Appadurai (ed.), *The Social Life of Things: Commodities in Cultural Perspective*. Cambridge: Cambridge University Press, pp. 3–63.

Bawden, G. 1989. The Andean State as a State of Mind. *Journal of Anthropological Research* 45: 327–333.

Bochkarev, V. S. 1991. Volgo-Uralskyi ochag kulturogeneza epokhi pozdnei bronzy. In V. M. Masson (ed.), *Sotsiogenez i kulturogenez v istoricheskom aspekte*. St. Petersburg: Institute for the History of Material Culture, pp. 24–27.

Budd, P., and T. Taylor. 1995. The Faerie Smith Meets the Bronze Industry: Magic versus Science in the Interpretation of Prehistoric Metal-Making. *World Archaeology* 27(1): 133–143.

Carneiro, R. 1981. The Chiefdom: Precursor of the State. In G. Jones and R. Kautz (eds.), *The Transition to Statehood in the New World*. Cambridge: Cambridge University Press, pp. 37–79.

Chernykh, E. N. 1970. *Drevneishaya metallurgiya Urala i Povolzh'ya*. Moscow: Nauka.

1992. *Ancient Metallurgy in the USSR*. Cambridge: Cambridge University Press.

2002. *Kargaly*. 2 vols. Moscow: Languages of Slavonic Culture.

Chernykh, E. N., and S. V. Kuz'minykh. 1987. *Drevniaia Metallurgiia Severnoi Evrazii Seiminsko-Turbinskii Fenomen*. Moscow: Nauka.

Chernykh, E. N., B. A. Prusakov, and L. V. Katkova. 2006. A Study of Ancient Copper. *Metal Science and Heat Treatment* 40(9): 368–373.

Childs, S. T. 1991. Transformations: Iron and Copper Production in Central Africa. In P. D. Glumac (ed.), *Recent Trends in Archaeometallurgical Research*. MASCA Research Papers in Science and Archaeology, vol. 8, part I. Philadelphia: University of Pennsylvania Museum of Archaeology and Anthropology, pp. 33–46.

Earle, T. K. 1991. *Chiefdoms: Power, Economy, and Ideology*. Cambridge: Cambridge University Press.

1997. *How Chiefs Come to Power*. Stanford: Stanford University Press.

Eliade, M. 1978. *The Forge and the Crucible*. Trans. S. Corrin. Chicago: University of Chicago Press.

2004 [1964]. *Shamanism: Archaic Techniques of Ecstasy*. Princeton: Princeton University Press.

Gellner, E. 1994. Foreword. In A. M. Khazanov, *Nomads and the Outside World*. 2 ed. Madison: University of Wisconsin Press, pp. ix–xxv.

Gening, V. F., G. B. Zdanovich, and V. V. Gening. 1992. *Sintashta*. Chelyabinsk: South Ural Press.

Goldstein, J. I., D. E. Newbury, D. C. Joy, L. C. Sawyer, E. Lifshin, C. E. Lyman, P. Echlin, and J. R. Michael. 2003. *Scanning Electron Microscopy and X-Ray Microanalysis*. New York: Kluwer.

Haaland, R. 2004. Technology, Transformation and Symbolism: Ethnographic Perspectives on European Iron Working. *Norwegian Archaeological Review* 37(1): 2–19.

Keller, C. M., and J. D. Keller. 1996. *Cognition and Tool Use: The Blacksmith at Work*. Cambridge: Cambridge University Press.

Kirch, P. V. 1984. *The Evolution of the Polynesian Chiefdoms*. Cambridge: Cambridge University Press.

Koryakova, L. 1996. Social Trends in Temperate Eurasia during the Second and First Millennia BC. *Journal of European Archaeology* 4: 243–280.

———. 2002. Social Landscape of Central Eurasia in the Bronze and Iron Ages: Tendencies, Factors, and Limits of Transformation. In K. Jones-Bley and D. G. Zdanovich (eds.), *Complex Societies of Central Eurasia from the 3rd to the 1st Millennium BC*, vol. 1. Washington, DC: Institute for the Study of Man, pp. 97–117.

Koryakova, L., and A. Epimakhov. 2007. *The Urals and Western Siberia in the Bronze and Iron Ages*. Cambridge: Cambridge University Press.

Kuznetsov, P. F. 1991. Unikal'noe pogrebenie epokhi rannei bronzy na r. Kutuluk. In *Drevnosti vostochno-evropeiskoi lesostepi*. Samara: Samara State Pedagogical University Press, pp. 137–139.

Kuznetsov, P. F., O. D. Mochalov, D. L. Peterson, L. M. S. Popova, A. P. Semenova, and D. Kormilitsin. 2005. Poisk sledov gornorudnogo delo epokhi pozdnei bronzy v srednom Povol'zhe (arkheologicheskie raboty v neissledovanikh raionakh Samaraskoi oblasti). *Izvestia Samarskogo nauchnogo tsentra Rossiskoi akademii nauk* 2: 332–343.

Kuznetsov, P. F., and A. P. Semenova. 2000. Pamyatniki potapovskogo tipa. In Yu. I. Kolev, A. E. Mamonov, and M. A. Turetskii (eds.), *Istoriya Samarskogo Povolzh'ya c drevneishikh vremen do nashikh dnei: bronzovyi vek*. Samara: Izdatel'stvo Samarskogo Nauchnogo Tsentra, pp. 122–151.

Lincoln, B. 1991. On the Scythian Royal Burials. In *Death, War, and Sacrifice*. Chicago: University of Chicago Press, pp. 188–197.

Matveeva, G. I., Yu. I. Kolev, and A. I. Korolev. 2004. Gorno-metallurgicheskii kompleks bronzovogo veka u c. Mikhailo-Ovsyanka na yuge Samarskoi oblasti. In *Voprosy arkheologii Urala i Povolzh'e*, vol. 2. Samara, pp. 69–88.

Mauss, M. 1990 [1950]. *The Gift*. London: Routledge.

McGuire, R. H. 1983. Breaking Down Cultural Complexity: Inequality and Heterogeneity. In M. Schiffer (ed.), *Advances in Archaeological Method and Theory*. New York: Academic Press, pp. 91–142.

McLellan, D. (ed.). 2000. *Karl Marx: Selected Writings*. 2nd ed. Oxford: Oxford University Press.

Meliksetian, Kh., E. Pernicka, and R. Badalyan. 2007. Compositions and Some Considerations on the Provenance of Armenian Early Bronze Age Copper Artefacts. Paper delivered at the Second International Conference on Archaeometallurgy in Europe, Aquileia, Italy, June.

Mochalov, O. D. 2008. *Keramika pogrebal'nykh pamyatnikov epokhi bronzy lesostepi Volgo-Ural'skovo mezhdurech'ya.* Samara: Samara State Pedagogical University.

Muhly, J. D. 1980. The Bronze Age Setting. In T. A. Wertime (ed.), *The Coming of the Age of Iron.* New Haven: Yale University Press, pp. 25–67.

Northover, P. 1985. The Complete Examination of Archaeological Metalwork. In P. Phillips (ed.), *The Archaeologist and the Laboratory.* CBA Research Reports, vol. 58. London: Council for British Archaeology, pp. 56–59.

1989. Non-ferrous Metallurgy in Archaeology. In J. Henderson (ed.), *Scientific Analysis in Archaeology.* UCLA Institute of Archaeological Research Tools, vol. 5. Los Angeles: University of California at Los Angeles, pp. 213–236.

1998. Analysis of Copper Alloy Metalwork from Arbedo TI. In M. P. Schindler (ed.), *Der Depotfund von Arbedo TI. Antiqua 30.* Basel: Schweizerische Gesellschaft für Ur- und Frühgeschichte, pp. 289–315.

2002. Analysis of Non-ferrous Metalwork from the Iron Age Cemetery at Pottenbrunn, NÖ. In P. C. Ramsl (ed.), *Das eisenzeitliche Gräberfeld von Pottenbrunn.* Forschungsansätze zu wirtschaftlichen Grundlagen und sozialen Strukturen der latènezeitlichen Bevölkerung des Traisentales, Niederösterreich, Materialheft A 11. Vienna: Fundberichte aus Österreich, pp. 251–263.

Pavlov, A. E. 1996. *Samara.* Samara: Samara dom pechati.

Peterson, D. L. 2007. Changing Technologies and Transformations of Value in the Eurasian Steppes and Northeastern Caucasus, circa 3000–1500 BCE. Ph.D. dissertation, University of Chicago.

Peterson, D. L., P. F. Kuznetsov, and O. D. Mochalov. 2006. The Samara Bronze Age Metals Project: Investigating Changing Technologies and Transformations of Value in the Eurasian Steppes (with an brief appendix by Emmett Brown and Audrey Brown). In D. L. Peterson, L. M. Popova, and A. T. Smith (eds.), *Beyond the Steppe and the Sown: Proceedings of the 2002 University of Chicago Conference on Eurasian Archaeology. Colloquia Pontica 13.* Leiden: Brill, pp. 322–342.

Renfrew, C. 1986. Varna and the Emergence of Wealth in Prehistoric Europe. In A. Appadurai (ed.), *The Social Life of Things.* Cambridge: Cambridge University Press, pp. 141–168.

Scott, D. 1991. *Metallography and Microstructures of Ancient and Historic Metals.* Malibu: J. Paul Getty Museum.

Shennan, S. 1993. After Social Evolution: A New Archaeological Agenda? In N. Yoffee and A. Sherratt (eds.), *Archaeological Theory: Who Sets the Agenda?* Cambridge: Cambridge University Press, pp. 53–59.

1999a. Cost, Benefit and Value in the Organization of Early European Copper Production. *Antiquity* 73: 352–363.

1999b. The Development of Rank Societies. In G. Barker (ed.), *Companion Encyclopedia of Archaeology,* vol. 2. London: Routledge, pp. 870–907.

Sherratt, A. 1997. *Economy and Society in Prehistoric Europe: Changing Perspectives.* Princeton: Princeton University Press.

Smith, A. T. 2003. *The Political Landscape: Constellations of Authority in Early Complex Polities.* Berkeley: University of California Press.

Taylor, T. 1999. Envaluing Metal: Theorizing the Eneolithic "Hiatus." In S. M. M. Young, A. M. Pollard, P. Budd, and R. A. Ixer (eds.), *Metals in Antiquity.* Oxford: Archaeopress, pp. 22–32.

Vasil'ev, I. B., P. F. Kuznetsov, and A. P. Semenova. 1992. Pogrebeniya znati epokhi bronzy b srednem Povolzh'e. *Arkheologicheskie vesti* 1: 52–63.

1994. *Potapovskii kurgannyi mogil'nik indoiranskikh plemen na Volge.* Samara: Samara State University.

Vasil'ev, I. B., P. F. Kuznetsov, and M. A. Turetskii. 2000. Yamnaya i Poltovkinskaya kultury. In Yu. I. Kolev, A. E. Mamonov, and M. A. Turetskii (eds.), *Istoriya Samarskogo Povolzh'ya c drevneishikh vremen do nashikh dnei: bronzovyi vek.* Samara: Izdatel'stvo Samarskogo Nauchnogo Tsentra, pp. 6–64.

Vinogradov, N. B. 2003. *Mogil'nik bronzovogo veka Krivoe Ozero v Yuzhnom Zaural'ye.* Chelyabinsk: Yuzhno-Ural'skoe knizhnoe izd-vo.

Wright, H. 1994. Prestate Political Formations. In G. Stein and M. Rothman (eds.), *Chiefdoms and Early States in the Near East.* Madison: Prehistory Press, pp. 67–84.

Yoffee, N. 1993. Too Many Chiefs? (or, Safe Texts for the '90s). In N. Yoffee and A. Sherratt (eds.), *Archaeological Theory: Who Sets the Agenda?* Cambridge: Cambridge University Press, pp. 60–78.

Zaikov, V. V., and D. G. Zdanovich. 2000. Kamennye izdeliya i mineral'no syr'evaya baza kamennoi industrii Arkaim. In G. B. Zdanovich, S. Ya. Zdanovich, and I. V. Predeina (eds.), *Arkheologicheskii istochnik i modelirovanie drevnikh tekhnologii: Trudy muzeya-zapovednika Arkaim.* Chelyabinsk: Arkaim Center, Institute of History and Archaeology, Ural Branch of the Russian Academy of Sciences, pp. 73–94.

Zdanovich, D. G. (ed.). 2002. *Arkaim: Nekropol'.* Kn. 1. Chelyabinsk: Uzhno-Ural'skoe knishnoe izd-vo.

Zdanovich, G. B., and D. G. Zdanovich. 2002. The "Country of Towns" of Southern Trans-Urals and Some Aspects of Steppe Assimilation in the Bronze Age. In K. Boyle, A. C. Renfrew, and M. Levine (eds.), *Ancient Interactions: East and West in Eurasia.* Cambridge: McDonald Institute, pp. 249–263.

CHAPTER 12

Early Metallurgy and Socio-Cultural Complexity
Archaeological Discoveries in Northwest China

JIANJUN MEI

THE ORIGIN of the use of metals and alloys in China is among the central issues in current studies of the emergence of early civilizations in China. Over the past 50 years, considerable scholarly interest has focused on a debate over whether metallurgy was introduced into China or invented independently (Loehr 1949, 1956; Barnard 1961, 1983; Barnard and Sato 1975: 1–75; Ho 1975: 177–221; Smith 1977; Sun and Han 1981; Jettmar 1981; Watson 1985: 335; Muhly 1988; Wagner 1993: 28–33; Linduff et al. 2000). Most of the studies have focused on typological and technological issues concerning early metals, with only a few paying attention to the role that the use of metals and the development of metallurgy played in the emergence of complex societies in China (Linduff 1998, 2002, 2004; Shelach 2001).

Currently, our understanding of the relationship between metallurgy and socio-cultural complexity in early China is poor, and many significant questions remain to be addressed. Recent archaeological discoveries in the western regions of Xinjiang, Gansu, and Qinghai in present-day northwest China have thrown new light on the beginnings of the use of metals and alloys in that region (Sun and Han 1997; Mei 2000, 2001, 2003a, 2003b; Li and Shui 2000; Qian et al. 2001; Li 2003, 2005). On the basis of this new archaeological evidence, this chapter offers some preliminary observations on the relationship between early metallurgy and socio-cultural complexity in northwest China. I first present an overview of the studies on the beginnings and early development of copper and bronze metallurgy in the region, with an emphasis on early contacts between this area and the Eurasian steppe. Then, I propose some social factors and symbolic dimensions of early metallurgy on the basis of a preliminary observation of recently excavated metal artifacts from the Xiaohe cemetery in Xinjiang. Finally, I examine socio-cultural factors that may have played a role in shaping the growth of early metallurgy in northwest China.

The Beginnings and Early Development of Metallurgy in Northwest China

OUR UNDERSTANDING of the emergence and early development of copper and bronze metallurgy in northwest China has been improved considerably over the past decade because of a number of new archaeological discoveries in the region, notably in the eastern part of Xinjiang (Fig. 12.1). Until the early 1990s, the predominant view concerning the beginnings of metallurgy in China was still in favor of the idea of independent invention, which was based mainly on archaeological finds of early copper and bronze objects in Gansu and Qinghai. These early metal objects, which include knives, axes, awls, earrings, mirrors, and spearheads (Figs. 12.2 and 12.3), are associated with such archaeological cultures as Majiayao, Machang, Qijia, and Siba and can be dated from the early third to the mid-second millennium BCE (Sun and Han 1981; Ko 1986: 2; Su et al. 1995: 48–49). On the basis of typological analogies, a number of scholars in the 1990s, including An (1993) and Louisa Fitzgerald-Huber (1995), began to claim that western stimulation was the impetus for the production of early metal artifacts found in Gansu and Qinghai. At the same time, a preliminary survey of archaeological discoveries made in Xinjiang, west of Gansu and Qinghai, offered new light on the issue of diffusion or independent invention (Shui 1993; Mei and Shell 1998, 1999).

The subsequent study of copper and bronze metallurgy in late prehistoric Xinjiang has demonstrated that Xinjiang lay on the boundary of early cultural interaction between this region and areas to the west (Mei 2000: 72–75). One of the major results of the study of this region has been to recognize the early dates of cultural intersection along a prehistoric "Silk Road" (i.e., early trade route). The use of painted pottery spread westward from Gansu into Xinjiang, but bronze technology was transmitted in the reverse direction (Debaine-Francfort 2001; Li 2002; Mei 2003a). However, this idea has not been widely accepted, as a number of scholars retain belief in the model of an indigenous origin for metallurgy in northwest China (Qian et al. 2001). This debate

Figure 12.1. A map of northern China and eastern Central Asia.

will continue until additional archaeological data from northwest China produces more decisive evidence. So far, the argument in favor of a stimulus from connections between northwest China and the Eurasian steppe, particularly in areas such as southern Siberia, is based primarily on analogous typological and metallurgical evidence.

Typological analysis is based on the observation that a variety of bronze forms associated with the Qijia, Siba, and Tianshanbeilu cultures, such as the socketed axe, handled knife with curved back, socketed spearhead, awl and knife with bone handle, and the circular mirror, all have parallels in Eurasian steppe cultures such as the Okunev, Seima-Turbino, and "Andronovo" (Figs. 12.2–6; Fitzgerald-Huber 1995; Takahama 2000; Mei 2001, 2003a). Metallurgical data were obtained from the examination of hundreds of early copper and bronze objects found in Gansu, Qinghai, and Xinjiang. This research revealed the use of tin bronze and arsenical copper, which is analogous to the composition of objects produced in the Eurasian steppe (Sun and Han 1997; Sun 1998; Mei 2000, 2003a; Li 2002, 2003, 2005). However, these two types of evidence have not been entirely sufficient to draw a clear and complete picture of the nature of the connections between northwest China and the Eurasian steppe and, more specifically, the consequences of those contacts.

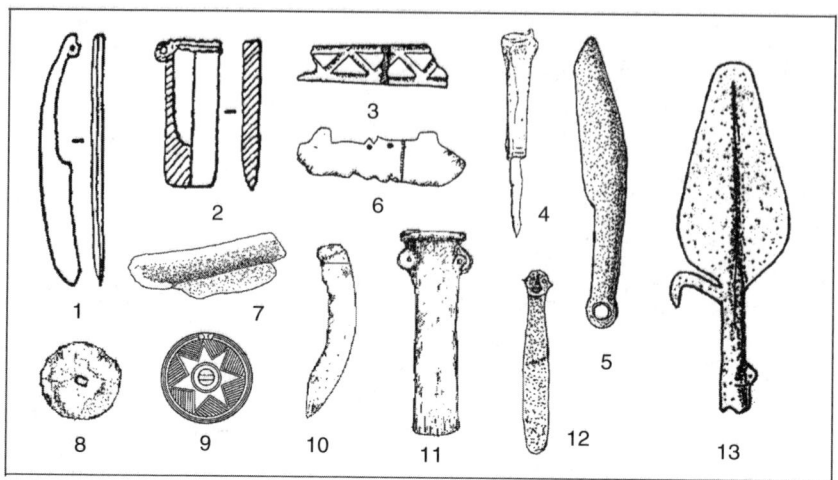

Figure 12.2. Copper and bronze artifacts recovered at various Qijia culture sites in Gansu and Qinghai: (1, 5) knives; (2, 11) socketed axes or celts; (3) handle; (4, 7) knives with bone handles; (6) knife with two protrusions on the back; (8, 9) mirrors; (10) sickle; (12) dagger-knife; (13) spearhead (7: drawing after Tian 1983: 76, photo; 5, 12: drawings after GM 1994: pls. 36 and 37; 13: drawing after Takahama 2000: 122, fig. 3; the rest after Debaine-Francfort 1995: 86, fig. 49; 104, fig. 61; and 119, fig. 71).

The nature of early contacts between northwest China and the Eurasian steppe has been suggested through observation of metallurgical and typological parallels, but many metal objects that are typical of the steppes are conspicuously absent in the excavated materials. These include socketed celts decorated with geometric patterns, knives and daggers with animal-headed handles (Seima-Turbino), and shaft-hole axes (Andronovo) (Fig. 12.4; Fitzgerald-Huber 1995; Mei 2003b). If the contacts were of a significant scale and included the movement of communities of peoples, we might expect to find a full set of metal objects from the steppe region in northwest China. What we have found, however, is the co-existence of various "steppe-style" artifacts and technology along with many local features of the prehistoric cultures of northwest China. This suggests that the contacts were most likely indirect, sporadic, and small scale. It has also become increasingly clear that the processes of interaction in this region were complex and carried out through intermediary links during different periods (Mei 2003b: 41–42; Li 2005: 266–267).

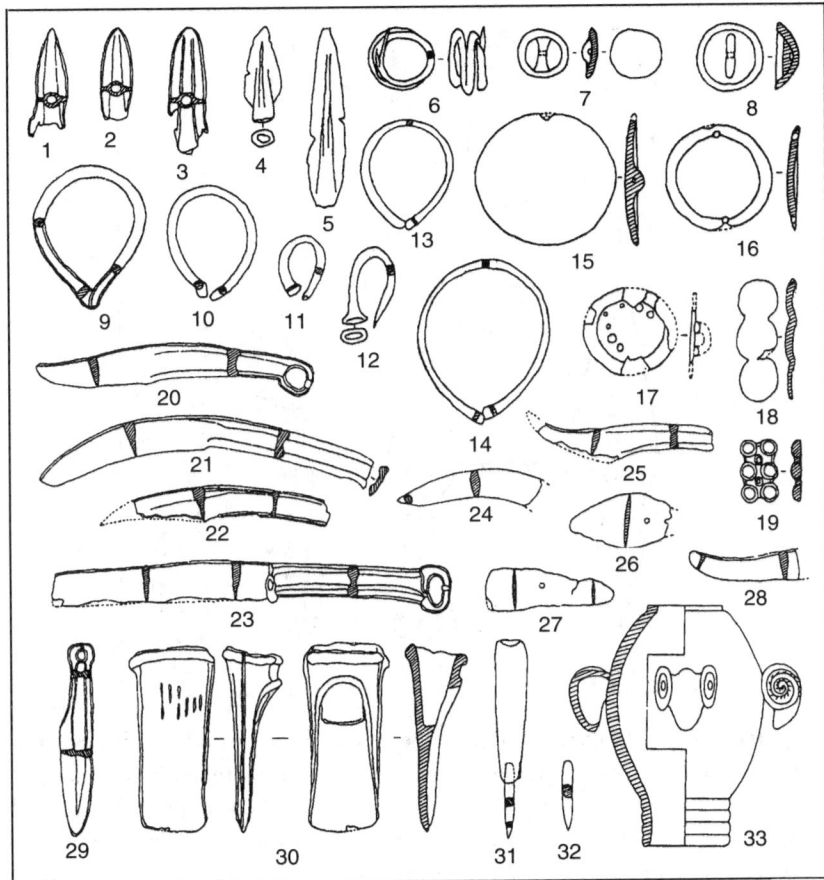

Figure 12.3. Copper and bronze artifacts of the Siba culture in Gansu: (1–5) arrowheads; (6) ring; (7–8) buttons; (9–14) earrings and rings; (15–19) ornaments; (20–29) knives; (30) socketed axe; (31–32) awls; (33) mace head (after Bai 2002: 29, fig. 3).

The role of local technological innovation is also an important issue. Although some metal objects found in northwest China parallel those from the Eurasian steppe, others exhibit local features. For example, circular mirrors with geometric decorations on the back excavated from the sites of Qijia and Tianshanbeilu are particularly noteworthy (Fig. 12.2: 9; Fig. 12.5: 29). Thus far, this kind of mirror has not been found in the Eurasian steppe, making its appearance hard to explain as a result of direct cultural transmission. It seems more likely that these objects were a local innovation (Mei 2003b), although how and where

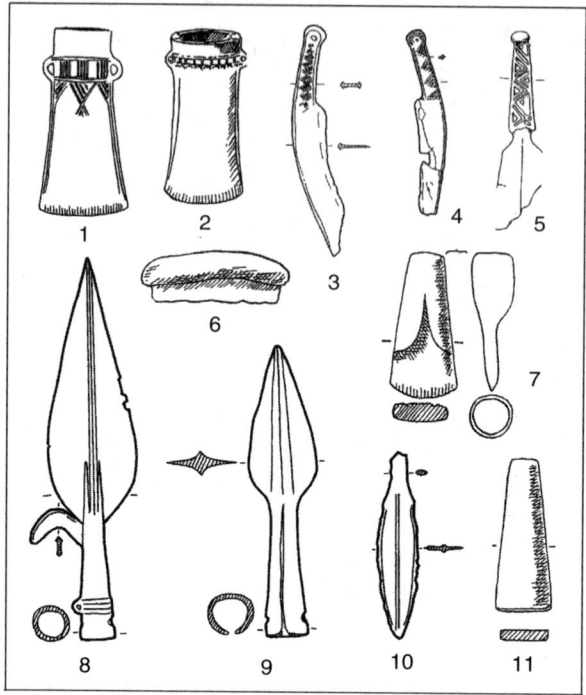

Figure 12.4. Copper and bronze artifacts ascribed to the Seima-Turbino transcultural phenomenon: (1, 2, 7) socketed celts or axes; (3, 4) knives; (5, 10) daggers; (6) knife with a bone handle; (8, 9) socketed spearheads; (11) flat axe (1, 2, 6, 7, 11: after Gimbutas 1956: 153, fig. 9, and 160, fig. 16; 3–5: after Fitzgerald-Huber 1995: 45, fig. 8; 8–10: after Chernykh 1992: 221, fig. 75).

this happened remains to be explained. The early appearance and use of arsenical copper in northwest China is also a vexing problem. It may be that arsenical copper was introduced into northwest China, but its use in local production was presumably due to local mineral resources, as suggested by a metallurgical examination of dozens of metal objects from eastern Xinjiang (Sun and Han 1997; Mei 2000; Qian et al. 2000, 2001; Li 2005). Nevertheless, to confirm such a suggestion, more substantial evidence of mines and metallurgical production sites in northwest China is needed.

A third question can be raised about the dating of metal finds in northwest China versus those dated in the Eurasian steppe. Currently, the discovery of four copper and bronze items from Majiayao and

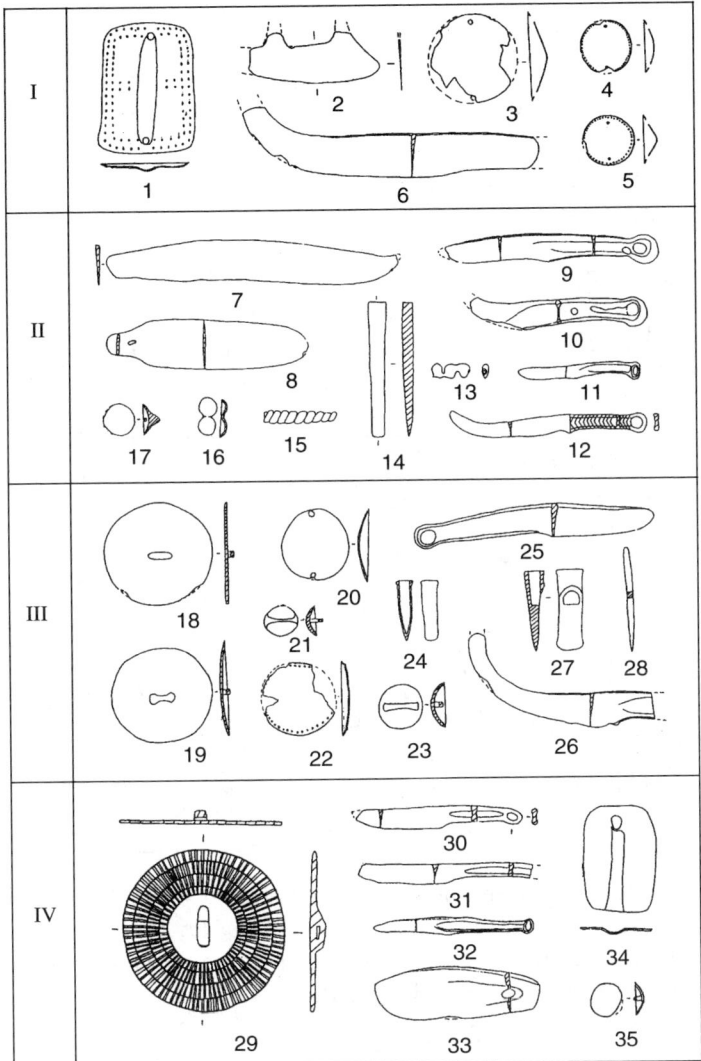

Figure 12.5. Copper and bronze artifacts found at the Tianshanbeilu cemetery, Hami, Xinjiang (Periods I to IV in chronological order): (1, 3–5, 13, 15–23, 34, 35) plaques, buttons, and other ornaments; (2, 6–12, 25, 26, 30–33) knives, sickles, and knife-dagger; (14, 24, 27, 28) awls and axes; (29) mirror (after Lü et al. 2001: 182–183, fig. 15–18).

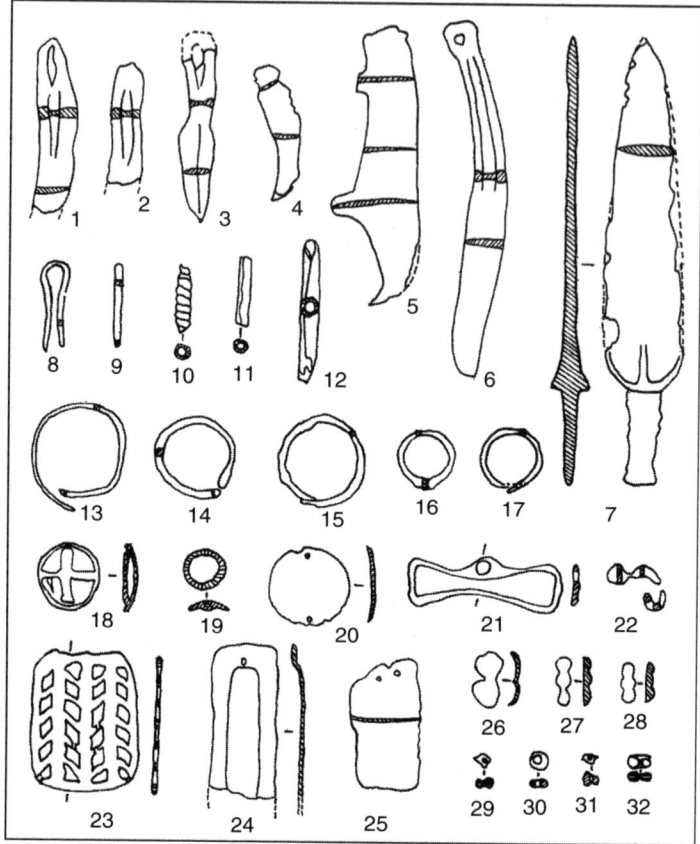

Figure 12.6. Copper and bronze artifacts found at the Tianshanbeilu cemetery, Hami, Xinjiang: (1, 2, 4–6) knives; (3, 7) daggers; (8–32) rings, plaques, buttons, tubes, and other small ornaments and implements (after Qian et al. 2001: 80, fig. 1).

Machang sites in Gansu, which date to the early and late third millennium BCE, respectively, present a major challenge because these items are still among the earliest metals found in northwest China (Sun and Han 1997; Mei 2001, 2003a). No metal objects found so far in Xinjiang can be dated earlier than 2000 BCE. Because of this current chronology, a western impetus for the beginnings of copper and bronze metallurgy in the Gansu region is still actively debated. Furthermore, the chronological sequence of early metal-using cultures in the Eurasian steppe, such as Okunev, Seima-Turbino, and Andronovo, is uncertain as well.

No general consensus has been reached, and a clearer understanding of the early relationship between northwest China and the Eurasian steppe awaits further systematic archaeological investigation.

Therefore, while we have a growing body of evidence to suggest connections between northwest China and the Eurasian steppe during the late third and early second millennia BCE, at this stage we may only hypothesize that the peoples of the Eurasian steppe played a significant role in the early development of copper and bronze metallurgy in northwest China. This is because there is insufficient evidence from the period before the late third millennium BCE, and so the question regarding diffusion versus independent invention remains an open one. Even for the late period, in which more archaeological evidence is available, many questions cannot be answered definitely. Obviously, the relationship between the emergence of metallurgy and social complexity in northwest China is one of them.

The Social and Symbolic Dimensions of Early Metallurgy in Northwest China

IN CONSIDERING the relationship between early metallurgy and socio-cultural complexity, the whole process from mining and production to the use and trade of metal products should be examined. In the case of early metallurgy in northwest China, little evidence has so far been found for mining and production, except for several stone casting mounds excavated at the Huoshaogou site of the Siba culture in Gansu (Sun and Han 1981: 289). Therefore, our understanding of the social dimensions of early metallurgy depends mainly on the products, namely the finished metal objects.

Hundreds of early metal objects have been found at various sites in northwest China. The social importance of these objects may vary considerably from one to another, or it may not be possible to define their importance with certainty. Because the context in which an object was found is crucial for understanding its social or symbolic dimensions, the recent archaeological investigation of the intact Xiaohe cemetery in the Tarim basin in southern Xinjiang offers a particularly significant site for study.

Located at the eastern edge of the Takalamakan desert, the Xiaohe cemetery site first became known to the world in the 1930s after Folke Bergman (1939), a Swedish archaeologist, visited it and later reported it in his book *Archaeological Researches in Sinkiang – Especially the Lop-Nor*

Region. Since then, no one has recorded a visit to the site until 2000 when it was "re-discovered" by an expedition team with the aid of GPS technology. Between 2002 and 2005, it was fully excavated by the archaeologists from the Xinjiang Institute of Cultural Relics and Archaeology. The excavation in 2002 has been briefly published, and a full report of the excavations during the past four years is in preparation (Xinjiang 2003).

According to the first report, the finds at the Xiaohe cemetery are in many ways comparable to those found at the Gumugou cemetery, which is radiocarbon-dated to the early second millennium BCE and located nearby (Mei 2000: 10). The Xiaohe cemetery most likely falls into a similar chronological range or is perhaps earlier. Once known to the local people as a place with a thousand coffins, the Xiaohe cemetery is very unusual and impressive because of the degree of preservation provided by the extremely dry environmental conditions. Not only are the organic materials, such as wooden coffins, grass baskets, woolen blankets, and felt hats well preserved, but the vertical and horizontal arrangements of coffins have survived in very good condition. The most unusual finds are two types of wooden poles, placed vertically at one end of the coffins. These poles, shaped like a male sex organ, are associated with coffins of females, while the poles shaped like a female organ are associated with the coffins for males. Neither ceramic nor metal implements were recovered, though the cutting marks on the coffins and poles strongly suggest the use of some sort of metal tools. The major metal items recovered at the Xiaohe cemetery were small, thin strips made of copper or copper alloys, measuring around 1 square centimeter. Such small copper strips were found at the Gumugou site as well, but their exact meaning or purpose has not been explained. Some important clues have appeared, however, at the Xiaohe site. For example, some small pieces of copper or copper alloys were found either at the top of the "male" poles as an inlay or on the chest of the deceased in the coffins (Liu and Li 2007: 17). Both uses suggest some sort of special meaning within the specific funeral context from which they were recovered.

In my opinion, the way that the ancient residents at Xiaohe employed metals provides us with sufficient evidence to see that these metals had some social or symbolic function, though I cannot define it precisely. We can speculate that the Xiaohe people valued these rare metal adornments and used them in an extraordinary way. It may be that metal was such a rare commodity that social practice prohibited the deposit of larger metal items within burials. If we take the desert environment into account, it may be reasonable to suggest that the metal items found at Xiaohe were imported from other places rather than locally produced. Because the metal pieces are very thin and soft and their color is that of pure copper, some scholars have inferred that they could be naturally formed copper rather than human smelted products (R. Han personal communication). However, recent research by the author and his student has indicated that three of these metal strips (kindly provided by the Xinjiang archaeologists) were not only smelted but are actually of a tin bronze composition. Metallographic examinations reveal the existence of copper sulfide inclusions, a clear sign of the smelting process, and evidence that hammering was the main technique used to shape the strips. Evidence for hot forging is also apparent (Chen et al. 2007). To get a better understanding of the social and symbolic dimensions of the early metals found at the Xiaohe cemetery, especially the role of trade in supplying metals, further research based on a complete archaeological report is needed.

Another example I wish to highlight is a bronze spearhead found at the Shenna site in Qinghai, considered by some scholars to be associated with the Qijia culture (2200–1700 BCE). The size of this spearhead is an unusually large 61 centimeters in length. It has a downward hook and a small loop on either side of the socket (Fig. 12.2: 13). Such a distinctive form parallels the spearheads excavated at the Rostovka cemetery site in southern Siberia (Fig. 12.4: 8), suggesting some cultural affinity (Takahama 2000). The Rostovka spearheads are all less than 40 centimeters in length, with a narrow body and a sharp point, indicating that they are functional weapons (Chernykh 1992: 221). By contrast, the Shenna spearhead is not just larger but has a blunt end, indicating that it is not an ordinary weapon but rather an object with some "ritual" or "ceremonial" meaning that was apparently made for burial (Mei 2003a: 10). It could have been used as a symbol for some sort of status or authority, perhaps for a military or political leader, or could have been an exotic object signifying wealth or power. The appearance of such an extraordinary metal object is presumably a rare indication of social inequality emerging in northwest China during the late third and early second millennia BCE.

Other unique or rare early metal objects found in northwest China could be seen to carry some social or symbolic dimensions, such as a bronze mace head (Fig. 12.3: 33), decorated bronze mirrors (Figs. 12.2: 9; 12.5: 29), a bronze knife decorated with a human head (Fig. 12:2: 12), a silver nose ring, and gold earrings (Li 2005: 241–245). Even the stone casting molds found at the Huoshaogou cemetery could convey some information concerning the presence of craft specialization. The practice of burying casting molds is comparable to an Iron Age example seen at the Yanghai cemetery in Turfan, Xinjiang, where a clay tuyère was found in a grave (Xinjiang 2004: 42). It seems reasonable to assume that burying casting molds or tuyeres is an indication that metallurgical practice played an important role in mortuary practice of the society and perhaps in the life of the deceased.

The Socio-Cultural Factors in the Development of Early Metallurgy

IN HER essay discussing metallurgy and adaptation in Gansu in the second millennium BCE, Linduff (2002) shows a sharp contrast between the Central Plains of China and northwest China in the development of early metallurgy. Whereas early bronze metallurgy in the Central Plain of China is characterized by the predominance of piece-mold casting and ritual bronze vessels (Fig. 12.7: 3–5), major bronze types seen in northwest China are still limited to knives, axes, awls, mirrors, rings, buttons and the like (Figs. 12.2–12.3; 12.5–12.6). The alloy types that were in use in these two regions also are somewhat distinct; tin bronze and arsenical copper are the two major types employed in northwest China; by contrast, lead bronze and leaded tin bronze predominate in the Central Plain (Sun and Han 1997; Jin 2000). Why do the two regions present such a sharp difference in the development of early metallurgy? It seems likely that the different socio-cultural backgrounds in the two regions may have played an important role in the emergence of such trajectories of development.

Although the process of the development of piece-mold casting in the Central Plain remains unclear, a "ritual" stage was already developed there when metallurgical production emerged during the third millennium BCE. In other words, institutionalized or regularized ritual practice including both jade and ceramics already existed before the appearance of bronze metallurgy. When bronze metallurgy was first introduced into the Central Plain, it was adapted into the existing socio-cultural customs that confirmed social hierarchies replacing jade as the exclusive commodity in elite burials. A high level of ceramic technology already had been achieved during early third millennium BCE, and an increased value placed on bronze vessels and weapons to accommodate ceremonies of ancestral reverence and to mark royal burials may have resulted in experimentation with casting including piece-mold casting technology.

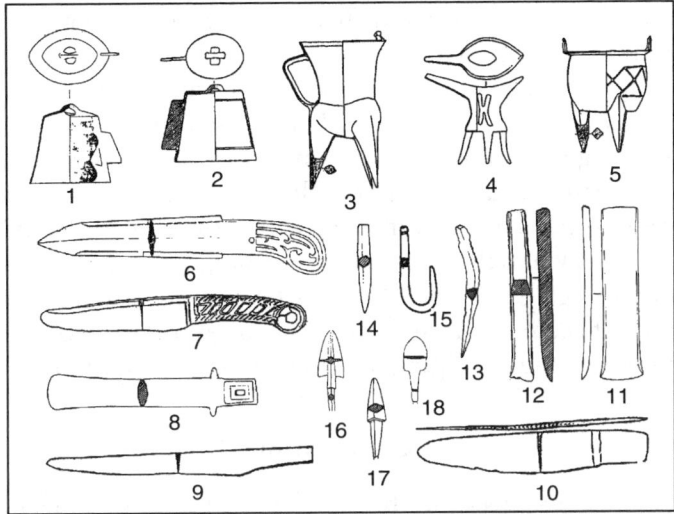

Figure 12.7. Copper and bronze artifacts of the Erlitou culture found in Yanshi, Henan, central China: (1–2) bells; (3–5) vessels; (6) dagger-axe; (7, 9, 10) knives; (8) axe; (11–15) small implements; (16–18) arrowheads (after Bai 2002: 33–36, fig. 7–9, 11).

By contrast, in northwest China, no such "ritual" need for bronze items existed to mark the elite, even though there were local needs of other sorts mentioned previously.

In the preceding discussion, I have raised the question of local technological innovations but did not touch on what would be the driving force behind such innovations. It appears that the existing cultural tradition in a given society or community may have played a crucial role in stimulating local innovation. Decorated bronze mirrors, for example, if accepted as a local invention, can be understood if we know the socio-cultural setting that produced them. The effect of local ideological systems can be interpreted from archaeological settings in Gansu and Qinghai, especially those associated with the Kayue culture (later than the Qijia), where it seems that the use of mirrors was related to some sort of "ritual" activity. For example, in one Kayue tomb at Dahuazhongzhuang cemetery in Qinghai, 36 small "mirrors" were found. This recalls the shamanistic practice of using dozens of small mirrors to embellish one's clothing (Liu 1993). Whether that sort of ritual practice existed during the Qijia period or whether its origins could be traced back to a period earlier than the Qijia culture is difficult to ascertain, although such possibilities should not be excluded.

Socio-cultural factors may explain or trail broader societal change or difference. Li (2005: 271–272) recently proposed that subsistence strategies affected regional varieties of early metallurgy in northwest China and the Central Plain. He also suggested that the geographic distribution of mineral resources was a factor in metal compositions in different regions, though he was open to the idea that steppe models may have played a part in the initial stage. Undoubtedly, subsistence and natural resources are two important economic factors that must be examined more systematically for a better understanding of these important questions.

Conclusion

OUR UNDERSTANDING of the development of early metallurgy in northwest China has been improved considerably over the past decade. It has been argued that both the steppe and local cultural traditions could have played an important role in stimulating the development of early metallurgy. However, the beginning stage of early metallurgy in northwest China remains rather unclear, and further archaeological evidence and research are needed. There is some evidence showing a correlation between the use of early metals and the emergence of social inequality among some societies in northwest China. This chapter has argued that the archaeological context is essential for understanding the social or symbolic dimensions of the use of early metals. It also has been suggested that socio-cultural factors could have resulted in local technological innovations and influenced the formation of the regional character of early metallurgy in northwest China.

Acknowledgments

I WOULD like to thank Bryan Hanks for kindly inviting me to participate in the symposium "New Research Directions in Eurasian Steppe Archaeology: The Emergence of Complex Societies in the Third to First Millennia BCE." I am also grateful to Professors Rubin Han, Sun Shuyun, Katheryn M. Linduff, Abodu Idiris, Li Wenying, Jessica Rawson, Colin Renfrew, Mayke Wagner, Wang Bo, Lü Enguo, Li Xiao, Zhang Yuzhong, Liu Xuetang, and Li Shuicheng, for their encouragement, support, and help with this research. Professors Abodu Idiris and Li Wenying kindly provided me with the three metal samples from their excavation at the Xiaohe cemetery. Mr. Chen Kunlong and Mr. Ling Yong helped me with the examination of the Xiaohe samples. To them, I wish to express my appreciation. Finally, I am deeply indebted to Bryan Hanks, Katheryn Linduff, and an anonymous reviewer for helping to improve this paper substantially in various ways. Any errors that still remain are entirely my own responsibility. The research described in this chapter received financial support under a project entitled "Exploring the Origins of Chinese Civilization" (Project number: 2006 BAK 21B03).

Abbreviations

KGXB   *Kaogu xuebao* (Acta Archaeologia Sinica)
WW   *Wenwu* (Cultural Relics)
XJWW   *Xinjiang wenwu* (Cultural Relics in Xinjiang)

References

An, Z. 1993. Shilun Zhongguo de zaoqi tongqi (On Early Copper and Bronze Artifacts in China). *Kaogu* (Archaeology) 12: 1110–1119.

Bai, Y. 2002. Zhongguo de zaoqi tongqi yu qingtongqi de qiyuan (The Origins of Early Chinese Copper and Bronze). *Dongnan Wenhua* (Southeast Culture) 7: 25–37.

Barnard, N. 1961. *Bronze Casting and Bronze Alloys in Ancient China*. Monumenta Serica Monography XIV. Tokyo: Australian National University and Monumenta Serica.

    1983. Further Evidence to Support the Hypothesis of Indigenous Origins of Metallurgy in Ancient China. In D. N. Keightley (ed.), *The Origins of Chinese Civilization*. Los Angeles: University of California Press, pp. 237–277.

Barnard, N., and T. Sato. 1975. *Metallurgical Remains of Ancient China*. Tokyo: Nichiosha.

Bergman, F. 1939. *Archaeological Researches in Sinkiang – Especially the Lop-Nor Region*. Sino-Swedish Expedition, 7. Stockholm: Bokförlags Aktiebolaget Thule.

Chen, K., Y. Ling, and J. Mei. 2007. Xiaohe mudi chutu sanjian tongpian de chubu fenxi (A Preliminary Analysis of Three Pieces of Copper Excavated at the Xiaohe Cemetery). *XJWW* 2: 125–128.

Chernykh, E. N. 1992. *Ancient Metallurgy in the USSR: The Early Metal Age*. Trans. S. Wright. Cambridge: Cambridge University Press.

Debaine-Francfort, C. 1995. *Du Néolithique à l'Âge du Bronze en Chine du Nord-Ouest: La culture de Qijia et ses connexions*. Paris: Éditions Recherche sur les Civilisations.

    2001. Xinjiang and Northwestern China around 1000 BC: Cultural Contacts and Transmissions. In R. Eichmann and H. Parzinger (eds.), *Migration und Kulturtransfer*. Bonn: Dr. Rudolf Habelt GmbH, pp. 57–70.

Fitzgerald-Huber, L. G. 1995. Qijia and Erlitou: The Question of Contacts with Distant Cultures. *Early China* 20: 17–67.

Gimbutas, M. 1956. Borodino, Seima and Their Contemporaries. *Proceedings of the Prehistoric Society* 22: 143–172.

GM (Gansu Museum). 1994. *Sichou Zhilu Gansu Wenwu Jinhua* (The Cultural Treasures of the Silk Road in Gansu). Lanzhou: Gansu Provincial Museum.

Ho, P. T. 1975. *The Cradle of the East: An Inquiry into the Indigenous Origins of Techniques and Ideas of Neolithic and Early Historic China, 5000–1000 B.C.* Hong Kong: Chinese University of Hong Kong.

Jettmar, K. 1981. Cultural and Ethnic Groups West of China in the Second and First Millennia B.C. *Asian Perspectives* 24(2): 145–162.

Jin, Z. 2000. Erlitou qitongqi de zjran kexue yanjiu yu Xia wenming tansuo (A Scientific Study of the Erlitou Bronzes and an Exploration of the Xia Civilization). *WW* 1: 57–64.

Ko, T. 1986. A Brief History of Metallurgy. In Beijing University of Iron and Steel Technology (ed.), *Zhongguo Yejin Shi Lunwenji* (Papers on the History of Metallurgy in China). Beijing, pp. 1–11.

Li, S. 2002. The Interaction between Northwest China and Central Asia during the Second Millennium BC: An Archaeological Perspective. In K. Boyle, C. Renfrew, and M. Levine (eds.), *Ancient Interactions: East and West in Eurasia.* Cambridge: McDonald Institute, pp. 171–182.

2003. Ancient Interactions in Eurasia and Northwest China: Revisiting J. G. Andersson's Legacy. *Bulletin of the Museum of Far Eastern Antiquities* 75: 9–30.

2005. Xibei yu Zhongyuan zaoqi yetongye de quyu tezheng ji jiaohu zuoyong (The Regional Features and Their Interactions of Early Metallurgy in Northwest and Central Plains of China). *KGXB* 3: 239–278.

Li, S., and T. Shui. 2000. Siba wenhua tongqi yanjiu (A Study of the Copper and Bronze Objects of the Siba Culture). *WW* 3: 36–44.

Linduff, K. M. 1998. The Emergence and Demise of Bronze-Producing Cultures outside the Central Plains of China. In V. H. Mair (ed.), *The Bronze Age and Early Iron Age Peoples of Eastern Central Asia*, 2 vols. Washington, DC: Institute for the Study of Man, pp. 619–643.

2002. At the Eastern Edge: Metallurgy and Adaptation in Gansu (PRC) in the 2nd Millennium BC. In K. Jones-Bley and D. G. Zdanovich (eds.), *Complex Societies of Central Eurasia from the 3rd to the 1st Millennium BC*, vol. 2. Washington, DC: Institute for the Study of Man, pp. 595–611.

2004. How Far Does the Eurasian Metallurgical Tradition Extend? In Katheryn M. Linduff (ed.), *Metallurgy in Ancient Eastern Eurasia from the Urals to the Yellow River.* Lewiston: Edwin Mellen Press, pp. 1–14.

Linduff, K. M., R. Han, and S. Sun (eds.). 2000. *The Beginnings of Metallurgy in China.* Lewiston: Edwin Mellen Press.

Liu, X. 1993. Xinjiang diqu zaoqi tongjing ji xiangguan wenti (Early Bronze Mirrors in the Xinjiang Region and the Relevant Issues). *XJWW* 1: 121–31.

Liu, X., and W. Li. 2007. Zhongguo zaoqi qingtong wenhua de qiyuan ji qi xiang-guan wenti xintan (A New Study on the Origin of Early Chinese Bronze Culture and Relevant Issues). *Zangxue Xuekan* (Journal of Tibetology) 3: 1–63.

Loehr, M. 1949. Weapons and Tools from Anyang and Siberian Analogies. *American Journal of Archaeology* 53: 26–144.

1956. *Chinese Bronze Age Weapons.* Ann Arbor: University of Michigan Press.

Lü, E., X. Chang, and B. Wang. 2001. Xinjiang qingtong shidai kaogu wenhua qianlun (On the Bronze Age Cultures in Xinjiang). In B. Su (ed.), *Su Bingqi Yu Dangdai Zhongguo Kaoguxue* (Su Bingqi and Archaeology in Contemporary China). Beijing: Kexue Chubanshe, pp. 172–193.

Mei, J. 2000. *Copper and Bronze Metallurgy in Late Prehistoric Xinjiang: Its Cultural Context and Relationship with Neighboring Regions.* BAR International Series 865. Oxford: Archaeopress.

2001. Guanyu Zhongguo yejin qiyuan ji zaoqi tongqi yanjiu de jige wenti (On the Origins of Metallurgy and the Studies of Early Copper Objects in China). *Tulufanxue Yanjiu* (Turfanological Research) 2: 57–68.

2003a. Cultural Interaction between China and Central Asia during the Bronze Age. *Proceedings of the British Academy* 121: 1–39.

2003b. Qijia and Seima-Turbino: The Question of Early Contacts between Northwest China and the Eurasian Steppe. *Bulletin of the Museum of Far Eastern Antiquities* 75: 31–54.

Mei, J., and C. Shell. 1998. Copper and Bronze Metallurgy in Late Prehistoric Xinjiang. In V. H. Mair (ed.), *The Bronze Age and Early Iron Age Peoples of Eastern Central Asia,* vol. 2. Washington, DC: Institute for the Study of Man, pp. 581–603.

1999. The Existence of Andronovo Cultural Influence in Xinjiang during the Second Millennium BC. *Antiquity* 73(281): 570–578.

Muhly, J. D. 1988. The Beginnings of Metallurgy in the Old World. In R. Maddin (ed.), *The Beginning of the Use of Metals and Alloys.* Cambridge, MA: MIT Press, pp. 2–20.

Qian, W., S. Sun, and R. Han. 2000. Gudai shentong yanjiu zongshu (A Review of the Studies of Ancient Arsenical Copper). *Wenwu Baohu Yu Kaogu Kexue* (Sciences of Conservation and Archaeology) 12(2): 43–50.

2001. Xinjiang Hami Tianshanbeilu mudi chutu tongqi de chubu yanjiu (A Preliminary Study of the Bronzes Unearthed from the Tianshanbeilu Cemetery in Hami, Xinjiang). *WW* 6: 78–89.

Smith, C. S. 1977. Review of Barnard and Sato 1975 and Ho Ping-ti 1975. *Technology & Culture* 18(1): 80–86.

Shelach, G. 2001. Interaction Spheres and the Development of Social Complexity in Northeast China. *Review of Archaeology* 22(2): 22–34.

Shui, T. 1993. Xinjiang qingtong shidai zhu wenhua de bijiao yanjiu (A Comparative Study of the Bronze Age Cultures in Xinjiang). *Guoxue Yanjiu* (Studies in Sinology) 1: 447–490.

Su, R., J. Hua, K. Li, and B. Lu. 1995. *Zhongguo Shanggu Jinshu Jishu* (The Metal Technology of Early China). Jinan: Shandong Keji Chubanshe.

Sun, S. 1998. Donghuishan yizhi Siba wenhua tongqi de jianding ji yanjiu (The Examination and Study of the Copper Objects from the Siba Culture Site at Donghuishan). In Gansu Provincial Institute of Cultural Relics and Archaeology and Jilin University (eds.), *Minle Donghuishan Kaogu: Siba wenhua mudi de jieshi yu yanjiu* (Excavation Report of the Donghuishan Site in Mingle – Excavation and Study on the Cemetery of Siba Culture). Beijing: Kexue Chubanshe, pp. 191–195.

Sun, S., and R. Han. 1981. Zhongguo zaoqi tongqi de chubu yanjiu (A Preliminary Study of Early Chinese Copper and Bronze Artifacts). *KGXB* 3: 287–301.

———. 1997. Gansu zaoqi tongqi de faxian yu yelian zhizao jishu de yanjiu (Discovery of Early Copper and Bronze Artifacts in Gansu and Studies of Their Smelting and Manufacturing Techniques). *WW* 7: 75–84.

Takahama, S. 2000. On Several Types of Copper Objects of the First Half of the Second Millennium BC from Central Eurasia. *Metals and Civilisation* [in Japanese]. Nara: Research Center for the Silk Road, pp. 111–23.

Tian, Y. 1983. Gansu Linxia faxian Qijia wenhua gubing tongren dao (A Copper Blade with Bone Handle Found in Linxia, Gansu). *WW* 1: 76.

Wagner, D. B. 1993. *Iron and Steel in Ancient China.* Leiden: Brill.

Watson, W. 1985. An Interpenetration of Opposites? Pre-Han Bronze Metallurgy in West China. *Proceedings of the British Academy* 70: 327–358.

Xinjiang (Xinjiang Institute of Cultural Relics and Archaeology). 2003. 2002 nian Xiaohe mudi kaogu diaocha yu fajue baogao (The Report of the Archaeological Investigation and Excavation at the Xiaohe Cemetery in 2002). *XJWW* 2: 8–64.

———. 2004. Shanshan xian Yanghai erhao mudi fajue jianbao (A Brief Report of the Excavation at the Cemetery II in Shanshan County). *XJWW* 1: 28–49.

PART THREE

FRONTIERS AND BORDER DYNAMICS

CHAPTER 13

Introduction

Thomas Barfield

THE TRANSFORMATION of the Eurasian steppe that took place from the end of the second to the middle of the first millennium BCE presents a host of questions. That such a transformation took place during that period is clear. The archaeological remains of fortified sites and settlements with mixed agro-pastoral economies of the Bronze Age disappear and are replaced by more-extensive pastoral economies where permanent structures are few and most archaeological material comes from tombs.

The three chapters here examine some of the most contentious questions in the field today from different times and places. The most hotly debated of these have been whether the introduction of a more extensive steppe pastoralism characteristic of the horse-riding nomads at the end of the period was a product of population migrations, invasions by elite groups bringing a new culture with them, or the adoption of a new way of life by an existing population. The next most contentious question is whether the growing size and complexity in these new societies' political structures was the product of an indigenous internal development or a reaction to external forces. This issue has led to debate about the nature of cross-cultural frontier relations with neighboring sedentary societies, particularly China. However, archaeologists have increasingly insisted on the importance of examining the steppe's northern frontier with the Eurasian forest zone and the foraging societies that inhabited it. Finally, there is the question of distinguishing differences (economic, cultural, ethnic, linguistic, political, etc.) among steppe pastoral societies that ancient (and indeed modern) historians have generally treated as uniformly similar.

Shelach examines steppe societies on China's northern frontier using data from excavated tombs to show the rise of militarism among societies there preceded the full-scale introduction of the bow-wielding cavalry armies that transformed their relation with China beginning in the fourth century BCE. That the steppe had become militarized before this date is not surprising. The use of new chariot technology in the Warring States period in China was an import from the steppe along with the horses needed to pull them. Because the frontier groups were intimately involved in warfare along China's frontier at this time, the grave goods reflect this. But Shelach's careful analysis of the distribution of goods throws doubt on the contention that a warrior elite from elsewhere had entered the area and ruled over existing peoples. Indeed the evidence from the tombs would indicate that there were a variety of local social status markers and many ways to achieve high rank. But if a new population or invading elite is not the source of the observed archaeological changes, then how are we to explain them?

I would suggest that one good option would be to look at a similar phenomenon: the rapid spread of horse-riding bison hunting in North America and the creation of a distinctive "Plains Indian" culture. Here the introduction of horses first taken from the Spaniards in New Mexico during the late seventeenth century had a profound impact on the societies that acquired them. Within a century, a wide variety of tribes with heterogeneous origins had all adopted similar cultural practices that included horse riding, bison hunting, and the use of mobile tents. These societies also valorized warfare and horse raiding. This cultural complex did not spread by the expansion of a single bison-hunting population or by the arrival of a new bison-hunting elite that dominated non-bison hunting people. Instead, parts of existing populations drawn from the region's different settled riverine horticultural societies, nomadic forest foragers, and existing pedestrian hunters in the mountain and plains all adopted a new cultural complex. They not only filled a new and highly productive economic niche but, through their military power, became dominant over their neighbors and formed the most formidable opposition to Euro-American expansion into the plain's ecosystem. Mobile horse-riding pastoralism would likely have had a similar impact on existing societies on the Eurasian steppe and could have been equally as rapid. Just as the horse-riding bison hunters became stereotyped as the quintessential Native Americans, so too the horse-riding steppe pastoralists seemed similarly uniform to outside observers despite their heterogeneous origins.

The question of frontier relations is the focus of Popova's long-term study focusing on the Volga-Urals region in the Bronze Age. Most of the debate about frontiers centers on the relationship between steppe societies and the agrarian states to their south. Popova notes that this debate has obscured an equally important steppe frontier, that of the forest belt to the north inhabited by societies engaged in hunting, fishing, or horticulture. It is clear that during the Bronze Age and afterward the steppe-forest interface was the source of considerable trade, particularly in high-value furs. It was a luxury trade that would have linked small foraging societies to larger networks of production and thereby had an impact on their social structure because it would require a different pattern of hunting from one designed to meet subsistence needs alone. We know from historical accounts in both Siberia and North America that traders using boats on rivers and lakes could penetrate deep into the forest zone to buy furs from native peoples. They might leave little physical evidence of their passing yet still link the forest peoples with centers of production thousands of miles distant.

The relationship between pastoralists or agro-pastoralists and foragers could be symbiotic because their production strategies were complementary. But as Late Bronze Age steppe peoples began specializing more in pastoral production, they pushed into regions formerly occupied exclusively by foragers, making the two competitors in some regions. Popova shows that depending on changing climatic conditions, the boundary between the two zones shifted or interpenetrated. One question that is difficult to answer from the available material is to what extent foragers moved onto the steppe to take up pastoralism. Pastoralism requires a relationship to its domesticated animals that is entirely different from the one hunters have to the undomesticated animals they pursue. Indeed, some of the archaeological evidence might be equally well explained by foragers taking up a new mode of production themselves and not the immigration of existing pastoralists. Certainly, the widespread use of reindeer motifs in the art of many early steppe societies would indicate a strong familiarity with forest hunting traditions, as Bunker's analysis here clearly shows. Her work would indicate that a "fur route" linking the steppe and forest zones served as a conduit for new cultural influences long before the emergence of a "silk route" that later linked sedentary agricultural societies to their northern steppe neighbors.

Popova also raises the issue of political complexity in steppe societies, complaining that too much emphasis has been put on their external relations. But the debate over whether political complexity on the
steppe is the result of internal or external factors has been clouded by
a conflation of a number of issues, particularly the old Soviet legacy of
treating descriptive labels such as "chieftainship" or "state" as evolutionary stages through which all societies historically evolve. But even
these later horse-riding steppe societies show an enormous diversity
of political organization, even though they all have similar economies.
The conclusion we need to draw from this is that political order and
pastoral economic structures are not directly linked in steppe pastoral
societies. The pastoral economy can run perfectly well either within a
vast imperial state structure or without any political hierarchy at all.
Cattle-raising pastoralists in East Africa historically lacked states and
even chiefdoms, relying on segmentary opposition or age sets to organize themselves. Even the region of Mongolia, home of the world's most
powerful nomadic states, has at times also run the same pastoral system
with no centralized political structure. Indeed, in only a single lifetime
the Mongols under Chinggis Khan moved from a political system close
to anarchy to the world's largest empire without changing the structure
of the underlying pastoral economy.

Even if the economies of later steppe pastoral societies did appear
uniform to outsiders, they did have significant differences. According
to Herodotus, the Scythians secluded their women while the neighboring Sarmatians would not allow a woman to marry until she had
killed an enemy in battle. But such direct written descriptions of nomad
life are rare and those which exist often treat the people of the steppe
negatively. Bunker argues that the steppe nomads did leave a record of
themselves, not in writing but through their arts. A close examination
of their artistic styles and motifs reveals the internal differentiation
among nomadic societies along China's northern border at the end of
the first millennium BCE into eastern and western zones. The eastern
zone of the Liaoxi steppe and Manchuria was ecologically mixed with
strong connections to the forest zone. Nomads here abutted hunters,
fishers, and farmers and engaged in a more complex economy than their
neighbors in the steppes to the west, whose economy was more purely
pastoral.

Bunker asserts that the analysis of the artistic complexes can stand
on their own without the need for texts to supplement them, and her
analysis confirms a picture of difference that remains consistent over

2,000 years. The east has had a more egalitarian social structure and more diversified economic base. With the exception of the Mongol Yuan dynasties, all the major foreign dynasties that ruled over China came from here. By contrast, the western zone was the home of all the major steppe empires, and its social structures were uniformly hierarchical. Although the Chinese considered them all barbarians and all alike, they were in fact culturally distinct. Bunker's non-literary evidence demonstrates that the roots of these differences were well established at the end of the Bronze Age.

Considering the vast area covered by the Eurasian steppe, it must be admitted that the amount of archaeological data is still very small. For the later periods when horse-riding pastoralists came to dominate the steppe, there are very few habitation sites available for analysis. We often substitute analogies with historical steppe nomadic societies to fill in the gaps, but without archeological data to confirm the similarities, we cannot be sure. This applies even more directly to the Bronze Age in which steppe peoples appear to have had more diverse economies and a wider variety of social organizations. Our ideas of social evolution, though more often implicit assumptions than explicit hypotheses, make it hard to accept the idea that the internal complexity of Bronze Age steppe societies was probably greater than a millennium later when horse-riding steppe pastoralism became more fully established. The Bronze Age agro-pastoral steppe economies encouraged the construction of fixed residences and internal social hierarchies that had much more potential for the development of social complexity than more extensive styles of pastoralism alone. The cultural and social differences that distinguish the various parts of the steppe in later periods, of which Bunker's analysis of artistic styles is an example, may well be the result of profound differences that existed in the Bronze Age. People may adopt a new economy and political organization, but they tend to haul a good deal of older cultural baggage with them.

The three chapters here also make major contributions to our understanding of frontier relations between peoples of the steppes and their near neighbors. This is a topic that has been extensively explored in later periods when Iron Age technology and horse-riding pastoralists enter the historical record, but it was obviously important in earlier periods as well when written documentation is lacking. Shelach's analysis of frontier warfare on the eastern steppe is a reminder that at this period the bi-polar frontier that appears so clearly in the form of the "Great Wall" dividing the nomads from China was by no means so stark earlier.

Indeed, assuming a sharp dichotomy between steppe nomads and sedentary Chinese may be an anachronistic projection upon a frontier relationship that was far more fluid. Archaeology can play an important role in determining whether these societies were as materially different at this period as they are portrayed in surviving historical texts.

On the eastern steppe, Bronze Age China in the Warring States period experienced intense military conflict among Chinese states as well as between frontier peoples and Chinese states. Sources of power on the steppe at this period may well have been related to relationships (or conflicts) with polities in China. However, too great an emphasis on military power for explaining frontier relationships may overshadow other types of frontier interactions. As Bunker's analysis shows, artistic motifs and modes of social organization that have their roots in the forest zones of the north are strong evidence for a long-term interaction in which cultural influences and trade transcended power relations. Similarly, Popova's observation that boundaries between foragers and pastoralists were so blurred in the Volga-Urals region implies that it may be misleading to even posit the existence of distinct groups with fixed territories. Here, archaeology takes on even greater importance, because it is one of the few tools available to identify the types of frontier relationships that took place between pastoralists and foragers. For generations "frontier relations" has been synonymous with nomad sedentary relations. The new focus on the steppe-forest frontier is a reminder that zones of interaction are always plural. Perhaps one day we can go further and tease out the internal archaeological frontiers that divided nomadic pastoralists from each other as well.

CHAPTER 14

Violence on the Frontiers?
Sources of Power and Socio-Political Change at the Easternmost Parts of the Eurasian Steppe during the Late Second and Early First Millennia BCE

GIDEON SHELACH

THIS CHAPTER focuses on the late second and early first millennia BCE, a period in which, according to many scholars, societies throughout the Eurasian steppe underwent meaningful changes (e.g., Hanks 2002: 183; Khazanov 1984: 92–93; Renfrew 2002: 4–7), and addresses models for social, political, and cultural change in frontier zones. Theories addressing socio-political change can be classified into two types: indigenous and exogenous. Indigenous theories see change as evolving through processes such as competition or cooperation among local individuals and groups and their interaction with the local environment. Exogenous theories attribute change, including the development of socio-political complexity, to forces outside the local communities. Such forces can be human-derived – large scale migrations, for example – but also natural, such as climatic changes. Although external and internal processes are not mutually exclusive, the intellectual traditions in which models evolved to explain socio-political change in prehistoric societies commonly make them seem that way.

Introduction

NOWHERE IS the blend of external and internal dimensions of change more evident than in the Eurasian steppe, where contacts among societies were frequent but where unique local cultures, adaptations, and hierarchies evolved since at least the third millennium BCE. This is especially true at the frontier zones of this large region: areas in which intensive interactions took place between societies with different economic strategies, ideologies, and cultural attributes. This chapter focuses on one such place – the easternmost fringes of the Eurasian steppe. Ecologically this area borders on the fertile loess plains of the Yellow River, where intensive agricultural adaptation is possible. This is a border between the cultures of the Chinese Bronze Age, with their shared ideology and material attributes, and the Bronze Age cultures of the steppes.

The chapter highlights and synchronizes exogenous and indigenous stimuli for change among societies located at this frontier zone during the beginning of the first millennium BCE, placing special emphasis on the analysis of data associated with wars, violence, and the control exerted by the elite over the population. Was the use of violence an important mechanism to achieve control in those areas? Was the change observed at the eastern part of the steppes associated with interactions between an already complex local society and a small group of elite warriors that migrated to the area from the west? Or was it the local elites who borrowed from the outside only symbols of war and violence to help establish themselves vis-à-vis the local population and the more complex state-level societies to the south?

The validity of these models is addressed through the analysis of burial data from the different sub-regions within North China and a systematic examination of survey data from the Chifeng region. Such a discussion, it is hoped, will contribute to a more refined understanding of the history of the eastern part of the Eurasian steppe. Moreover, although this chapter does not attempt to develop a new, all-inclusive model, it should stimulate novel thinking about processes of change. The discussion is relevant for the understanding of other socio-political trajectories of societies in the Eurasian steppe and in similar environments elsewhere.

The Northern Zone of China: Environmental Conditions and Cultural-Historical Background

CHINA'S "NORTHERN ZONE"[1] is the easternmost part of the Eurasian steppe – a crescent-shaped area that covers the northern parts of Hebei, Shanxi, and Shaanxi provinces, most of Gansu and Qinghai, Inner Mongolia and Liaoning, and the western parts of Jilin and Heilongjiang. During the second and first millennia BCE, this area lay beyond the Shang and Zhou polities that developed in the Yellow, Wei, and Yangzi River basins. It lies more or less between 40 and 46 degrees north latitude (down to almost 35 degrees in the Ordos region) and between 95 and 128 degrees east longitude. On the north it borders on Mongolia and southern Siberia, and on the west the Gobi and Takalamakan deserts.

Today, the area receives an average annual precipitation of 500 millimeters or less. Most areas in the east receive today 350 to 500 millimeters, and in the west 100 to 400 (Liu 1998). Average temperatures also drop as one moves north and west. In the Chifeng (赤峰) region, which is located in the heart of the eastern part of the Northern Zone, average temperatures in January are minus 9.7 degrees Celsius, and an average of 23.2 degrees in July. In Huhhot (呼和浩特), located toward the central-western part of this region, average degrees in January are minus 13.1 and in July 21.9 Celsius (Liu 1998).

Although climatic conditions during the second and first millennia BCE were probably warmer and moister, the area was then, as now, less suitable for agriculture than areas in the main river basins to the south. The soils and natural vegetation of most of the Northern Zone are classified as that of a steppe (*caoyuan* 草原 in Chinese) (Hou 1982). The productivity of the areas so defined varied greatly owing to their specific location, altitude, and soil type. For example, the sands that cover large areas along the Xilamulun River (西拉木伦河) in the east have poor capacity to hold water and therefore provide less favorable conditions for the growing of wild and domesticated vegetation than the loess soils of other parts of this region.

Human presence in this area goes back to the Pleistocene, and sedentary village life, probably associated with agriculture, is dated to circa 6200 BCE (Shelach 2001). By the late third and early second millennia BCE, the local population had multiplied (Drennan et al. 2003), and in different regions within the Northern Zone settlement hierarchies developed (Linduff et al. 2004; Shelach 1999; Tian and Han 2003). Some sites are larger than 20 hectares and show substantial investment in both permanent structures, such as stone walls that enclosed the sites or parts of a site (known from the eastern part of the Northern Zone), and domestic structures (Liaoning 2001; Neimenggu 2000; Shelach 1999). Stone tools, a substantial number of storage pits, and the remains of grains and animal bones suggest that societies throughout the Northern Zone relied heavily on agricultural production (Debaine-Francfort 1995; Huang 2000; Gansu sheng 1990; Li and Gao 1985; Linduff 1995: 138; Shui 2001; H. Wang 1992; L. Wang 2004; Zhongguo 1996: 362–409). High-quality and complex ceramic vessels and evidence for incipient bronze production (Debaine-Francfort 1995; Li and Han 2000; Li and Shui 2000;

Linduff 1998; Shelach 1999: 101–105; Tian and Han 2003; Zhongguo 1996: 188–190) point to at least some degree of craft specialization.

Transition to the First Millennium BCE: Continuity or Change?

THE LATE second and early first millennia BCE were periods of social, political, and economic change for societies throughout the Eurasian steppe. However, the locus and magnitude of change among societies of the Northern Zone has rarely, if ever, been systematically addressed. A cursory look at the archaeological record may suggest a dramatic change. In the eastern part of this region, for example, fortified sites, typical of the second millennium, all but disappear, and investment in permanent structures is minimal (Shelach 1999: 146–148; Linduff et al. 2004). In the Ordos region, part of the western section of the Northern Zone, the number of known sites drops dramatically (Tian and Han 2003: 236; M. Wang 1994: 399). Throughout the Northern Zone new styles of artifacts and burial practices are introduced during the late second and early first millennia BCE. Change is most dramatically seen in the flowering of the local bronze industry (Linduff 1997: 33–73; Tian and Guo 1986; Zhu 1987).

These may seem like isolated and unconnected trends, but a systematic analysis of the archaeological data suggests a complex combination of continuity and change. Direct continuity from the second to the first millennium BCE is suggested, for example, by the shapes and styles of ceramic vessels. Regional styles, which had developed in the Northern Zone since at least the third millennium BCE, continued to evolve, without any significant break, during the second and first millennia, indicating the area supported centers of production, retail, and transportation, an entire network of associations (cf. Shui 2001; Tian and Guo 1988; Zhu 1987) (Fig. 14.1). In contrast to the intuitive impression that the settlement system collapsed in this period, results from the only systematic and large-scale regional survey conducted so far in the Northern Zone – the Chifeng International Collaborative Archaeological Research Project – suggest that population levels remained high from the late second to the first half of the first millennium BCE. Analyzing the data from the 765 square kilometers surveyed to date and taking into account such factors as the total occupied area for each period, the density of artifacts, and the length of the period, we have estimated that both the Lower Xiajiadian (夏家店下层 second millennium BCE) and the Upper Xiajiadian

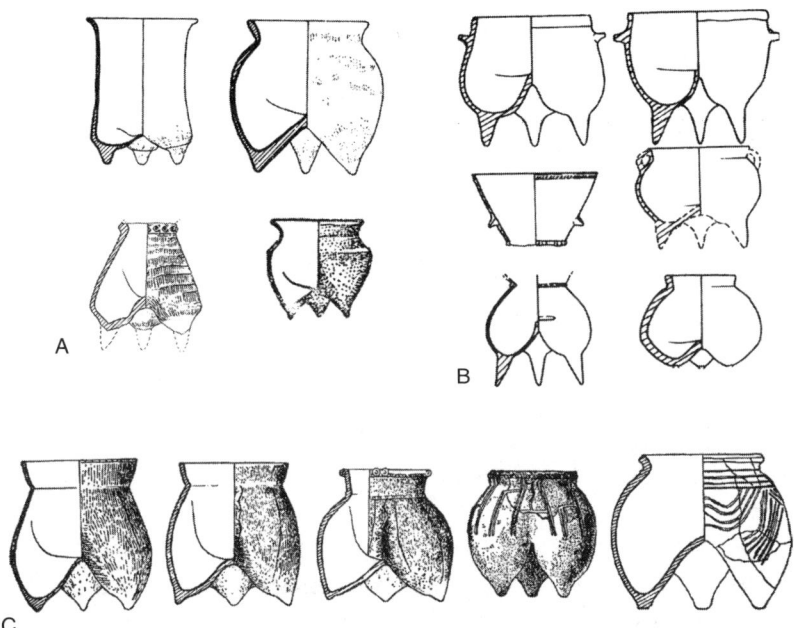

Figure 14.1. Continuation of local ceramic production in different parts of the Northern Zone. In the Chifeng region: (A) Lower Xiajiadian; (B) Upper Xiajiadian (after Zhang et al. 1987: 71; Qi 1991: 63). In the western part of the Northern Zone: (C) From the third phase at Zhukaigou (*left*) to Maoqinggou (*right*) (after Tian and Guo 1988, fig. 7).

(夏家店上层 late second and early first millennia BCE) are eras of high population density in the Chifeng region with no apparent collapse or dramatic loss of population between them (Drennan et al. 2003).

Economic changes are fundamental for our understanding of the transition from the second to the first millennium BCE in the Northern Zone. As elsewhere in the Eurasian steppe, according to the accepted model, in this period pastoralism became the main subsistence mode, and individuals and societies adopted a more mobile way of life. Indeed, at many sites, bones of herd animals dominate the faunal assemblages. At the Dahuazhongzhuang cemetery (大华中庄) in Qinghai, for example, among the 525 animal bones found at graves associated with the Kayue culture (卡约, circa 1600–600 BCE), 64% are bones of sheep/goat, 15.2% are cattle bones, and 20% are horse bones. Only one pig skeleton and one dog bone, as well as two bones of wild animals have been recovered (Qinghai 1985: 28–33). In the Ordos region, among the graves of

Table 14.1. Percentages of animals at different sites from the first millennium BCE

	Dahuazhongzhuang[a]	Maoqinggou[b]	Guandongche[c]
Sheep/goat	64	80.3	32.2
Cattle	15.2	14.5	19.2
Horse	20	3.9	3.2
Pig	0.19	0	25
Dog	0.19	1.3	16.2
Wild animal	0.38	0	3.2

[a] Gansu region, cemetery.
[b] Ordos region, cemetery, early graves.
[c] Chifeng region, habitation.
Sources: Data taken from Qinghai 1985: 28–33; Tian and Guo 1986: 306–313; and Zhu 2004: 43.

the first two phases of the Maoqinggou cemetery (毛庆沟) – representing perhaps the period between the seventh and fifth centuries BCE – altogether 61 sheep/goat skulls (80.3%), 11 cattle skulls (14.5%), 3 horse skulls (3.9%), and 1 dog skull (1.3%) have been recovered (Tian and Guo 1986: 306–313). Somewhat different patterns are emerging, however, from the analysis of faunal remains excavated from sites in the eastern part of the Northern Zone. While sheep/goat bones are more numerous as compared to their percentages in assemblages from the third and second millennia BCE, they are not as numerous as in the western part of the Northern Zone, and the percentages of bones of animals that are not usually associated with a pastoralist way of life, such as pig, remain quite high. At the Guandongche (关东车) site, in Keshiketeng banner (克什克腾旗), among the 31 animal skeletons identified, 32.2% are sheep/goats, 25% are pigs, 19.2% cattle, 16.2% dog, and only one horse bone and one wild animal bone were found (Zhu 2004: 433).

The data from Table 14.1 seem to indicate that while reliance on herding animals became perhaps the dominant mode of subsistence of societies in the western part of the Northern Zone, those in the eastern part were engaged in a more mixed economy. This conclusion is supported by analysis of the Chifeng survey data. Using GIS methods, I compared the location of Lower and Upper Xiajiadian sites in relation to current land-use categories digitized from a map published in the 1980s (Neimenggu 1988). While I realize that the four categories into which our survey area is divided (irrigated agricultural land,

Table 14.2. Occupation indexes of Lower Xiajiadian and Upper Xiajiadian sites and of modern villages according to different land-use categories

	Lower Xiajiadian (second millennium)	Upper Xiajiadian (first millennium)	Modern villages
Irrigated agriculture	0.48002	0.316702	0.541236
Non-irrigated agriculture	0.214325	0.230301	0.129944
Pastureland	0.175165	0.446608	0.059573
Forests	0.129269	0.006085	0.269276
TOTAL	0.9987788	0.9996955	1.0000292

non-irrigated agricultural land, pastureland, and forests) do not necessarily reflect prehistoric land-use patterns, I think that they are a good heuristic device because they do reflect the relative agricultural productivity of the land.[2] For each period, I calculated the ratio of occupied to non-occupied area in each land category. If sites had been randomly established, the relative part of occupied area in each land category would be the same; deviation from this random selection mode will reflect the attractiveness of each type of land to the prehistoric populations. To compare between the periods, I standardized these numbers to an occupation index that sums up to 1 for each period. For comparative purposes, and as a kind of "control group," I analyzed in the same way the location of the current villages in our survey area, whose occupation, we know, is based on intensive agriculture.

Table 14.2 and Figure 14.2 demonstrate close similarities between the occupation patterns of the Lower Xiajiadian and the modern villages. Both periods show significant preference for land currently under irrigation and much less to dry agriculture and pastureland – a pattern that can be associated with intensive agriculture economy. The Upper Xiajiadian pattern is more complex; while land used today for both wet and dry agriculture was then occupied to a significant degree, the land most preferred then is that currently used for herding. Such a pattern may suggest a mixed economy in which both pastoralism and agriculture are important.

The data presented here suggest that an economic change occurred between the second and the first millennia BCE. The magnitude of this change seems to have been felt more severely in the western part of

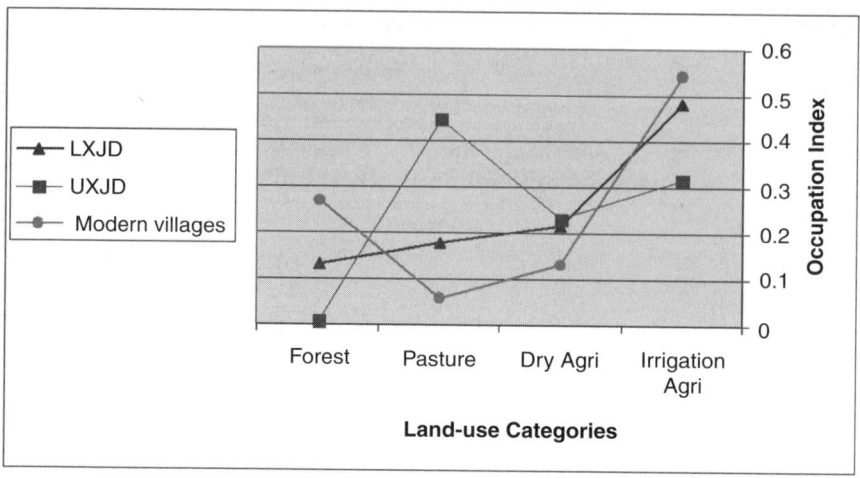

Figure 14.2. Graph of occupation indexes of different land-use categories of Lower Xiajiadian, Upper Xiajiadian, and modern villages.

the Northern Zone, where a more rapid and complete transition from agricultural to a pastoral-based economy is suggested. However, many crucial types of data are missing, especially for societies that inhabited the western part of the Northern Zone. Should systematic regional survey data become available for this region, it will be interesting to re-evaluate my conclusions. In the eastern part of the Northern Zone, comparable changes also occurred, but the pace and magnitude were more constrained. New kinds of domestic animals, such as sheep, goats, and horses are found at sites, but their relative numbers are not very high. Although a pastoralist economy is indicated by site structure and location, the archaeological evidence also suggests continuity of intensive agriculture. What emerged, it seems, even by the end of the period addressed here, is not a purely pastoralist economy but a mixed economy in which agriculture and pastoralism existed side by side and had a roughly similar importance.

Coming from Afar? Evidence for Inter-regional Interactions

As POINTED out at the beginning of this chapter, exogenous models, such as population replacement and the deterioration of climatic conditions, are probably the most common explanation for the transition to a pastoralist economy, not only in the Northern Zone but in other parts of the Eurasian steppe as well (cf. Bashilov and Yablonsky 2000; Chernykh et al. 2004; Christian 1998; Dergachev 2002; Hiebert and Shishlina 1996; Li et al. 2003; Lin 2003; Qiao 1992; Suo 2003; Xia et al. 2000).[3] These external prime movers have been evoked by many researchers trying to solve the conundrum. For example, why, in areas where agriculture had evolved over thousands of years and where agricultural-based communities had been stable and prosperous, was this tradition abandoned in favor of what seems a less productive and stable economy? A recent review article by Lin Yun (2003), one of the leading Chinese scholars of the Bronze Age of northern China, evokes both explanations: the need to cope with deteriorating climatic conditions pushed the local population to adopt a new way of life. At the same time, Lin also identified in the archaeological record evidence for the intrusion into the area of new people who pushed away the local population and established their foreign way of life.

I have argued elsewhere that although current paleoclimatic research suggests that moderate changes – a decrease in average yearly temperatures and precipitation – occurred during the late second and early first millennia BCE, this alone cannot explain contemporaneous social, political, or economic processes in the Northern Zone (Shelach 2005: 18–20). Here, I focus on the effects of contacts between societies in the Northern Zone and in areas of the Eurasian steppe to the north and west of it. However, before such effects are addressed, I briefly present evidence for such interactions.

Contacts among societies within the Northern Zone and between them and societies to their south in the Yellow and Wei river basins were probably quite common during the third millenium and the first half of the second millennium BCE (Li and Han 2000: 423–425; Shelach 2001: 24–25; Xu Yulin 1993: 328–329; Zhongguo 1996: 82–84, 191–194). However, no concrete evidence for contact with societies in the Eurasian steppe precedes the second half of the second millennium BCE. A few artifacts have been labeled as "Andronovo," implying earlier contacts with cultures of the Eurasian steppe, most notably bronze-socketed axes (Fig. 14.3), found mostly in Xinjiang and the western part of the Northern Zone (Mei and Shell 1998: 587; Peng 1998) but also in the east (Liaoning 1989). However, the identification of these artifacts with the Andronovo culture, rather than with the Karasuk or other cultures dated to the second part of the second millennium and to the early first millennium BCE, is debated even among those who support the western origins of such artifacts (cf. Fitzgerald-Huber

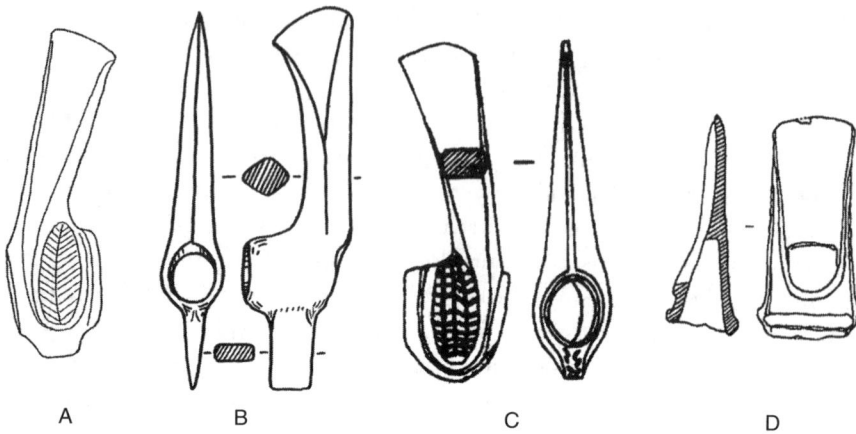

Figure 14.3. "Andronovo" axes from Central Asia, Xinjiang, and the Northern Zone: (A) Andronovo culture (after Kuz'mina 2004: fig. 2.2); (B) Wanliu, Liaoning (after Liaoning and Tieling 1989: 12, fig. 13); (C) Gongliu, Xinjiang (after Mei 2000: 94, fig. 2.22); (D) Siba culture (after Mei 2003: 53, fig. 4).

1995; Mei 2003; Peng 1998). Mace heads, most of them made of stone but some made of bronze, have recently been suggested as further evidence for early contacts with cultures to the west as far away as Egypt and Mesopotamia (S. Li 2002: 180–181; 2003: 15–16). Such an interpretation is, however, problematic because most of the mace heads that Shuicheng Li (2003: 29, fig. 9) presents have a generic oval shape that could easily have been developed locally and is not tied directly to any specific external model.[4] Moreover, given the wide range of dates provided by Li (2003: 16) for the mace heads found in northwest China – circa 3000–500 BCE – it is difficult to relate them to the early phase of contacts with the west.

By circa 1300 BCE we have much more concrete evidence for long-range contacts. Numerous artifacts, mostly bronzes, whose style and shape are correlated with those known from areas further to the north and west, are found in graves in both the western and eastern parts of the Northern Zone. It is possible to suggest at least two avenues of interaction with artifacts found most typically in the eastern part of the Northern Zone not commonly found in the west and vice versa.

In the western part of the Northern Zone, the artifacts most clearly associated with contacts toward the west and the north are socketed weapons. It is perhaps possible to date a few of those objects to circa 1500 BCE. For example, Mei (2003: 33–35) identifies two such objects as belonging to the Qijia (齐家) culture and an unspecified number to the slightly later Siba (四坝) culture. Both Mei (2003) and Fitzgerald-Huber (1995) associate these artifacts with the Seima-Turbino complex of the west and east of the Ural Mountains in Central Asia (cf. Chernykh 1992: 218–233). Similar objects found at different sites in Xinjiang (Mei 2000: 59–61, figs. 22.4 and 3.3; 2003: 35) suggest an interaction route that passed through the Altai Mountains, northern Xinjiang (the Hami basin), and the Hexi corridor (S. Li 2003: 18, fig. 11).

In the eastern part of the Northern Zone, artifacts of the so-called Northern Bronze Complex are among the earliest manifestations of the intensification of long-range interactions. The artifacts associated with this phenomenon include knives with animal, ring, or jingle heads, socketed weapons, the enigmatic "bow-shaped" artifacts, and helmets, as well as smaller ornaments such as earrings (Lin 1986; Guo 1995; Wu 1985; Yang 2002). As previous research has already noted, one of the interesting aspects of this phenomenon is the close affiliation that some of the bronze artifacts bear to artifacts found among cultures of the Eurasian steppe as far west as the Volga region. Artifacts similar to the knives and socketed weapons of the "Northern Bronze Complex" are known from the Karasuk culture of southern Siberia and the Seima-Turbino complex further west (Chernykh 1992: 215–233; Chernykh et al. 2004; Legrand 2004; Lin 1986; Linduff 1997: 29–32; Shelach 2005; Wu 1985). The accepted chronologies of the Karasuk, from circa 1400 to 1000 BCE (Legrand 2004: 139), and the Seima-Turbino, from around the fifteenth to fourteenth century BCE (Chernykh 1992: 194), synchronize with that of the "Northern Bronze Complex" of the Northern Zone. This suggests that regardless of the direction of the exchange, the spread of the styles and techniques associated with this phenomenon was relatively rapid.

According to the distribution of one diagnostic artifact of the "Northern Bronze Complex" – namely, knives with animal heads – the largest concentration in China is in the eastern part of the Northern Zone in Chifeng and Liaoning (Guo 1995; Yang 2002). Some have been found in the Zhangjiakou (张家口) area in northwest Hebei (Tao 1994: 18), in sites such as Chaodaogou (抄道沟) in Hebei Province (Hebei 1962; Ma 1995: 1, 11) and Yantoucun in north Shaanxi (Lin 1986: 245, fig. 51, no. 7; Shaanxi 1979: fig. 90), but no such artifacts are known from the area further to the west in the Northern Zone. Very similar objects were found in Shang graves, such as in the famous Fuhao tomb at Yinxu (Zhongguo 1980, photo 66) and in other Late Shang and Western Zhou contexts, but they seem to have been brought to this Central Plain context from the outside (Guo 1995; Linduff 1997: 32; Wu 1985; Yang 2002). Very similar knives are known from areas in northern Mongolia (Linduff 1997: 29), the Karasuk culture of the Minusinsk basin (Legrand 2004: 148, fig. 5.4), and even west of the Ural Mountains, in cemeteries of the Seima-Turbino complex (Chernykh 1992: 219, figs. 74, 75, 77; Chernykh et al. 2004: 27, figs. 1.10 and 1.11). It is interesting to note that the distribution of the artifacts currently known connect the eastern part of the Northern Zone through a very northern route that passes along the ecotone between the steppe and the forest zones of Central Asia. Other kinds of evidence, such as the common construction of slab graves in the eastern part of the Northern Zone (Shelach 1999: 148–150), Mongolia and Central Asia (Askarov et al. 1992: 460; Erdenebaatar 2004: 189; Legrand 2004: 142), but not in the western part of the Northern Zone, suggest the existence of this northeastern route of interactions.

During the same time that societies in the Eurasian steppe and the Northern Zone came into contact, other interactions reached so far as to include societies in the Yellow River basin. Most notably, the chariots that made their first appearance in China at the late Shang center of Yinxu (殷墟) near Anyang reflect the transmission of new technology from the west (Anthony and Vinogradov 1995; Barbieri-Low 2000; Shaughnessy 1988). The association of some of these chariots, and especially the people buried with them – presumably the drivers of the chariots and the horse grooms – with artifacts of the "Northern Bronze Complex" (e.g., C. Li 1977: 112, fig. 17) may suggest that chariots were introduced to China via the eastern part of the Northern Zone and not via its western part, as is commonly assumed (cf. Barbieri-Low 2000: 45–47).

Violence on the Frontiers? Militarism as a Source of Real and Symbolic Power

EVIDENCE FOR contacts between societies on the Eurasian steppe and in the Northern Zone are even more numerous at sites of the first half of the first millennium BCE (Mei 2000: 66–71; Shelach 2005), but the type of data associated with such contacts is more or less the same – mainly similarities among movable metal objects and the symbols found on them as well as on other media. However, the nature and magnitude of the interactions that produced such patterns of long-range similarities are hotly debated.

Research during most of the second half of the twentieth century, both in China and in the West, rejected the emphasis placed by early researchers on external contacts and focused on indigenous cultures and processes. Among European and North American archaeologists, this emphasis on local trajectories and the adaptation of the local society coupled with the almost complete disregard for external contacts – an intellectual trend that one critique termed "the premise of calorific priority" (Sherratt 1995: 7) – was a reaction against the hyper-diffusionist models common during the first half of the twentieth century (see discussion in Schortman and Urban 1987 and 1992). In China, it was a reaction against what was conceived as Eurocentric and imperialistic models of scholars such as the Swedish geologist and archaeologist Johan Gunnar Andersson (S. Li 2003: 10).

In recent years this attitude has changed, and from the crises of area studies during the 1990s emerged a new interest in cross-cultural interactions (Bentley 2005: 1–2). Archaeologists, like other social scientists, are no longer satisfied with viewing past societies as existing in isolation from each other, and contacts among prehistoric societies are no longer seen as meaningless incidents (cf. Anthony 1990; Cusick 1998; Kohl 1987; Renfrew 2002; Sherratt 1995). In China as well, this topic is no longer taboo. One notable example is the research on the "mummies" discovered in the Tarim basin – some of them identified as Caucasian and associated with Indo-European people – and the great attention they received in academic and non-academic circles, both in China and in the West (cf. Mair 1998; B. Wang 1999; and references in S. Li 2003: 13).[5]

Perhaps because of the nomadic nature of current and historic populations of the Eurasian steppe, explanations related to the movement of people and "influences" were, and to a certain degree remain, very popular among scholars working in these areas. Similarities in style and technique were commonly interpreted as evidence for waves of eastward migrations from the Black Sea area.[6] Reactions against such simplistic models ranged from a wholesale rejection of "diffusionistic" models to arguments about the direction of the diffusion. Among the first type is research that focuses on the unique, indigenous nature of different cultures in this vast area (cf. Linduff 1997; Yablonksy 2000); researchers of the second type debate the origins of selected traits, such as, for example, the recent voices advocating the eastern origins of the so-called Scythian-Siberian motifs (cf. Bokovenko 1996; Wu 2002).[7]

Table 14.3. Factor analysis of 45 graves from Xiaobaiyang cemetery

	Factor 1 (27.6%)	Factor 2 (17%)	Factor 3 (13.2%)	Factor 4 (10.4%)
Coffin	**0.802**	0.015	-0.213	-0.125
Animal bones	**0.722**	0.207	0.308	-0.321
Grave size (M2)	**0.720**	0.203	-0.265	0.194
Stone artifacts	**0.617**	0.365	0.353	0.134
Bone artifacts	**0.587**	**-0.656**	-0.031	0.166
Bronze weapons	**0.578**	**-0.686**	0.003	0.247
Bronze plaques	0.308	0.535	0.360	0.453
Ceramics	-0.056	-0.237	**0.718**	0.120
Bronze ornaments	-0.044	-0.127	**0.637**	-0.481
Beads	0.391	0.414	-0.256	**-0.505**
Small bronze artifacts	0.211	**-0.467**	-0.021	**-0.412**

Note: In each factor, 1 and –1 represent respectively the highest positive and negative correlation and 0 the lowest. Boldface indicates a close association.

A minority of the research on the Northern Zone subscribes to the notion of large-scale migration waves as the one mechanism that brought about cultural, economic, and social change in this region (cf. Fitzgerald-Huber 2003: 63; L. Wang 2004; Zhu 2004). Although I concur with Anthony (1990) that we should not dismiss such explanations out of hand, I think that the evidence does not support such a notion. The very continuation of the most mundane aspects of life, such as ceramic tradition or population levels, would argue against models of large population replacement. Having shown that there was long-range interaction, we must also show its impact and synthesize external inputs and indigenous processes.

As pointed out at the beginning of this chapter, one such model suggests the migration of a small group of elite warriors and their powerful imposition on the local society. This model is attractive because, while integrating an important external stimulus for change, it does not preclude local developments, such as the aggregation of population and intensification of economic production. The "migratory elite" model presumes that during the early first millennium BCE, socio-political status in the Northern Zone was associated with symbols of militaristic power such as bronze weapons and horseback riding.

To test this idea more systematically, I analyzed burial data from different parts of the Northern Zone. Rather than analyzing all the data

Table 14.4. Factor analysis of 37 early graves from Maoqinggou

	Factor 1 (23.3%)	Factor 2 (20.1%)	Factor 3 (15.3%)	Factor 4 (11.9%)
Bone artifacts	**0.675**	-0.221	0.303	0.136
All animals	**0.649**	-0.353	**0.405**	-0.154
Grave size (M2)	**0.606**	-0.500	-0.227	-0.233
Bronze weapons	**0.560**	**0.657**	0.066	-0.208
Horse gear	0.050	**0.718**	0.067	-0.552
Small bronze ornaments	-0.072	0.316	**0.762**	**0.442**
Ceramics	0.372	0.213	-0.502	**0.448**
Bronze belt parts	**0.472**	**0.469**	-0.274	0.336

Note: In each factor, 1 and −1 represent respectively the highest positive and negative correlation and 0 the lowest. Boldface indicates a close association.

available from this region, much of which comes from accidental finds or is partial, I concentrated on data from three cemeteries: Xiaobaiyang (小白阳) in northern Hebei in the eastern part of the Northern Zone; Maoqinggou (毛庆沟) from the Ordos region in the central part; and Dahuazhongzhuang (大华中庄) in northeastern Qinghai, located in the western part of this region. Each of these sites represents the most complete set of data from their respective regions, and for such analysis it is better, I think, to work with coherent datasets rather than with a conglomeration of data from different sites.

The 45 well-preserved graves at the Xiaobaiyang cemetery are dated to the Upper Xiajiadian (Zhangjiakou 1987: 50–51), making it the largest coherent burial dataset of this period. From the Maoqinggou cemetery I used data collected from the description and tabulation of the 36 well-preserved graves that have been assigned to periods I and II, dated from the eighth to the fifth century BCE (Tian and Guo 1986: 306–313). From the 117 graves assigned to the Kayue period at the Dahuazhongzhuang cemetery, I analyzed the data pertaining to the 95 undisturbed graves (Qinghai 1985: 28–33).

It is taken for granted here that death rituals, including funerals, are not completely divorced from the socio-political and economic reality of the society that conducted them. While, for example, artifacts placed in graves were intended for the use of the deceased in his or her afterlife, they are used, at the same time, to display the social position of the deceased family and affirm its position vis-à-vis other participants in the funeral.

Table 14.5. Factor analysis of 95 well-preserved graves from Dahuazhongzhuang cemetery

	Factor 1 (20.9%)	Factor 2 (12%)	Factor 3 (9.8%)	Factor 4 (9.2%)
Animal bones	**0.734**	0.239	0.023	0.012
Grave size (M3)	**0.705**	0.178	0.312	-0.258
Bronze weapons	**0.640**	-0.208	-0.349	**0.390**
Coffin	**0.607**	-0.033	**0.417**	0.077
Bronze mirrors	**0.549**	0.171	-0.459	0.117
Bronze plaques	**0.453**	**-0.669**	0.036	0.098
Bronze beads	0.069	**-0.619**	0.393	-0.099
Beads	**0.351**	**-0.531**	-0.245	0.206
Spindle whorls	0.003	**-0.500**	0.336	-0.096
Stone artifacts	0.156	0.340	0.349	**0.644**
Other small bronzes	0.072	0.260	**0.362**	**0.520**
Other large bronzes	0.315	0.227	**0.329**	-0.483
Bone artifacts	**0.491**	0.104	-0.322	-0.265
Ceramics	**0.496**	0.169	0.130	-0.208
Shells	0.270	0.053	-0.236	-0.185

Note: In each factor 1 and −1 represent respectively the highest positive and negative correlation and 0 the lowest. Boldface indicates a close association.

Therefore, the patterns we observed in the mortuary data can inform us on analogous (though not necessarily identical) social patterns.

Factor analysis of burial data from the three cemeteries supports the notion that prestige is closely associated with militaristic symbols. In all three analyses (Tables 14.3–14.5), bronze weapons are correlated in factor 1, which accounts for the greatest portion of the variability, with aspects of investment in the construction of the grave and its furnishing. In the Xiaobaiyang data (Table 14.3), factor 2, accounting for the 17% of the variability, which correlates with bronze weapons, bone artifacts (many of them arrowheads and spearheads) and small bronze artifacts may further represent an identity based on military activity or hunting. Similar associations are seen also in factor 2 of the Maoqinggou analysis (Table 14.4).

While it is clear that weapons are associated with socio-political status, the results of the factor analysis may be seen as a reflection of the

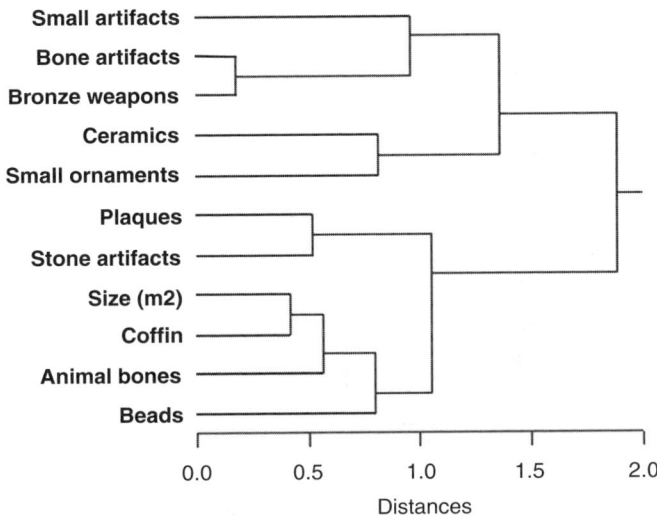

Figure 14.4. Hierarchical clustering of burial data from the Xiaobaiyang cemetery.

accumulation of wealth (the ability to construct larger and more expensive graves and accumulate different types of artifacts, including weapons) rather than the construction of a well-defined militaristic identity (or ideology), as one may expect from the "migratory elite" model. A complementary way of looking for symbolic expressions of identity in burial data is through the clustering of the mortuary variables.[8] Such a clustering may suggest a "package" of markers, which tend to appear together and symbolize a certain kind of identity.

Using the hierarchical clustering method (with Pearson's r correlation coefficient) on the standardized dataset of the three cemeteries produced suggestive grouping patterns. Experimenting with the Xiaobaiyang data, I found that the different types of bronze weapons were clustered close together. I therefore collapsed them into one category. As can be seen in Figure 14.4, the mortuary variables from Xiaobaiyang seem to be clustered into four groups:[9] (1) all bronze weapons and bone artifacts, with small bronze artifacts added higher up; (2) grave size and coffin, with animal bones and beads added to this group higher up; (3) stone artifacts and bronze plaques; and (4) ceramic vessels and small bronze ornaments. Interestingly, those four groups are clustered into two broader divisions, groups 1 and 4 and groups 2 and 3. We can hypothesize that those groups are associated with four different dimensions of identity.

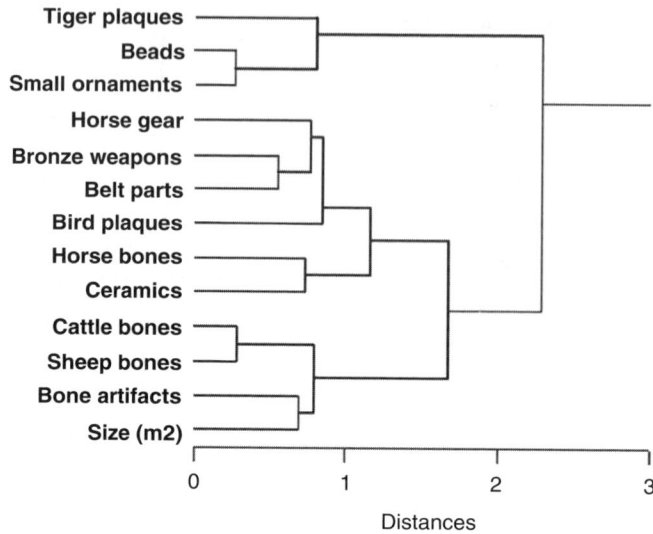

Figure 14.5. Hierarchical clustering of burial data from the Maoqinggou cemetery.

Group 1, correlating bronze weapons and bone artifacts (many of them bone arrowheads) can be seen as reflecting the warrior (masculine?) identity we were looking for. It is important to note that such an identity is clearly separated from variables directly associated with investment in the construction of the grave and the burial ceremony (group 2), suggesting that they could be independent identities.

The clustering of the Maoqinggou results in three main groups (Fig. 14.5). Of these, as in the Xiaobaiyang data, two represent investment (or accumulation of wealth) and militaristic identity. One interesting aspect of the Maoqinggou analysis is the grouping of sacrificial animal bones; whereas sheep and cattle bones correlate with grave size, reflecting investment in the construction of the grave and the burial ceremony (presumably the animals were consumed during the ceremony), horse bones are clustered with ceramics and more broadly with belt plaques, weapons, and horse gear, suggesting that horses were symbols of "militaristic" identity.

Although the analysis of the Dahuazhongzhuang mortuary variables produced less clear-cut results (Fig. 14.6), we can still see the clustering, in the central group, of variables such as grave size, coffin, sacrificial animals, and ceramic vessels, which are associated with direct investment

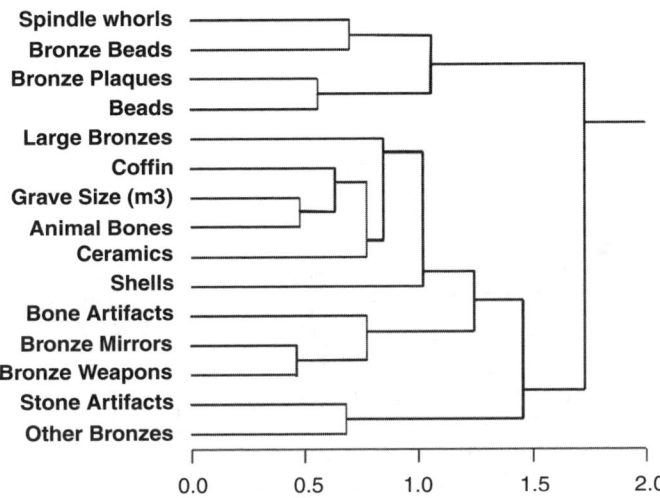

Figure 14.6. Hierarchical clustering of burial data from the Dahuazhongzhuang cemetery.

in the construction of the grave and its general furnishing (with foods?). A second group, correlating with bronze mirrors, bronze weapons, and bone artifacts, is, again, associated with "militaristic" identity.

Both the clustering of the graves and the factor analysis of burial variables suggest the importance of symbols of social and economic prestige in the construction of identity in graves. As the hierarchical clustering of mortuary data demonstrates, however, socio-economic prestige is not the only type of identity that was symbolized in the mortuary practices of societies from the Northern Zone. Significantly, more or less the same patterns recur in the analysis of the three cemeteries, suggesting that similar socio-political forces were involved in the three regions they represent. In all three regions, we can identify three or four symbolic "packages." One is clustered around the construction of the grave and its furnishings and funerary ceremony, while the other is associated with weapons and hunting. The third cluster, which I have not discussed here, may be associated with more personal or with "ethnic-like" identities.

To demonstrate how a "militaristic" identity was constructed and how it was symbolized, it is worthwhile to examine a few individual graves. A prime example is Grave M45 from the Zhoujiadi (周家地) cemetery, located in the Aohan banner of the Chifeng region. The deceased placed

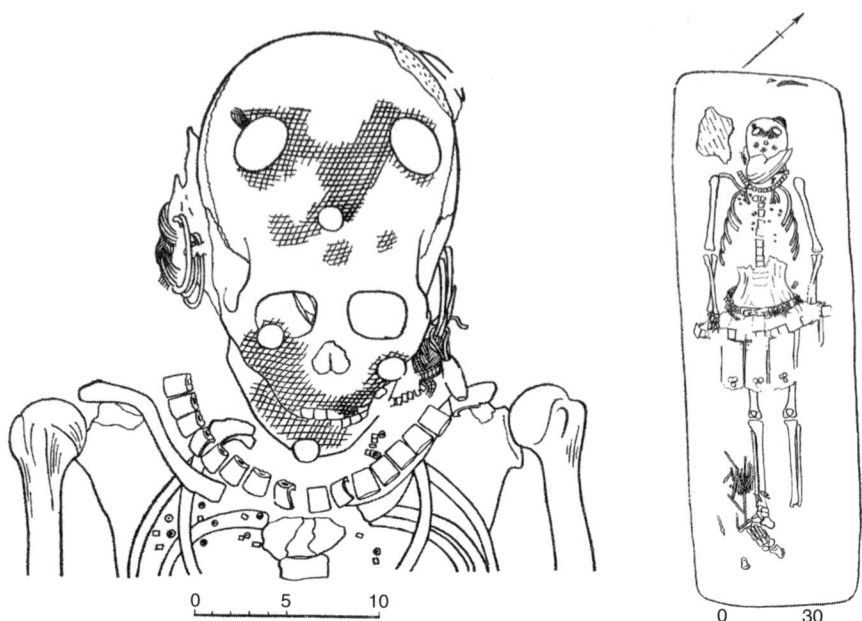

Figure 14.7. Drawings of Grave M45 from the Zhoujiadi (周家地) cemetery, in Aohan banner of the Chifeng region (after Zhongguo 1984: 418).

in this grave, a male[10] who was 12 to 13 years old, probably died as a result of an arrow piercing the upper left side of his skull. There were fractures in at least two more places on the skull and his right foot seems to have been cut off as well (Zhongguo 1984: 418). The burial treatment includes the placement of bronze plaques on the arrow wound and other fractures to the skull. Bronze buttons were placed on the eyes and perhaps on the mouth of the deceased, and remains of fabric found on the skull may suggest that these bronzes were attached to a cloth or that it was placed over them. On this a large freshwater shellfish was placed, which covered the lower part of the deceased's face (Fig. 14.7). The head and neck of the deceased was adorned with small bone and bronze ornaments, and there were two belts placed around the waist. One belt was thin and square, made of leather with bronze buttons, and had a bronze belt buckle. The other belt was wider and made of bronze parts to which larger leather parts were attached. Bronze buttons were fixed to the leather parts in decorative patterns, and a bronze knife was attached to

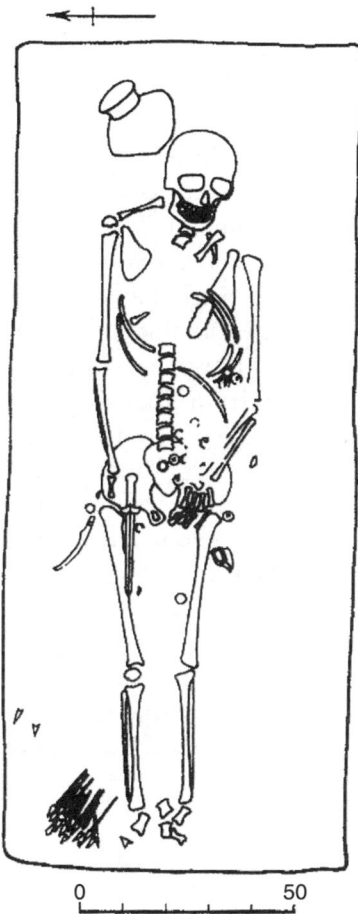

Figure 14.8. Drawing of Grave M37 from Xiaobaiyang (after Zhangjiakou 1987: 43).

the side of this belt. Close to the right foot of the deceased was placed a batch of 39 bone arrows and one bronze arrow.

The treatment of the head of the occupant of Grave M45 is to the best of my knowledge unique to this grave. We can assume that it is associated with the fact that he died violently, presumably in war. Although graves that are more richly furnished are known from the eastern part of the Northern Zone (cf. Xiang and Li 1995), none of them belongs to such a young occupant. The relatively lavish furnishing of a grave of a juvenile, which includes more than 20 bronze artifacts (not including small buttons) and more than 60 bone artifacts and beads, may be associated not with the accumulation of wealth during his short lifetime but rather with the special circumstances of his death. The large number of weapons included in this grave suggests that the prestige of its occupant was associated with military or quasi-military activity. It is possible that the deceased wore some of the artifacts found in the grave during his lifetime; however, some, such as the bronze buttons placed on his skull and the two belts around the waist, suggest a funerary display aimed at emphasizing a "militaristic" identity.

While a few aspects of the funerary treatment of Grave M45 are unique, the general patterns of display are seen in other graves as well. For example, Grave M37, also from Xiaobaiyang, shows similar patterns of artifact arrangements (Fig. 14.8). A knife and a dagger are placed near the left side of the deceased, and a bronze axe is placed slightly lower. These artifacts may have been attached to a belt, which disintegrated. Near his right foot a batch of eight bone arrows and 15 bronze arrowheads were placed. These positions are similar to Grave M45 and are found in other graves as well (e.g., Tian and Guo 1986: 239, 248). Although Grave M37 is not the most lavishly furnished at Xiaobaiyang cemetery, it places an emphasis on weapons not seen in some of the graves with more numerous artifacts. This may suggest, again, the construction of a "militaristic" identity, which is related to, but not identical with, social or economic status.

Conclusions

THE ANALYSIS presented here supports the existence of complex patterns of continuity and change of different aspects of the socio-political, economic, and cultural makeup of societies in the Northern Zone during the late second and early first millennia BCE. During the same time, long-range interactions between these societies and their counterparts in the Eurasian steppe intensified. Although we lack direct evidence for the mode of these interactions, we can observe that the artifacts associated with them are either bronze weapons or bronze plaques, belt buckles, and other personal ornamentation. While most of those artifacts were probably produced locally (Shelach 2005: 24–28), the origin of their style is exogenous. Why such external symbols were adopted is an intriguing question. Because most of these symbols were displayed as ornaments and attached paraphernalia, including weapons that were attached to the belt and visible during day-to-day interactions, we can speculate that the elite and perhaps other segments of the society found them as useful instruments in constructing their identities. Displaying such symbols during funerals may be seen as a way to secure the social hierarchy and emphasize the militaristic identity of the deceased and his kin. We can further speculate that the "non-Chinese" (or non-Zhou) nature of those symbols made them a useful marker of regional identity that was different from that of the Zhou people.

As we have seen, one way, and certainly not the only one, of achieving a prominent position in these early societies was associated with the display of and association with elements of military power. The fact that the main tools associated with such force, such as the bronze knives, daggers, and axes, are all also associated with long-range contacts, may lend support to a militaristic "migratory elite" model. It is as likely, however, that it was the local elite who adopted external symbols and techniques to enhance their local power and prestige. In either case, it is important to evaluate how external elements are used and manipulated in the context of local societies.

Weapons found in graves of the late second and early first millennia BCE suggest that violence and coercion were important means to achieve status in those societies. With the exception of the skeleton interred in Grave M45, however, actual evidence for violence is very rare. This may be partly due to the scarcity of research on the human skeletons found in graves, but it also may suggest that "militaristic" identity was more symbolic than real. Moreover, most of the bronze weapons found in graves from the Northern Zone are short daggers, knives, and axes more suitable for hand-to-hand combat than for mounted warfare. Some of them, like the knife, may not in effect be weapons at all but rather tools associated with herding. Even the arrowheads, found in relatively large numbers, could be used for hunting as well as for fighting. This all suggests that such grave goods reflect the affiliation of a person with a social group rather than his or her common participation in combat or in larger-scale wars.

The developments in the Northern Zone discussed here are important not only for our better understanding of the local processes but also for a more comprehensive understanding of the larger regional history. During the first millennium BCE, interactions of different types between societies in the Northern Zone and their Zhou (Chinese) neighbors to the south increased dramatically. These interactions became an important aspect of the development of the Chinese empire and continued to play an important role throughout the imperial era. Because most of the historical records pertaining to these relations were written by the Chinese side, it is important to find other types of data and develop novel approaches to understand the developments and motivations of the individuals and groups who lived in the Northern Zone.

It is possible to argue that militaristic symbols, but probably not frequent combat, were among the core elements in the construction of the antithetic identity of the two groups. Although the people of the Northern Zone are frequently described in early Chinese texts as having "beasts' hearts" (Pines 2005), conflicts with them were much less ferocious than during later periods. Chinese texts describing conflicts between the Zhou (Chinese) states and their northern neighbors mention clashes against infantry armies and chariots (Di Cosmo 2002: 109), but cavalry, the hallmark of later nomadic warfare, is never mentioned before the fourth century BCE. The same may be learned perhaps from the type of weapons found in the Northern Zone cemeteries analyzed here, wherein most weapons are designed for hand-to-hand combat that may be associated with personal prestige but are not the type of weapons that are later associated with the militaristic powers of the northern nomads.

Notes

1. This name, which in different variations is used by Chinese and non-Chinese scholars (cf. Di Cosmo 1999; Lin 1986; Wu 1985; Linduff 1997), can be seen as Sinocentric: the "Northern Zone," or Beifang (北方) in Chinese, reflecting its geographic position vis-à-vis China. However, in the absence of a better term, I use it here in reference to the position of the region in China without implying any predetermined dependency on areas to the south.
2. During prehistoric times it is almost certain that people did not engage in irrigation agriculture, certainly not on the scale practiced today in the Chifeng area. However, these areas, located mostly on relatively flat land close to the main rivers, should have been fertile even without irrigation. The land currently labeled as forests represents recent planting of trees, but it occupied a relatively small portion of the survey area and does not impact the results of my analysis.
3. For a critical overview of such models for Central Asia see Frachetti 2004: 245–252.
4. The one mace head that is unique – bronze, to which four cast sheep heads are attached, excavated at the Huoshaogou cemetery – was cast using the piece-mold technique, suggesting influence from the Central Plains of China rather than from the west (Mei 2003: 43).
5. A journal devoted entirely to the history of interactions with the Eurasian continent – called *Ouya xuekan* (歐亞學刊) or *Eurasian Studies* – has been published in China since 1999. This attitude may be associated with the new socio-cultural atmosphere in which global trade and mutual enrichments are celebrated.

6. For a critical overview of such models, especially among Russian archaeologists, see Frachetti 2004, chap. 5.
7. Similar sweeping models of westward – rather than traditional eastward – migratory waves can be also found in the writings of Evgenii N. Chernykh (1992; Chernykh et al. 2004).
8. For theoretical discussion on identity construction among ancient societies and especially the question of how identity is symbolized, constructed, and contested in funerals and burials, see my earlier publications (Shelach 2008a, 2008b).
9. I did not include shells and spindle whorls because these two variables each appeared in only one grave, so their grouping may be misleading.
10. Because of the young age of this individual, absolute sexing may not be certain.

References

Anthony, D. W. 1990. Migration in Archaeology: The Baby and the Bathwater. *American Anthropologist* 92: 895–914.

Anthony, D. W., and N. B. Vinogradov. 1995. Birth of the Chariot. *Archaeology* 48(2): 36–41.

Askarov, A., V. Volkov, and N. Ser-Odjv. 1992. Pastoral and Nomadic Tribes at the Beginning of the First Millennium B.C. In A. H. Damo, and V. M. Masson (eds.), *History of Civilizations of Central Asia*, vol. 1: *The Dawn of Civilization: Earliest Times to 700 B.C.* Paris: Unesco, pp. 459–475.

Barbieri-Low, A. J. 2000. Wheeled Vehicles in the Chinese Bronze Age (c. 2000–741 B.C.). *Sino-Platonic Papers*. 99: 1–99.

Bashilov, Vladimir A., and Leonid T. Yablonsky. 2000. Some Current Problems concerning the History of Early Iron Age Eurasian Steppe Nomadic Societies. In J. Davis-Kimball, E. M. Murphy, L. Koryakova, and L. T. Yablonksy (eds.), *Kurgans, Ritual Sites, and Settlements: Eurasian Bronze and Iron Age.* BAR International Series 890. Oxford: Archaeopress, pp. 9–12.

Bentley, J. H. 2005. Regional Histories, Global Processes, Cross-Cultural Interactions. In J. H. Bentley, P. Bridenthal, and A. A. Yang, (eds.), *Interactions: Transregional Perspectives on World History.* Honolulu: University of Hawai'i Press, pp. 1–13.

Bokovenko, N. A. 1996. Asian Influence on European Scythia. *Ancient Civilization from Scythia to Siberia* 3: 97–122.

Chernykh, E. N. 1992. *Ancient Metallurgy in the USSR: The Early Metal Age.* Cambridge: Cambridge University Press.

Chernykh, E. N., Evgenii V. Kuz'minykh, and L. B. Orlovskaia. 2004. Ancient Metallurgy of Northeast Asia: From the Urals to the Sayan-Altai. In K. Linduff (ed.), *Metallurgy in Ancient Eastern Eurasia from the Urals to the Yellow River.* Lewiston: Edwin Mellen Press, pp. 15–36.

Christian, D. 1998. *A History of Russia, Central Asia and Mongolia.* Oxford: Blackwell.

Cusick, James G. 1998. Introduction. In J. Cusick (ed.), *Studies in Culture Contact: Interaction, Culture Change, and Archaeology.* Carbondale: Center for Archaeological Investigations, Southern Illinois University, pp. 1–20.

Debaine-Francfort, C. 1995. *Du Neolithique à l'Age du Bronze en Chine du nord-ouest: la culture de Qijia et ses connexions.* Paris: Editions Recherche sur les Civilisations.

Dergachev, V. 2002. Two Studies in Defence of the Migration Concept. In K. Boyle, C. Renfrew, and M. Levine (eds.), *Ancient Interactions: East and West in Eurasia.* Cambridge: McDonald Institute, pp. 93–112.

Di Cosmo, N. 1999. The Northern Frontier in Pre-imperial China. In M. Loewe and E. Shaughnessy (eds.), *The Cambridge History of Ancient China.* Cambridge: Cambridge University Press, pp. 885–966.

2002. *Ancient China and Its Enemies.* Cambridge: Cambridge University Press.

Drennan, R. D., C. E. Peterson, G. G. Indrisano, Mingyu Teng, G. Shelach, Yanping Zhu, K. M. Linduff, and Guo Zhizhong. 2003. Approaches to Regional Demographic Reconstruction. In Chifeng International Collaborative Project Team (eds.), *Chifeng International Collaborative Archaeological Project: Regional Archaeology in Eastern Inner Mongolia; A Methodological Exploration.* Beijing: Kexue chubanshe, pp. 152–165.

Erdenebaatar, D. 2004. Burial Materials Related to the History of the Bronze Age in the Territory of Mongolia. In K. Linduff (ed.), *Metallurgy in Ancient Eastern Eurasia from the Urals to the Yellow River.* Lewiston: Edwin Mellen Press, pp. 189–221.

Fitzgerald-Huber, L. 1995. Qijia and Erlitou: The Question of Contacts with Distant Cultures. *Early China* 20: 17–68.

2003. The Qijia Culture: Paths East and West. *Bulletin of the Museum of Far Eastern Antiquities* 75: 55–78.

Frachetti, M. D. 2004. Bronze Age Pastoral Landscapes of Eurasia and the Nature of Social Interaction in the Mountain Steppe Zone of Eastern Kazakhstan. Ph.D. dissertation, University of Pennsylvania.

Gansu sheng wenwu kaogu yanjiusuo 甘肃省文物考古研究所. 1990. Gansusheng wenwu kaogu gongzuo shi nian 甘肃省文物考古工作十年 (Ten Years of Archaeological Work in Gansu Province), *Wenwu kaogu gongzuo shi nian* 文物考古工作十年. Beijing: Wenwu Chubanshe, pp. 316–326.

Guo, D. 1995. "Northern Type" Bronze Artifacts Unearthed in the Liaoning Region, and Related Issues. In S. Nelson (ed.), *The Archaeology of Northeast China.* London: Routledge, pp. 182–205.

Hanks, B. K. 2002. The Eurasian Steppe "Nomadic World" of the First Millennium BC: Inherent Problems within the Study of Iron Age Nomadic Groups. In K. Boyle, C. Renfrew, and M. Levine (eds.), *Ancient Interactions: East and West in Eurasia.* Cambridge: McDonald Institute, pp. 183–197.

Hebei sheng wenhuaju wenwu gozuo dui 河北省文化局文物工作队. 1962. Hebei qinglong xian chaodaogou faxian yi pi qingtongqi 河北青龙县抄道沟发现一批青铜器 (A Group of Bronzes Found at Chaodoagou, Qinglong County, Hebei). *Kaogu* 12: 644–645.

Hiebert, F. T., and N. I. Shishlina. 1996. Introduction. *Anthropology and Archaeology of Eurasia* 34 (4): 5–12.

Hou, Xueyu 候学煜 (ed.), 1982. *Zhonghua renmin gongheguo zhibei tu* 中华人民共和国植被图 (Vegetation Coverage Map of the PRC). Beijing: Ditu chubanshe.

Huang, Yunping 黄蕴平. 2000. Zhukaigou yizhi shougu de jianding yu yanjiu 朱开沟遗址兽骨的鉴定与研究 (Identification and Analysis of Animal Bones from the Zhukaigou Site). In Neimenggu zizhiqu wenwu kaogu yanjiusuo 内蒙古自治区文物考古研究所, E'erduosi bowuguan 鄂尔多斯博物馆 (eds.),

Zhukaigou – Qingtong shidai zaoqi yizhi fajue baogoa 朱开沟-青铜时代早期遗址发掘报告 (Zhukaigou – A Report on the Excavation of an Early Bronze Age Site). Beijing: Wenwu chubanshe, pp. 400–421.

Khazanov, A. M. 1984. *Nomads and the Outside World*. Cambridge: Cambridge University Press.

Kohl, P. L. 1987. The Ancient Economy, Transferable Technologies and the Bronze Age World-System: A View from the Northeastern Frontier of the Ancient Near East. In M. Rowlands, M. Larsen, and K. Kristiansen (eds.), *Center and Periphery in the Ancient World*. Cambridge: Cambridge University Press, pp. 13–24.

Kuz'mina, E. 2004. Historical Perspectives on the Andronovo and Early Metal Use in Eastern Asia. In K. M. Linduff (ed.), *Metallurgy in Ancient Eastern Eurasian from the Urals to the Yellow River*. Lewiston: Edwin Mellen Press, pp. 37–84.

Legrand, S. 2004. Karasuk Metallurgy: Technological Development and Regional Influence. In K. M. Linduff (ed.), *Metallurgy in Ancient Eastern Eurasia from the Urals to the Yellow River*. Lewiston: Edwin Mellen Press, pp. 139–156.

Li, Chi. 1977. *Anyang*. Seattle: University of Washington Press.

Li, Gongdu 李恭笃, and Meixuan Gao 高美璇. 1985. Xiajiadian xiacang wenhua ruogan wenti yanjiu 夏家店下层文化若干问题研究 (Research of Certain Questions Concerning the Lower Xiajiadian Culture). *Liaoning daxue xuebao* 1985(5): 154–161.

Li, Shuicheng 李水. 2002. The Interaction between Northwest China and Central Asia during the Second Millennium BC: An Archaeological Perspective. In K. Boyle, C. Renfrew, and M. Levine (eds.), *Ancient Interactions: East and West in Eurasia*. Cambridge: McDonald Institute, pp. 171–182.

2003. Ancient Interactions in Eurasia and Northwest China: Revisiting J. G. Andersson's Legacy. *Bulletin of the Museum of Far Eastern Antiquities* 75: 9–30.

Li, Xiaoqiang, Jie Zhou, and Johan Dodson. 2003. The Vegetation Characteristics of the "Yuan" Area at Yaoxian on the Loess Plateau in China over the Last 12000 Years. *Review of Palaeobotany and Palynology* 124: 1–7.

Li, Xiuhui 李秀辉, and Rubin Han 韩汝玢. 2000. Zhukaoigou Yizhi chutu tongqi de jinxiangxue yanjiu 朱开沟遗址出土铜器的金相学研究 (A Metallurgical Study of Bronze Artifacts from Zhukaigou). In *Zhukaigou – Qingtong shidai zaoqi yizhi fajue baogoa* 朱开沟-青铜时代早期遗址发掘报告 (Zhukaigou – A Report on the Excavation of an Early Bronze Age Site). Neimenggu zizhiqu wenwu kaogu yanjiusuo 内蒙古自治区文物考古研究所, E'erduosi bowuguan 鄂尔多斯博物馆. Beijing: Wenwu chubanshe, pp. 422–446.

Liaonning daxue lishixi kaogu yanjiushi 辽宁大学历史系考古研究室 and Tieling shi bowuguan 铁岭市博物馆. 1989. Liaoning Faku xian Wanliu yizhi fajue 辽宁法库县湾柳遗址发掘 (Excavations at the Wanliu Site, Faku County, Liaoning). *Kaogu* 1989(12): 1076–1086.

Liaoning sheng wenwu kaogu yanjiusuo 辽宁省文物考古研究. 2001. Liaoning Beipiaoshi Kangjiatun chengzhi fajue jianbao 辽宁北票市康家屯城址发掘简报 (Preliminary Report on the Excavations at the Kangjiatun Fortified Site, Beipiao City, Liaoning). *Kaogu*. 2001(8): 31–44.

Lin, Yun 林云. 1986. A Reexamination of the Relationship between Bronzes of the Shang Culture and the Northern Zone. In K. C. Chang (ed.), *Studies of Shang Archaeology*. New Haven: Yale University Press, pp. 237–273.

2003. Zhongguo beifang changcheng didai youmu wenhua dai de xhingcheng guocheng 中国北方长城地带游牧文化带的形成过程 (The Formation of Northern Nomads Belt along the Great Wall Area of China). *Yanjing xuebao* 燕京学报 14: 95–145.

Linduff, K.M. 1995. Zhukaigou: Steppe Culture and the Rise of Chinese Civilization. *Antiquity*. 69: 133–145.

1997. Archaeological Overview. In E. C. Bunker (ed.), *Ancient Bronzes of the Eastern Eurasian Steppes: The Arthur M. Sackler Collection*. New York: Sackler/Freer Gallery, Smithsonian Institution, Abrams, pp. 18–98.

1998. The Emergence and Demise of Bronze-Producing Cultures Outside the Central Plain of China. In V. Mair (ed.), *The Bronze Age and Early Iron Age Peoples of Eastern Central Asian*. Philadelphia: University of Pennsylvania Museum Publications, pp. 619–643.

Linduff, K.M., R. Drennan, and G. Shelach. 2004. Early Complex Societies in NE China: The Chifeng International Collaborative Archaeological Research Project. *Journal of Field Archaeology* 29(1–2): 45–73.

Liu, Mingguang 刘明光 (ed.). 1998. *Zhongguo ziran dili tuji* 中国自然地理图集. 2nd ed. Beijing: Zhongguo ditu chubanshe.

Ma, Chengyuan 马承源 (ed.). 1995. *Zhongguo Qingtongqi Quanji: Beifang Minzu* 中国青铜器全集：北方民族. (Collection of Chinese Bronzes: The Northern Minorities). Beijing: Wenwu chubanshe.

Mair, Victor H. (ed.). 1998. *The Bronze Age and Early Iron Age Peoples of Eastern Central Asia*. Philadelphia: University of Pennsylvania Museum Publications.

Mei, Jianjun. 2000. *Copper and Bronze Metallurgy in Late Prehistoric Xinjiang: Its Cultural Context and Relationship with Neighboring Regions*. Bar International Series 863. Oxford: Archaeopress.

2003. Qijia and Seima-Turbino: The Question of Early Contacts between Northwest China and the Eurasian Steppe. *Bulletin of the Museum of Far Eastern Antiquities* 75: 31–54.

Mei, Jianjun, and Colin Shell. 1998. Copper and Bronze Metallurgy in Late Prehistoric Xinjiang. In V. Mair (ed.), *The Bronze Age and Early Iron Age Peoples of Eastern Central Asia*. Philadelphia: University of Pennsylvania Museum Publications, pp. 581–603.

Neimenggu caochang ziyuan yaogan yingyong kaocha dui 内蒙古草场治源遥感应用考察队. 1988. *Neimenggu zizhiqu Chifeng shi ziran tiaojian yu caochang ziyuan ditu* 内蒙古自治区赤峰市自然条件与草场资源地图 (Map of Natural and Steppe Resources of the Chifeng City Area, Inner Mongolia Autonomous Region). Beijing: Kexue chubanshe.

Neimenggu zizhiqu wenwu kaogu yanjiusuo 内蒙古自治区文物考古研究所, E'erduosi bowuguan 鄂尔多斯博物馆. 2000. *Zhukaigou – Qingtong shidai zaoqi yizhi fajue baogoa* 朱开沟-青铜时代早期遗址发掘报告 (Zhukaigou – A Report on the Excavation of an Early Bronze Age Site). Beijing: Wenwu chubanshe.

Peng, Ke. 1998. The Andronovo Bronze Artifacts Discovered in Toquztara County in Ili, Xinjiang. In V. Mair (ed.), *The Bronze Age and Early Iron Age Peoples of Eastern Central Asia*. Philadelphia: University of Pennsylvania Museum Publications, pp. 573–580.

Pines, Yuri. 2005. Beasts or Humans: Pre-imperial Origins of the "Sino-Barbarian" Dichotomy. In R. Amitai, and M. Biran (eds.), *Mongols, Turks and Others: Eurasian Nomads and the Outside World*. Leiden: Brill, pp. 59–102.

Qi, Xiaoguang 齐晓光. 1991. Neimeggu Keshiketeng qi Longtoushan yizhi fajue de zhuyao shouhuo 内蒙古克什克腾旗龙头山遗址发掘的主要收获 (The Main Results of the Excavations at Longtoushan Site, Keshiketeng Banner, Inner Mongolia). In Neimenggu wenwu kaogu yanjiusuo (ed.), *Neimenggu dongbu qu kaoguxue wenhua yanjiu wenji* 内蒙古东部区考古学文北研究文集. Beijing: Haiyang chubanshe, pp. 58–72.

Qiao, Xiaoqin, 乔晓勤. 1992. Guanyu beifang youmu wenhua qiyuan de tantao 關於北方遊牧文化起源的探討 (Inquiry into the Origin of the Northern Nomadic Culture). *Neimangu wenwu kaogu* 1992(1–2): 21–25.

Qinghai sheng Huangyuan xian bowuguan 青海省湟源县博物馆, Qinghai sheng wenwu kaogu dui 青海省文物考古队, and Qinghai sheng shehui kexueyuan lishi yanjiushi 青海省社会科学院历史研究室. 1985. Qinghai Huangyuan xian Dahuazhongzhuang Kayue wenhua mudi fajue jianbao 青海湟源县大华中庄卡约文化墓地发掘简报 (Preliminary Report on the Excavation of the Kayue Culture Cemetery of Dahuazhongzhuang, Huangyuan County, Qinghai). *Kaogu yu wenwu*. 1985(5): 11–34.

Renfrew, C. 2002. Pastoralism and Interaction: Some Introductory Questions. In K. Boyle, C. Renfrew, and M. Levine (eds.), *Ancient Interactions: East and West in Eurasia*. Cambridge: McDonald Institute, pp. 1–10.

Schortman, E. M., and P. A. Urban. 1987. Modeling Interregional Interaction in Prehistory. In M. B. Schiffer (ed.), *Advances in Archaeological Method and Theory*, vol. 11. San Diego: Academic Press, pp. 37–95.

1992. The Place of Interaction Studies in Archaeological Thought. In E. M. Schortman and P. A. Urban (eds.), *Resources, Power, and Interaction*. New York: Plenum Press, pp. 3–15.

Shaanxi sheng kaogu yanjiusuo, 陕西省考古研究所. 1979. *Shaanxi sheng chutu Shang Zhou qingtongqi*. 青海省出土商周青铜器, vol. 1. Beijing: Wenwu chubanshe.

Shaughnessy, E. L. 1988. Historical Perspectives on the Introduction of the Chariot into China. *Harvard Journal of Asiatic Studies* 48(1): 189–237.

Shelach, G. 1999. *Leadership Strategies, Economic Activity, and Interregional Interaction: Social Complexity in Northeast China*. New York: Plenum Press.

2001. Interaction Spheres and the Development of Social Complexity in Northeast China. *Review of Archaeology* 22(2): 22–34.

2005. Early Pastoral Societies of Northeast China: Local Change and Interregional Interaction during c. 1100 to 600 BC. In R. Amitai, and M. Biran (eds.), *Mongols, Turks and Others: Eurasian Nomads and the Outside World*. Leiden: Brill, pp. 15–58.

2008a. He Who Eats the Horse, She Who Rides It? Symbols of Gender Identity on the Eastern Edges of the Eurasian Steppe. In Katheryn M. Linduff and Karen S. Rubinson (eds.), *Are All Warriors Male? Gender Roles on the Ancient Eurasian Steppe*. Lanham: Altamira, pp. 93–110.

2008b. *Prehistoric Societies on the Northern Frontiers of China: Archaeological Perspectives on Identity Formation and Economic Change during the First Millennium BCE*. London: Equinox.

Sherratt, A. 1995. Reviving the Grand Narrative: Archaeology and Long-Term Change. *Journal of European Archaeology* 3(1): 1–32.

Shui, Tao 水涛. 2001. Gansu diqu qingtong shidai wenhua jiegou he jingji xingtai yanjiu 甘肃地区青铜时代的文化结构和经济形态研究 (The Cultures and Economic Adaptation in Gansu during the Bronze Age). In Shui Tao (ed.), *Zhongguo xibei diqu qingtong shidai kaogu lunji* 中国海北地区青铜时代考古论集 (Collection of Papers on the Bronze Age of Northwest China). Beijing: Kexue Chubanshe, pp. 193–327.

Suo, Xiufen 索秀芬. 2003. Neimenggu nongmu jiaocuodai kaoguxue wenhua jingji xingtai zhuanbian ji qi yuanyin 内蒙古农牧交错带考古学文化经济形态转变及其原因 (Economic Change Reflected by Archaeological Cultures of the Agriculture-Pastoralism Transition Zone of Inner Mongolia). *Neimenggu wenuw kaogu* 1: 62–68.

Tao, Zongye 陶宗冶. 1994. Shilun Zhangjiakou diqu Zhanguo yiqian de kaogu wenhua yicun 试论张家口地区战国以前的考古文化遗存 (Discussing the Archaeological Cultures of the Pre-warring States Period at the Zhangjiakou Area). *Beifang wenwu* 1994(2): 14–22.

Tian, Guangjin, 天广金, and Suxin Guo 郭素新. 1986. E'erduosi shi qingtongqi 鄂尔多斯式青铜器 (Ordos Style Bronzes). Beijing: Wenwu chubanshe.

——— 1988. E'erduosi shi qingtongqi de yuanyuan 鄂尔多斯式青铜器的渊源 (The Origins of the Ordos Style Bronzes). *Kaogu Xuebao* 1988(3): 257–274.

Tian, Guangjin 天广金, and Jianye Han 韩建业. 2003. Zhukaigou wenhua yanjiu 朱开沟文化研究 (Research of the Zhukaigou Culture). In Beijing daxue kaogu wenbo xueyuan 背景大学考古文博学院 (ed.), *Kaoguxue yanjiu* 考古学研究, vol 5. Beijing: Kexue chubanshe, pp. 226–259.

Wang, Binghua 王炳华 (ed.), 1999. *Xinjiang gu shi* 新讲古尸 (The Ancient Corpses of Xinjiang). Urumqi: Xinjiang renmin chubanshe.

Wang, Huide 王惠德. 1992. Guanyu Xiajiadian xiaceng wenhua gucheng de ji ge wenti 关于夏家店下层文化古城的几个问题 (Some Questions concerning the Cities of the Lower Xiajiadian Culture). *Zhaowuda mengzu shizhuan xuebao* 13: 82–88.

Wang, Lixin 王立新. 2004. Shilun Changcheng didai zhongduan qingtong shidai wenhua de fazhang 试论长城地带中段青铜时代文化的发展 (Discussing the Development of Bronze Age Cultures in the Central Area of the Great Wall Zone). In Jilin daxue bianjiang kaogu yanjiu zhongxin 吉林大学边疆考古研究中心 (ed.), *Qingzhu Zhang Zhongpei xiansheng qishi sui lunwen ji* 庆祝张忠培先生七十岁论文集. Beijing: Kexue chubanshe, pp. 365–385.

Wang, Ming-ke 王明珂. 1994. E'erduosi ji qi linjin diqu zhuanhua youmuye de qiyuan 鄂爾多斯及其鄰近地區專化遊牧業的起源 (The Origins of Specialized Pastoralism in the Ordos and Neighboring Areas). *Zhong yang yan jiu yuan li shi yu yan yan jiu suo ji kan.* 65(2): 375–434.

Wu, En 乌恩. 1985. Yin zhi Zhouchu de beifang qingtongqi 殷至周初的北方青铜器 (Northern Bronzes from Yin to Early Zhou). *Kaogu xuebao* 1985(2): 135–156.

——— 2002. Ouya dalu caoyuan zaoqi youmu wenhua de jidian sikao 欧亚大陆草原早期游牧文化的几点思考 (Some Ideas on the Early Nomadic Culture in the Eurasian Steppe). *Kaogu Xuabao* 2002(4): 437–470.

Xia, Zhengkai 夏正楷, Hui Deng 邓辉, and Honglin Wu 武弘麟. 2000. Neimenggu Xilamulun he liuyu kaogu 内蒙海拉木伦河流域考古文化演变的地貌背景分析 (Geomorphologic Background of the Prehistoric Cultural Evolution in the Xilamulun River Basin, Inner Mongolia). *Dili xuebai* 地理学报 55(3): 329–336.

Xiang, Chunsong 项春松, and Yi Li 李义. 1995. Ningcheng Xiaoheishigou shiguomu diaocha qingli baogao. 宁城小黑石沟石椁墓调查清理报告 (Report of the Research of the Stone Coffin Grave Found at Xiaoheishigou, Ningcheng). *Wenwu* 文物 5: 4–22.

Xu, Yulin 许玉林. 1993. Liaoning Shang Zhou sheqi de qingtong wenhua 辽宁商周时期青铜文化 (The Bronze Culture of Liaoning during the Shang and Zhou Periods). In Su Bingqi 苏秉琦(ed.), *Kaoguxue wenhua lunji* 考古学文化论集, vol. 3. Beijing: Wenwu chubanshe, pp. 311–334.

Yablonksy, L. T. 2000. "Scythian Triad" and "Scythian World." In J. Davis-Kimball, E. M. Murphy, L. Koryakova, and L. T. Yablonksy (eds.), *Kurgans, Ritual Sites, and Settlements: Eurasian Bronze and Iron Age*. Oxford: BAR International Series, pp. 3–8.

Yang, Jianhua 杨建华. 2002. Yanshan nan bei Shang Zhou zhi ji qingtongwi yicun de fenqun yanjiu 燕山南北商周之际青铜器遗存的分群研究 (Classification of Bronze Objects Dated to the Shang-Zhou Transition Period from the Area North and South of the Yan Mountains). *Kaogu xuebao* 2002(2): 157–174.

Zhang, Zhongpei 张忠培, Zhesheng Kong 孔哲生, Wenjun Zhang 张文军, and Young Chen 陈雍. 1987. Xiajiadian xiaceng wenhua yanjiu 夏家店下层文化研究 (Research of the Lower Xiajiadian Culture). *Kaoguxue wenhua lunji*, 58–78. Beijing: Wenwu chubanshe.

Zhangjiakou shi wenwu shiye guanlisuo 张家口市文物事业管理所. 1987. Hebei Xuanhuaxian Xiaoyang mudi fajue baogao 河北宣化县小白杨墓地发掘报告 (Report on the Excavations at the Xiaobaiyang Graveyard, Xuanhua County, Hebei). *Wenwu* 文物 1987(5): 41–51.

Zhongguo shehui kexueyuan kaogu yanjiusuo 中国社会科学院考古研究所. 1980. *Yinxu Fuhao mu* 殷墟妇好墓. Beijing: Wenwu chubanshe.

Zhongguo shehui kexueyuan kaogu yanjiusuo Neimenggu gongzuodui 中国社会科学院考古研究所内蒙古工作队. 1984. Neimenggu Aohanqi Zhoujiadi mudi fajue jianbao 内蒙古敖汉旗周家地墓地发掘简报 (Brief Report on the Excavation of the Zhoujiadi Graveyard Site at Aohan Banner, Inner Mongolia). *Kaogu* 1984(5): 417–426.

1996. *Dadianzi: Xiajiadian xiaceng wenhua yizhi yu mudi fajue baogao* 大甸子—夏家店下层文化遗址与墓地发掘报告. (Dadianzi: Excavations of the Domestic Site and Cemetery of the Lower Xiajiadian Period). Beijing: Kexue chubanshe.

Zhu, Yonggang 朱永刚. 1987. Xiajiadian shangceng wenhua de chubu yanjiu 夏家店上层文化的初步研究 (Research of the Upper Xiajiadian culture). Su Bingqi 苏秉琦 (ed.), In *Kaoguxue wenhua lunji* 考古学文化论集, vol. 1. Beijing: Wenwu chubanshe, pp. 99–128.

2004. Chaganmulun he gu yizhi wenhua leixing ji xiangguan wenti 查干木伦河流域古遗址文化类型及相关问题 (The Archaeological Cultures of the Chaganmulun River Basin and Related Issues). *Kaogu wenwu* 考古文物 2004(3): 40–47.

CHAPTER 15

First-Millennium BCE *Beifang* Artifacts as Historical Documents

EMMA C. BUNKER

ANCIENT CHINESE texts often mention pastoral peoples living in the *beifang*, but the textual descriptions range from misleading to inaccurate, portraying them as wily, needy, ruthless people without honor who constantly harassed their Chinese neighbors. Modern scholars tend to base their opinions on the ancient Chinese records, viewing the *beifang* as a reservoir of "barbarian" activities that have periodically contaminated Eastern Zhou, Qin, and Han Chinese culture. As a result, the *beifang* "folk" have been consistently given poor scholarly press over the years without any recognition of their cultural achievements. This lacuna is being blamed, in part, on the supposed lack of written records provided by the *beifang* groups themselves.

In lieu of written records from the *beifang*, I intend to demonstrate that excavated *beifang* artifacts can serve as encoded historical records, as has been suggested for various other "prehistoric" cultures (Hodder 1986; Pruecel 2006), revealing details of daily life, regional characteristics, metallurgical techniques, and exotic cultural contacts about which we have much to learn. The artifacts belong to archaeological cultures that are markedly specific to geographic locations. Only very recently has any part of this region been studied systematically with the goal of understanding the socio-political background of the peoples who appear to be represented by these artifacts (Linduff 1997a; Linduff et al. 2002–2004; Indrisano and Linduff forthcoming). If we can decode the information, long hidden from our understanding, that such artifacts provide, we will see that the view is significantly different from what we have traditionally associated with the *beifang*. This area is culturally far more complex than has been acknowledged in the past (see Shelach, Chapter 16 in this volume). A closer study of these artifacts will confirm the existence of distinctive regional sub-cultures within the *beifang* region that must be considered individually if we are to understand the relationships between various *beifang* peoples and early China. A practical knowledge of pastoral lifestyles is also useful to ensure that our decoding of the information provided by the various artifacts is accurate.

Late Prehistoric Beifang

DURING THE first millennium BCE, China's northern frontier zone, known as the *beifang* (Lin Yun 1986: 248), was inhabited by semi-mobile agro-pastoral groups with animal-oriented economies based on varied combinations of herding, hunting, and stockbreeding, occasionally supplemented by minimal agriculture (Bunker et al. 2002: 7–14; Linduff et al. 2002–2004; Shelach 1999). The *beifang* is a vast arc of marginal land extending from Heilongjiang in the northeast to Ningxia and Gansu in the west that has never been culturally or climatically homogeneous because it is divided ecologically by the Taihang mountain range that runs along the western border of Hebei Province. To the east, the mountainous land covered with forests is suitable for hunting, trapping, and fur trading, and further east the fertile soil of the Liao River valley sustains hunting, fishing, fur trading, settled stockbreeding, and limited agriculture (Shelach 1999). By contrast, the grassy plains west of the Taihang are more conducive to large-scale herding. The lifestyles and beliefs of the pastoral groups living in these two significantly different areas reflect the diverse ecology of their specific geographic location, with occasional evidence of exotic contacts with other outside cultures, both direct and indirect, through trade.

The various regional *beifang* groups, or at least some part of each community, led lives organized around prescribed seasonal activities, including daily excursions as well as longer migrations from fixed home bases to well-established destinations that provided water, pasture, and rich hunting grounds dictated by their environmental needs. The horse, both ridden and driven, provided their primary means of transportation, making them famous for their equestrian prowess, as are all people of the steppe (Olsen et al. 2006). They built no permanent architecture until late in the first millennium BCE under the Xiongnu (Allard and Erdenebaatar 2005: 561–62) but lived in portable felt-covered trellis tents that accommodated their semi-mobile lifestyles (Bunker et al.

2002: 12–14). These tents are the famous dome-shaped yurts, known as *gers* in Mongolian, that appear to have developed farther west early in the first millennium BCE, and are still in use today (Stronach 2005: 190–94).

Early steppe administrative centers were seldom permanent and certainly were not great architectural centers. The Song Dynasty Chinese official, Bi Zhongyou, described the Qidan Liao court's winter encampment as a "felt city" (Wittfogel and Feng 1949: 134), and in the thirteenth century William of Rubruck noted that the imperial capital of the Mongol Empire at Kharkhorum was about the size of a large French village (Rogers et al. 2005: 816). This lack of permanent settlements was bewildering to the settled Chinese who considered their semi-mobile pastoral neighbors culturally inferior, not understanding the pastoral desire to preserve nomadic ideals while fulfilling the necessary requirements of administration and control with a mobile court system.

The *beifang* groups themselves left no written records documenting their customs and beliefs, while ancient Chinese literature abounds with references to their existence (Di Cosmo 2002; Prusek 1966, 1971). By contrast, the dynastic Chinese to the south led sedentary lives based chiefly on agriculture, and bequeathed an immense legacy of written records and architectural remains to posterity that attests to their history and cultural achievements. The chief interests for the Chinese were the geographic locations, political alliances, potential danger, and possible value as trading partners of their *beifang* neighbors, and not their cultural achievements (So and Bunker 1995).

Most modern scholars have little real knowledge of regional *beifang* lifestyles, blaming their ignorance on the lack of available *beifang* documents. As a result of this supposed lacuna, modern scholars based their opinions on information gleaned from ancient Chinese literature, which ranges from misleading to inaccurate, lumping all the regional *beifang* groups under one "barbarian" umbrella. Modern scholars' chief interests are the roles played by the *beifang* and other steppe peoples in the transmission of technology and artistic styles between east and west, and north and south.

Many scholars today still discuss ancient China's early *beifang* neighbors from a distinctly pejorative Sinocentric point of view. As Nicola Di Cosmo (2002: 314) observes, "Of what was happening on the Inner Asian side, … little is known." When the only sources consulted are Chinese, a Chinese perspective would seem inevitable when writing about the *beifang* peoples. But, are we sure that the *beifang* groups themselves left no historical records that can be consulted?

I intend to demonstrate that motifs on first-millennium BCE *beifang* artifacts can reveal poorly understood regional characteristics, including identification of local fauna, subsistence practices, metallurgical technology, mythological beliefs, and exotic cultural contacts. Comparative studies of the various regional groups reveal that the artifacts marking the graves often reveal individual social status, military prowess, gender, age, and wealth, as well as individual leadership roles within the community, as they do in the cemetery at Maoqinggou in south central Inner Mongolia (Hollmann and Kossak 1992; Wu 2004).

Beifang Culture during the First Millennium BCE

THE ADVENT of riding astride and the accompanying transition to a more mobile and animal-oriented economy during the early first millennium BCE apparently fueled an artistic flowering among the various regional *beifang* groups that reflects these developments (Fig. 15.1a,b). We can retrieve some part of their various regional histories and beliefs by using their artifacts as historical texts, referring primarily to excavated examples in the Peoples' Republic of China, Mongolia, and Siberia, as well as a few undocumented but stylistically similar examples in public and private collections around the world. We must discard the earlier scholarly approach, especially the term "animal style" that ignores and/ or obscures the existence of regional groups and visual styles that define cultures individualized through their distinctions (Rostovtzeff 1929; So and Bunker 1995: 26–27).

Among the pastoral people, the need for personal identification, group recognition, and status was of paramount importance, because they built no great urban centers and moved seasonally (Barfield 1993). Their mobile lifestyle is reflected in their choice of artifacts, which are typical of a mobile way of life – small, portable objects that have specific functions and were decorated with artistic symbols that provide clues to their specific geographic locations, individual identification, and cultural beliefs. Their artifacts consist primarily of personal items such as ornaments or tools and include belt plaques, necklaces, earrings, garment plaques, small tools and weapons designed to be suspended from belts, and horse and chariot ornaments. These are all apparently worn or used in life but are commonly placed in tombs to designate both the position of the individual and his or her role within

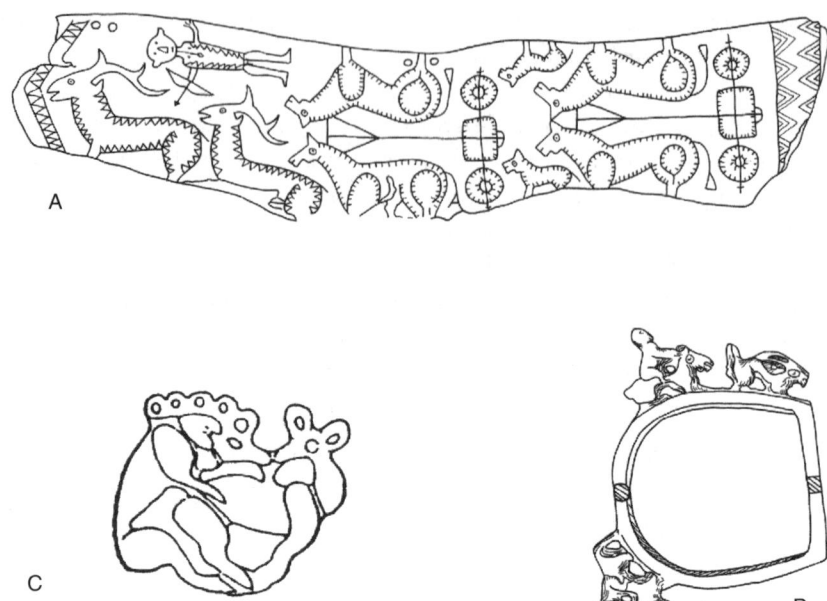

Figure 15.1. Dongbei artifacts: (A) Hunting scene, bronze fitting, Nanshan'gen, Ningcheng Xian, Inner Mongolia Autonomous Region, eighth century BCE (after So and Bunker 1995: 49); (B) Bone fragment, Nanshan'gen, Ningcheng Xian, Inner Mongolia Autonomous Region, eighth century BCE (after So and Bunker 1995: 49); (C) Copulating deer, illustration of bronze plaque, Chifeng region, Inner Mongolia Autonomous Region, fifth century BCE (after Bunker et al. 1997: 166).

the community (X. Wu 2004; J. Wu 2004). The artifacts produced throughout the *beifang* are typologically similar, but the visual symbolism that distinguishes these objects and the metallurgical techniques employed to make them are regionally specific and reflect distinct regional characteristics (Bunker 2006).

The *beifang* can be divided geographically into two significantly different cultural regions: northeast China, known as the *dongbei*, consisting of territory east of the Taihang; and northwest China, known as the *xibei*, consisting of regions west of the Taihang. Within each region, people developed an artistic vocabulary that stems from the specific economies that governed their lives. The iconography of each region was apparently based on the local fauna, and that dictated the zoomorphic motifs and presumably both ideological and mythological visual language of each group. A correct identification of the birds and animals represented is essential to an understanding of the information supplied by each artifact, as zoomorphic images are usually area specific and help to suggest a provenience for the object.

Every motif is a recognizable image, apparently specific to the region and the peoples who lived in it. Those using and viewing such items displayed or recognized the regional person; as such, objects communicated messages that were probably defined (and confined for use) by local lifestyles and ecology (Shennan 1994: 1–32). In a society without written records, where much depended on complex hierarchal relationships within clans and on recognition by outside groups, such artifacts could serve as identifiers of the group. From these items, viewers could learn history, value systems, clan identification, religious beliefs, and heroic codes of ethics, all of which were thought previously to have been transmitted through oral accounts.

The Dongbei

THE *DONGBEI* (northeast) encompasses Heilongjiang, Jilin, and Liaoning provinces, as well as southeast Inner Mongolia and parts of northern Hebei (Nelson 1995: 1–4). The land is fertile, watered by many rivers, and is often mountainous. The economy was and still is based on hunting (Fig. 15.1b), fishing, fur trading, stockbreeding, and agriculture. The ecology did not require long-distance migrations in order to support large-scale herding economies.

Dongbei artifacts include belt ornaments, garment plaques, equestrian gear, knives, and daggers, and all are embellished with images of wild animals and birds that reflect the local fauna. Deer, gazelles, ibex, wild boars, leopards, tigers, bears, and game birds were the most popular quarry, chiefly valued for their pelts, hides and feathers. Images of copulating deer (Fig. 15.1c) and ibexes abound and probably served as amulets to ensure proliferation among the wild herds on which the hunters' economic success depended. Such images never appear, unless through inter-regional trade, on artifacts created for the herding tribes living west of the Taihang range. Animal reproduction in domestic herds was presumably supervised, as it is today, so grazing rights, water sources, predators, disease, and weather would have been more important issues than fertility symbols for herders living west of the Taihang range. The lifestyle of the herder is significantly different from that of the hunter, and although not exclusive of each other, one activity was favored over the other to represent each area.

Belts, traditionally formed by multiple plaques strung on cord or leather, were signature artifacts among the *dongbei* people. Such plaques were adorned with a limited variety of zoomorphic symbols, including cervid images with backswept antlers formed by tangent circles, an image that had a long history throughout the Eurasian steppe (Fig. 15.1c). The history of this image reveals much about trans-Eurasian contacts not found in ancient literature, if one traces its visual evolution (Tchlenova 1963: 66–67, tables 1, 2; Bunker et al. 2002: 17–18).

Based originally on the reindeer, this specific cervid motif evolved first in the Eurasian steppe and was altered regionally to reflect the local cervid in the various regions to which it was diffused. Subsequently, the motif was transmitted eastward to *dongbei* territory and westward through Central Asia to the Black Sea region where, altered through contact with Near Eastern and Greek artistic conventions, it became the leitmotif of the Scythians (Amandry 1965). The wide-ranging territory in which this cervid image was portrayed suggests a common source in northeastern Eurasia from which it must have diffused east and west, rather than being the result of a simple east-west transmission. The manner in which the cervid's legs are folded, the back legs overlapping the front, is a speed pose that developed early in the first millennium BCE but was subsequently replaced late in the first millennium BCE by the so-called flying gallop. Although ubiquitous in the *dongbei*, such cervid motifs are seldom seen on artifacts found west of the Taihang, unless the artifact is an exotic intrusion.

Dongbei belts were frequently hung with knives and scabbards, typical hunters' tools needed for butchering and skinning the animals taken (Bunker et al. 2002: 74). The outward curving blades on some knives are especially designed to skin an animal without damaging its pelt, the same way skinning knives are designed today. Daggers, useful for stabbing in combat at close quarters, are less frequent in *dongbei* burials than knives.

Knife hilts and scabbards display realistically portrayed wild animals and game birds that derive from a strong pictorial tradition traceable to representations on early first millennium BCE "deer stones" and petroglyphs found throughout Mongolia and southeastern Siberia (Novgorodova 1980). In ancient times, the people of the *dongbei* had close contact with the far north via the "Fur Road," a complex trade route leading north through the Amur Valley and crossing Eurasia north of the fiftieth parallel, roughly the same as the route traveled today by the trans-Siberian railroad (So and Bunker 1995: 65; Kuz'mina 2007). The Fur Road bypassed northwest China (*xibei*), perhaps explaining why horseback riding occurred earlier in the *dongbei* than in the *xibei* regions, and why certain visual motifs from the far north represented in *dongbei* vocabularies never appeared in those of the herding people living west of the Taihang.

Dongbei artifacts are chiefly piece-mold cast, but a few, such as jingles, were cast by the lost-wax process to accommodate the complexities of the form. Lost-wax casting occurs earlier in the *dongbei* than in the Chinese heartland. Remnants of mining and foundry debris indicate that artifacts created for the *dongbei* inhabitants were probably cast locally and not commissioned from foundries in dynastic China (Linduff 1997b: 72). The few actual Chinese vessels found among the local grave goods represent exotic status possessions that must have been acquired through trade and tribute.

Dongbei zoomorphic motifs are far more naturalistic than the stylized zoomorphic images depicted on artifacts belonging to the herding tribes living west of the Taihang range, and they record entirely different concerns and visual expressions. The *dongbei* artifacts and their decoration document activity resulting from hunting and marketing animal products, especially fur, to their Chinese clients. Their hunting was presumably seasonal, forcing them to travel short distances from one suitable hunting ground to another, depending on the season and the location of their quarry, returning in winter to a base where tanning, leatherwork, and metalwork were accomplished.

Northern Hebei Province

Northern Hebei belongs geographically to the *dongbei*, but the artifacts lack certain *dongbei* features, such as the emphasis on fertility exemplified by animal copulation scenes. The people inhabiting the rugged Jundushan and Yinshan ranges were chiefly horse-riding hunters and fur traders with nearby home bases.

Excavations at several cemetery sites reveal elite burials in which the dead were literally covered with metal ornaments: necklaces, pectorals, belt plaques, garment plaques, and a plethora of weapons and tools (Fig. 15.2). Comparative studies of the grave goods indicate the important role of artifacts in acknowledging status, prestige, wealth, and power (So and Bunker 1995: 46–50; Linduff 1997b; X. Wu 2004; J. Wu

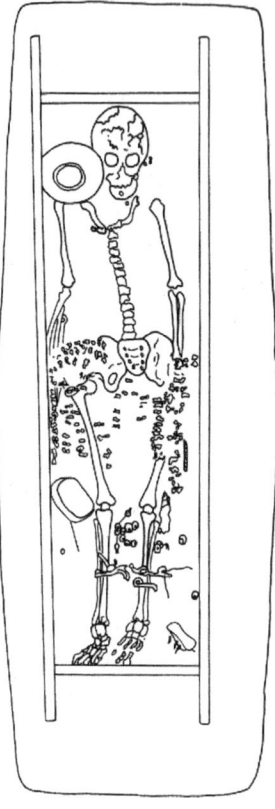

Figure 15.2. Tomb 156, Jundushan, Yanqing Xian, Beijing, sixth to fifth century BCE (after So and Bunker 1995: 161).

2004). Daggers with integrally cast hilts and blades, often with a raised midrib, are signature Eurasian steppe weapons, designed to hang from their owner's belt. Pectoral plaques worn high on the chests were conspicuous and were formed frequently in the shape of a crouching leopard with slightly open jaws and sharp teeth. Most pectoral plaques are made of bronze and piece-mold cast, a manufacturing technique confirmed by metal flashing within the design openings. A few plaques made of gold were cast by the lost-wax process, which indicates a deliberate casting choice for precious metal that reveals a sophisticated understanding of the technical properties of gold and bronze (So and Bunker 1995: 50, fig. 18; Bunker et al. 2002: 171–172, no. 157). Gold cast by lost wax results in a better-quality object than if it were cast by the piece-mold process. The technological sophistication, large size, rarity, restrictive use, and area-specific iconography of these casts allow us to speculate on the power of their message in life and death – they identified association with a particular group or community and probably status within that group (Brumfiel 2000; Jones 1997).

Wild animals and birds form the most popular zoomorphic motifs. Images of raptors, such as the steppe eagle (*Aquila rapax*), must document the practice of falconry in antiquity, because it is still a popular activity in the Eurasian steppe that enables men to display their hunting skills and demonstrate their mastery over nature. The cervid with back-swept antlers and folded legs that had developed earlier on deer stones also plays a role in the artistic vocabulary of northern Hebei (Bunker et al. 1997: 171; Bunker et al. 2002: 158–59), as does a coiled carnivore image.

The motif of a carnivore coiled into a ring has an East Asian *dongbei* priority, occurring first on Neolithic period jades found at Hongshan, and later on Western Zhou Chinese bridle ornaments (ca. eleventh century–771 BCE) (Bunker 2002: 24). The coiled-carnivore motif also occurs in Mongolia on deer stones and, in southern Siberia, on a bronze breastplate excavated at Arzhan in the Republic of Tuva (Graiznov 1984; Cugunov 2006). Later the motif was transmitted west to Central Asia, where it appears in Saka art, and then to northwest Eurasia, where it occurs in Scythian art (Tchlenova 1967: 31). Finials surmounted by ungulates indicate the practice of elaborate funerary rituals during which they marked the corners of the canopy that was erected over the deceased. Such funerary finials also occur at Arzhan in the Republic of Tuva (Graiznov 1984: 52, fig. 25) and in northwest Eurasia among the Scythians. Ultimately, these funerary customs may be traceable to third-millennium BCE practices in south Russia, such as those found at Maikop (Phillips 1961: 320). The northern Hebei and Arzhan finials lack mold marks, indicating that they were cast by the lost-wax process, rather than in a piece mold.

The inhabitants of northern Hebei had long-established access to the Fur Road through present-day Zhangjiakou, a major fur-trading and leather-tanning center that served as a gateway to Mongolia and other northern destinations in the past, and which is still in use today. In the late second millennium CE, Zhangjiakou also became part of the famous tea-trading road that led north to the Eurasian steppe and beyond. This route bypassed the *xibei*, so it is not surprising that many *beifang* motifs from northern Hebei traceable to southern Siberia and beyond did not appear on artifacts found in the *xibei* until much later. Only in the fourth century BCE did coiled-carnivore motifs occur in the *xibei*, west of the Taihang, having been re-introduced into the Far East by newly arrived pastoral groups fleeing the late fourth-century BCE Central Asian campaigns of Alexander the Great (Bunker 1992: 99–115; Kuz'mina 2007).

The non-Chinese inhabitants of northern Hebei appear to have been extremely self-sufficient, active traders, working on an equal basis with their nearby Chinese neighbors; their major commodities were exotic furs and leather. The presence of daggers in the graves has inspired scholars to attribute a warlike character to the northern Hebei hunters, but there is no evidence of violence so far in the archaeological record in the region. Knives are hunting tools, and daggers (*duan jian*, or short swords, in Chinese archaeological literature) are primarily small personal weapons. However, daggers were invaluable to the hunter for thrusting and stabbing at close range, rather than as a battle-ready, slashing weapon (Bunker et al. 2002: 74; and P. R. S. Moorey personal communication).

The Xibei

The *beifang* territory located west of the Taihang range, known as the *xibei*, encompasses large patches of grasslands separated geographically and ecologically into three separate regional spheres with similar herding strategies but with significant local variations. South-central Inner Mongolia formed one region. Northern Shaanxi, the Ordos, and adjacent areas in southwest Inner Mongolia formed a second region. A third included the Ningxia Hui Autonomous Region, southeastern Gansu, and parts of the Xinjiang Uyghur Autonomous Region. The pastoral groups inhabiting these regions were engaged in large-scale herding and livestock trading, resulting in a more mobile lifestyle than that of the *dongbei* groups living east of the Taihang.

In general, *xibei* artifacts are significantly different in appearance, symbolism, and function from those belonging to the hunting inhabitants of the *dongbei*. Zoomorphic-shaped pectorals worn as group identifiers in northern Hebei are non-existent in burials west of the Taihang. Instead, belt plaques and buckles decorated with zoomorphic combat and predation motifs reflect the herder's constant concern with protecting his herds from the ever-present danger of predators, as well as probably conferring equal prowess on him that he might need to defend his grazing rights (Fig. 15.3) (Bunker 1990b: 295). Such themes derive ultimately from third-millennium BCE motifs developed in the ancient Near East that were slowly transmitted eastward by the steppe peoples during the second and first millennia BCE (Basilov 1989: 4–6) and were

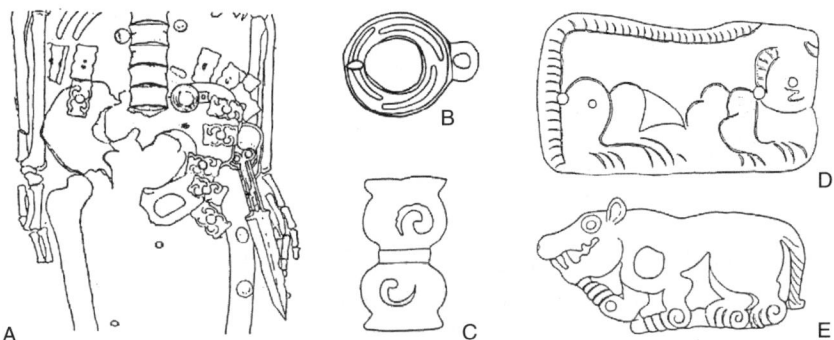

Figure 15.3. Belt ornaments from Maoqinggou, Liangcheng Xian, Inner Mongolia (after Bunker 1990b: 295): (A) M60; (B) Hook buckle from M60; (C) Belt plaque; (D) Tinned bronze belt plaque from M5; (E) Bronze belt plaque from M55.

altered from region to region to reflect the local fauna (So and Bunker 1995: 54–56).

South-Central Inner Mongolia

LUSH GREEN pastures abound south of the Yinshan in central Inner Mongolia, making the region ideal for large-scale livestock herding. Heads and hooves of horses, cattle, and goats among the grave goods suggest the composition of the herds (Linduff 1997b: 56–57).

Belt plaques depicting animal predation and raptor heads embellished by intaglio linear curves and angles are most frequently found in this region as burial debris. The linear curves and angles are similar to surface patterns found on Eastern Zhou carved jades and bronzes cast at Houma, the Eastern Zhou period Jin-state foundry in southern Shanxi Province (Figs. 15.3b–e). Mold marks around each attachment loop on the backs of the plaques show clearly that they were cast in multi-piece molds. These casting traditions recall those practiced at Houma (Weber 1973: 232; Li and Liang 1997).

There is no gold or evidence of amalgam gilding among the grave goods, but some bronze plaques were intentionally tinned and thereby indicative of status (Han and Li, Chapter 10 in this volume). Tinning, a surface enrichment developed by Chinese Zhou craftsmen to produce a pleasing silvery appearance, never occurs on *beifang* artifacts created east of the Taihang (Bunker 1990b: 295–97; Han and Bunker 1993). It appears that herders in south-central Inner Mongolia relied heavily on nearby Chinese foundries, such as Houma, and itinerant Chinese crafts-men to produce their artifacts, suggesting the development of Sino-Steppe commercial relationships. Such trade dynamics have just begun to be investigated (So and Bunker 1995; Li and Liang 1997; Linduff forthcoming).

Northern Shaanxi and the Ordos Region of Southwest Inner Mongolia

IN ANTIQUITY, the loess plateau located in the Ordos region of south-western Inner Mongolia was covered by grass, bushes, and trees and was sufficiently watered by numerous rivers and streams to produce rich grazing lands. The open pastures, watered terrain, and remains of large animals such as horses and cattle as sacrificial burial gifts and visual rep-resentations on most artifacts suggest that the local peoples practiced a mobile way of life in which long-distance herders moved seasonally in search of pasture and water, because the pasture in the area was not suf-ficient to sustain sedentary grazing.

The most informative artifacts characterize the customs of this region in the form of zoomorphically shaped hollow bronze sculptures in the round created to embellish the yokes of burial carts (Fig. 15.4a). Such artifact types do not occur east of the Taihang, documenting different funerary rituals in the Ordos from those already encountered in the *dongbei*. The custom of decorating a burial cart in this way also occurs in burials associated with the Qin, the nearby Chinese Eastern Zhou state that occupied most of present-day Shaanxi and whose inhabitants were known to be of non-Chinese ancestry (So 1995: 230).

The few belt accessories attributed to the greater Ordos region appear to be Chinese-made specifically for their *beifang* neighbors, as such orna-ments are not found in Chinese tombs of the period (So and Bunker 1995: 171–76; Li and Liang 1997). Some buckles have a fixed tongue of steppe invention, but the designs, especially those of a crouching feline, are stylistically Chinese (Bunker et al. 1997: 220–221). Instances of inter-marriage between the dynastic Qin Chinese and their pastoral neighbors are noted in ancient Chinese literature, further evidence for the increasingly close relationship between the steppe and the sown that developed in this region (Bunker 1990b: 296).

Sometime late in the fourth century BCE, mounted warriors from farther west appeared in the Ordos region and introduced new types

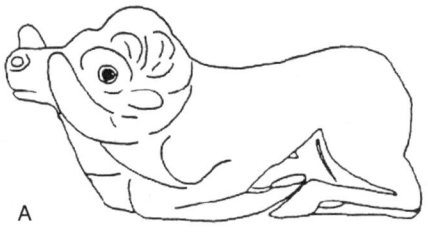

Figure 15.4. Grave goods from the Ordos Desert Region, Inner Mongolia, fourth century BCE: (A) Yoke ornament from Waertougou, Ordos Region (after Linduff 1997a: 53); (B) Gold inlaid belt plaque, Aluchaideng, Hangjin banner (after Tian and Guo 1986: 90).

of artifacts adorned with symbolic systems, including fantastic animal imagery with raptor-headed appendages not seen before in the region (Fig. 15.4b). These new arrivals have been tentatively associated with the Rouzhi (Yuezhi) (Watson 1961: 2:160–161, 168, 267–268), who are thought to have been an Inner Asian peoples with an Indo-European cultural heritage distantly related to the Saka and Scythian tribes to the west (Bunker 1990b: 298–299). A horned ungulate with tines that terminate in eared raptor heads is one of their major iconographic motifs, echoing earlier fantastic zoomorphs found on the tattooed man discovered at Pazyryk in the Altai mountains (Rudenko 1970: pls. 165, 168; So and Bunker 1995: 56–59) and on Scythian and Saka artifacts in the west (Bunker 1989: 87; Bunker et al. 2002: 26–28). A hierarchy of metals based on gold and silver also accompanied the appearance of these newcomers in the Ordos region (Bunker 1992).

Burials associated with the newcomers contain fewer funerary cart ornaments than did the earlier burials (So and Bunker 1995: 56–59). Instead, numerous riding bridle ornaments highlight a preference for riding astride that neatly coincides with the traditional late fourth-century BCE date for the adoption of mounted archery from its nomadic neighbors by the Chinese state of Zhao (Watson 1961: 2:159; Bunker et al. 2002: 26–27).

Eastern Gansu and Southern Ningxia

Numerous herding groups inhabited the loess plateau centered in eastern Gansu and southern Ningxia, a vast sea of grasslands watered by rivers descending from the Longshan. Animal-attack themes distinguish the plethora of belt plaques found among the grave goods there, probably revealing concerns with predators that attacked the herds and/ or carried off the young (Fig. 15.5, left). Most of the belt buckles are decorated with steppe subject matter in the form of animal predation motifs but are cast in Qin Chinese Warring States style (Fig. 15.5, right).

Burial vehicles were decorated with hollow-cast recumbent cervid figures designed to fit over the cart yokes, a custom noted earlier in the Ordos region and also followed by their Qin Chinese neighbors, non-Sinitic people themselves who were closely inter-married with their herding neighbors (Mair 2005: 72). Such burial customs appear to have been discontinued by the end of the fourth century BCE after the arrival of mounted warrior-herdsmen from farther west with a more mobile culture firmly based on riding astride (Watson 1961: 2:159).

Wolves, the local predators, are featured on artifacts in this region but not on examples cast in the *dongbei*, where wolves have not been documented. Tinning and lost-wax casting are responsible for creating artifacts decorated with raised spirals on a pseudo-granulated ground in typical Qin style. Fourth-century BCE lupine images display raptor heads at the tips of their tails and crests, iconography that had to have been adopted in the Ordos region from the west, ultimately from the Saka and Scythians (Bunker 2006: 112). The appearance of camels' heads on some artifacts, often zoologically misidentified, clearly indicates their

Figure 15.5. Belt plaques from Qingyang (after Bunker 1990b: 297).

provenience in the northwest, where camels graze to this day and through which ancient caravans passed on their long trek across Eurasia.

The general impression gleaned from the style and iconography of the artifacts and the dependence on Chinese technology found in this area suggests a close symbiotic relationship between the Chinese state of Qin and herding groups on their western frontier. Ample evidence for commercial contacts between the two groups would explain the similarities between the decoration on the artifacts of the pastoral groups and that on the artifacts of the Qin during the sixth through fourth century BCE (Bunker 1990b: 296–298; So 1995: 230–33; Shelach and Pines 2006; Linduff forthcoming). The picture revealed by the artifacts and their decoration suggests a busy herding lifestyle in which the major concerns were for the safety of the herds, the territorial protection of grazing rights, and a constant need for access to water sources.

The artifacts reveal extensive contact between the herders and their Qin state neighbors, who apparently fashioned their luxury goods and personal ornaments (Linduff forthcoming). Such a symbiotic arrangement suggests that the herding groups located in southern Ningxia and eastern Gansu lacked the know-how to make their own ornaments. The fact that no foundry debris has been found that can be associated with the local groups lends support for this theory. Further evidence for the Chinese manufacture of artifacts for their pastoral neighbors is furnished by the recent discovery of a bronze-caster's burial in the northern suburbs of Xian. The Warring States period tomb includes models for both Chinese and nomadic-style paraphernalia among the tools of his trade (Institute of Archaeology 2003; the author is grateful to Francis Allard for this reference).

Xiongnu Period, Late Third Century to First Century BCE

BY THE end of the third century BCE, the Xiongnu had formed a vast steppe empire by conquering the diverse hunting and herding peoples inhabiting the *beifang* during the third quarter of the first millennium BCE. The cultural diversity exhibited by the conquered groups absorbed into the Xiongnu confederacy resulted in a variety of artifacts adorned with a mixed iconography found at Xiongnu sites throughout the eastern Eurasian steppe.

The most characteristic artifacts attributed to the Xiongnu are belt buckles formed by mirror-image and identical pairs of plaques, with status being indicated by a hierarchy of metals based on gold, silver, and bronze supplemented by surface enrichment, such as gilding, tinning, and silver plating (Fig. 15.6). Several buckles made during the early years of Xiongnu dominance still display iconographic elements associated with the subjugated Rouzhi group, including zoomorphic images with raptor-headed appendages (Linduff 1997b: 79–86; Bunker et al. 2002: 29–32).

The unification of the eastern Eurasian steppe by the Xiongnu and the subsequent expulsion around 160 BCE of the Rouzhi (Yuezhi) were accompanied by more changes in the visual iconography and an increase

Figure 15.6. Xiongnu period bronze belt plaques, second to first century BCE: (A) Kexingzhuang, tomb 140 (after Linduff 1997: 86); (B) Confronted Camels (after Tian and Guo 1986: 77); (C) Eastern Mongolia (after Bunker et al. 1997: 261).

of Chinese objects in the graves. The importance of fantastic zoomorphs with raptor-headed appendages, associated earlier with the Rouzhi, slowly decreased in favor of a new emphasis on real animals and human activities frequently represented in abbreviated landscape settings (Fig. 15.6A). The only fantastic animal associated with the Xiongnu is a sinuous four-legged creature of lupine ancestry with sharp claws that ultimately evolved into the characteristic Han Dynasty Chinese dragon image (Rudenko 1958: 119–121; Bunker 1990b: 302–303).

Animals in combat are rarely represented on Xiongnu artifacts. Instead, two ungulates posed antithetically on either side of a tree or vegetal form are popular subjects (Fig. 15.6B). Such a design is ultimately traceable to the ancient Near East, suggesting some early Xiongnu connection with western traditions, presumably somewhere in Inner Asia long before the Xiongnu appeared on the northern frontier of dynastic China (Miniaev 1995: 123–136; Bunker et al. 2002: 30–31).

A serious interest in the quality of their herds and an acute sense of observation are reflected in certain scenes portrayed on belt plaques. Images include violent fights between two rut-crazed stallions or two male camels vying for females to breed, representing annual contests that must have been a common occurrence in late spring as the females came into heat (Fig. 15.6C). These scenes are later versions of compositions that developed farther west in Inner Asia, adding further credence to the suggestion that the Xiongnu may have migrated into the eastern Eurasian steppe world from farther west (Bunker et al. 2002: 132).

Other belt plaques produced for the Xiongnu portray a limited number of stereotyped zoomorphic and human scenes and are found in the burials of men, women, and children alike, as if each scene had some significance related to familial recognition and identity. According to references in Han texts, family continuity among the Xiongnu was of utmost importance (Watson 1961: 2:156; Bunker 1990b: 302), and this affiliation and one's position in the group were apparently conferred by the choice of metal and surface enrichment that formed and adorned the plaques. A systematic study of such features and their significance would be possible now that an abundance of data from burials in the region and from this period has been provided in the past two decades.

Although ancient Chinese texts make mention of the Xiongnu along China's northern frontier, the actual Xiongnu heartland seems to have been located further north (Miniaev 1995). Certain belt plaques include zoomorphic images in their symbolism that were not found among the local fauna in the *beifang* – for example, swans, well known in southern Siberia, and vultures, known chiefly in Mongolia and Buryatia. Archaeological evidence for settled Xiongnu communities in Mongolia and Buryatia indicates that the Xiongnu no longer practiced transhumance there but rather were semi-sedentary. Many northern Xiongnu metal ornaments appear to be Chinese-made, probably by itinerant or captive Chinese artisans in Xiongnu settlements, as it is doubtful that the Xiongnu themselves ever learned the complex metallurgical techniques necessary to amalgam-gild bronze and silver (So and Bunker 1995: 62–3; Bunker 1990: 47–8; Miniaev 1995).

The Xiongnu were symbiotically associated with their Chinese neighbors beyond the Great Wall to the south, both commercially through trade and treaty or tribute arrangements and politically through marriage alliances (Hulsewe 1993: 129–130; 405–406). The visual symbolism of the Xiongnu confederacy leadership seems to indicate a trend toward concern with human affairs, brought on by a closer association with Han China and an increasing trend toward a more settled rather than a mobile way of life (Yu 1967). This relationship is reflected in the visual arts of both the Han Chinese and the Xiongnu, an unbiased testament to their close association hitherto not always understood by historians.

Conclusion

THE PERCEIVED lack of written records from the first-millennium BCE *beifang* groups is no longer an excuse for basing references to them exclusively on ancient Chinese texts. It should now be clear that artifacts belonging to the *beifang* peoples can convey far more information than has been traditionally acknowledged. As a result, they can be "read" for an understanding of a pastoral world that has remained largely unknown to us and about which we still have much to learn. The artifacts reveal a carefully structured lifestyle, social hierarchy, and belief system that governed each regional group and can now serve as visual evidence of their sophisticated culture. The artifacts also reveal the local economies, belief systems, and the extent of the relationships with dynastic China – as these features can all be reflected in the designs and decoration. The archaeological and anthropological models developed to explain the pastoral societies of the Eurasian steppe have tended to obscure the individual regional and sub-regional groups and omit an evaluation of their artifacts (Barfield 1989, 1993; Di Cosmo 2002). In the past, the pastoral people of the *beifang* and their artifacts have been discussed more often by art historians focusing primarily on their artistic impact on early Chinese art rather than by anthropologists interested in the Beifang peoples themselves and their particular regional lifestyles and cultural achievements. In the present chapter, the older art-historical method has been replaced with a more anthropological approach, allowing us to understand more clearly how and why the many inter-cultural contacts between the Beifang people and ancient China occurred. On the basis of their artifacts, it would appear that the semi-sedentary *dongbei* pastoral groups living east of the Taihang mountain range during the first millennium BCE are more important for the study of Chinese history than the herding groups located west of the Taihang. The seasonal migratory patterns of the *dongbei* people set the stage for the later pastoral groups in the first and second millennia CE. The Xianbei, associated with the Wei Dynasty (fourth to fifth century CE), the tenth-century Qidans who founded the Liao Dynasty, and the Manchu of seventeenth-century Qing leadership, continued to define their heritage in terms of seasonal royal hunts, long after they had become rulers of north China. Falconry, so prominent in the *dongbei*, continued to be a major sport among the Xianbei, the Qidan, and the Manchu, a feature that becomes readily apparent when their artifacts are studied (Bunker 1999: 25, 26, 160, 185; Saunders 2005: 82).

According to historians, the Liao and Qing courts traveled throughout their territories from one hunting ground to another with large entourages of retainers and baggage trains, living in tents and seeking the appropriate seasonal quarry. "The tradition of the mobile royal camp as a symbol of the ruler's martial virility" was maintained by later Dongbei inhabitants for the preservation of their heritage and not for the harassment of their Chinese subjects and neighbors (Mair 2005: 67, 210–211, n. 32). The long tradition of a mobile administrative center among the Dongbei peoples allowed the Qidan and the Manchu to rule China from horseback and stay culturally "in the saddle" for several centuries.

It is time to stop viewing the ancient peoples of the *beifang* only through a Chinese lens and start learning to decode the historical legacy provided by their artifacts. As Di Cosmo (2002: 314, 317) has pointed out, the early segments of the Great Wall were built by the dynastic Chinese not as defense against the steppe people but as defensive measures between the various Warring States Chinese polities themselves. It would be more appropriate if the early *beifang* groups were not presented in a book entitled *Ancient China and Its Enemies* (Di Cosmo 2002). The surviving *beifang* artifacts, when used as cultural documents, expose a more constructive and symbiotic relationship between ancient China and its northern neighbors than Di Cosmo's book title infers. Were it not for the survival of the remarkable *beifang* artifacts in archaeological contexts and in scattered private and public collections significant links in the history of ancient northeastern Eurasia and dynastic China might otherwise have been lost. The nomadic lifestyle has been romanticized and difficult for "city folk" to understand, whether Chinese or westerners. In closing, perhaps it would be well to consider Tolkien's words in *The Fellowship of the Ring*, that "not all those that wander are lost."

References

Allard, Francis, and Diimaajav Erdenebaatar. 2005. Khirigsuurs, Ritual and Mobility in the Bronze Age of Mongolia. *Antiquity* 79: 547–563.

Amandry, Pierre. 1965. Un motif "Scythen" en Iran et an Greece. *Journal of Near Eastern Studies* 24(4): 149–159.

Barfield, Thomas J. 1989. *The Perilous Frontier: Nomadic Empires and China*. Cambridge: Blackwell.

1993. *The Nomadic Alternative*. Englewood Cliffs, NJ: Prentice-Hall.

Basilov, Vladimir N. 1989. *Nomads of Eurasia*. Seattle: Natural History Museum of Los Angeles County in association with University of Washington Press.

Brumfiel, Elizabeth. 2000. On the Archaeology of Choice: Agency Studies as a Research Strategem. In M. A. Anne Dobres and J. Robb (eds.), *Agency in Archaeology*. London: Routledge, pp. 251–255.

Bunker, Emma C. 1989. Dangerous Scholarship: On Citing Unexcavated Artifacts from Inner Mongolia and North China. *Orientations*, June, pp. 87–94.

1990a. Ancient Ordos Bronzes with Tin-Enriched Surfaces. *Orientations*, January, pp. 78–80.

1990b. Ancient Ordos Bronzes. In J. Rawson and E. Bunker (eds.), *Ancient Chinese and Ordos Bronzes*. Hong Kong: Oriental Ceramic Society of Hong Kong, pp. 287–362.

1992. Significant Changes in Iconography and Technology among Ancient China's Northwestern Pastoral Neighbors from the Fourth to First Century B.C. *Bulletin of the Asia Institute*, n.s., 6: 99–115.

1999. In Search of Liao Culture. In E. Bunker (ed.), *Adornment for the Body and Soul: Ancient Chinese Ornaments from the Mengdiexuan Collection*. Hong Kong: University Museum and Art Gallery, pp. 11–38.

2006. Northern China in the First Millennium BC, *Beifang* Artifacts as Historical Documents. *Arts and Cultures* 6: 90–123.

Bunker, Emma C., with Trudi S. Kawami, Katheryn Linduff, and Wu En. 1997. *Ancient Bronzes of the Eastern Eurasian Steppes from the Arthur M. Sackler Collections*. New York: Arthur M. Sackler Foundation.

Bunker, Emma C., with contributions by James C.Y. Watt and Zhixin Sun. 2002. *Nomadic Art of the Eastern Eurasian Steppes: The Eugene V. Thaw and Other New York Collections.* New York: Metropolitan Museum of Art.

Cugunov, K.V., H. Parsinger, and A. Nagler. 2006. *Der Goldschatz von Arzhan: Ein Fürstengrab der Skythenzeit in der südsibirischen Steppe.* Munich: H. C. Beck.

Di Cosmo, Nicola. 2002. *Ancient China and Its Enemies: The Rise of Nomadic Power in East Asian History.* Cambridge: Cambridge University Press.

Graiznov, Mikhail Petrovich. 1984. *Der Grosskurgan von Arzan in Tuva, Sudsibirien.* Trans. A. von Schebek. Materialien zur allegemeinen und vergleichenden Archaologie, vol. 23. Munich: C. H. Beck.

Han, Rubin, and Emma C. Bunker. 1993. Biaomian fuxi de E'erduosi qingtong shipin de younjiu. *Wenwu* 9: 80–96.

Hodder, Ian. 1986. *Reading the Past: Current Approaches to Interpretation in Archaeology.* Cambridge: Cambridge University Press.

Hollmann, Thomas O., and George W. Kossak. 1992. *Maoqinggou.* Mainz am Rhein: Verlag Philipp von Zabern.

Hulsewe, A.F.P. 1993. *Shi Chih* and *Han Shu.* In Michael Loewe (ed.), *Early Chinese Texts.* Berkeley: University of California Press, pp. 129–136, 405–414.

Indrisano, Gregory, and Katheryn Linduff. Forthcoming. Imperial Expansion in the Late Warring States and Han Dynasty Periods: A Case Study from South Central Inner Mongolia. In Gregory E. Areshian (ed.), *Archaeological Histories and Anthropological Interpretations of Imperialism.* Los Angeles: Cotson Institute of Archaeology, UCLA.

Institute of Archaeology, Shaanxi Province. 2003. Preliminary Excavation Report of the Warring States–Period Tomb of a Bronze-Caster in the Northern Suburbs of Xian. *Wenwu* (Cultural Relics) 9: 4–14.

Jones, Sian. 1997. *The Archaeology of Ethnicity.* London: Routledge.

Kuz'mina, Elena. 2007. *The Prehistory of the Silk Road.* Philadelphia: University of Pennsylvania Press.

Li, Xiating 李夏廷, and Liang Ziming 梁子明 (eds.). 1997. *The Art of the Houma Foundry.* Institute of Archaeology of Shanxi Province. Princeton: Princeton University Press.

Lin, Yun. 1986. A Re-examination of the Relationship between Bronzes of the Shang Culture and of the Northern Zone. In K.C. Chang (ed.), *Studies in Shang Archaeology.* New Haven: Yale University Press, pp. 237–273.

Linduff, Katheryn M. 1997a. An Archaeological Overview: Section 1. Reconstructing Frontier Cultures from Archaeological Evidence. In E. C. Bunker et al., *Ancient Bronzes from the Eastern Eurasian Steppes in the Arthur M. Sackler Collections.* New York: Arthur M. Sackler Foundation, pp. 1–98.

1997b. The Emergence of Nomadic Pastoralism, 9th–3rd Centuries BC. In E. C. Bunker et al., *Ancient Bronzes from the Eastern Eurasian Steppes in the Arthur M. Sackler Collections.* New York: Arthur M. Sackler Foundation, pp. 33–98.

Forthcoming. Chinese Production of Signature Artifacts for the Nomad Market in Zhou China. In *Metallurgy and Civilization.* London: Archetype Publications.

Linduff, K. M., Robert D. Drennan, and Gideon Shelach. 2002–2004. Early Complex Societies in Northeast China: The Chifeng International Collaborative Archaeological Research Project. *Journal of Field Archaeology.* 29, nos. 1–2: 45–73.

Mair, Victor. 2005. The North (West)ern Peoples and the Recurrent Origins of the "Chinese" State. In J. A. Fogel (ed.), *The Teleology of the Modern Nation-State, Japan and China*. Philadelphia: University of Pennsylvania Press, pp. 46–84, 205–217.

Minaiev, Sergei S. 1995. Noveishchie nakhodki khudozhestvennoi bronzi i problema formirovaniia, "geometricheskogo stilia" v iskusstve syunnu (The Newest Bronze Art Discoveries and the Formation of the "Geometric Style" in the Art of the Xiongnu). *Arkheologicheskii vestnik* 4: 123–136.

Nelson, Sarah Milledge. 1995. *The Archaeology of Northeast China*. London: Routledge.

Novgorodova, Eleanora A. 1980. *Alte Kunst der Mongoliya*. Leipzig: E. A. Seemann.

Olsen, Sandra L., Susan Grant, Alice M. Choyke, and Laszlo Bartosiewicz. 2006. *Horses and Humans: The Evolution of Human-Equine Relationships*. Bar International Series 1560. Oxford: Archaeopress.

Phillips, E. D. 1961. The Royal Hordes: The Nomad Peoples of the Steppes. In S. Piggott (ed.), *The Dawn of Civilization: The First World Survey of Human Cultures in Early Times*. London: Thames and Hudson, pp. 301–328.

Preucel, Robert W. 2006. *Archaeological Semiotics*. Malden: Blackwell.

Prusek, Jaroslav. 1966. The Steppe Zone in the Period of the Early Nomads and China of the 9th–7th Centuries BC. *Diogenes* 54: 23–46.

    1971. *The Chinese Statelets and the Northern Barbarians, 1400–300 BC*. Dordrecht: Reidel.

Rawson, Jessica, and Emma Bunker. 1990. *Ancient Chinese and Ordos Bronzes*. Hong Kong: Oriental Ceramic Society of Hong Kong.

Rogers, J. Daniel, Erdenebat Ulambayar, and Mathew Gallon. 2005. Urban Centres and the Emergence of Empires in Eastern Inner Asia. *Antiquity* 79(306): 801–818.

Rostovtzeff, Mikhail. 1929. *The Animal Style in South Russia and China*. Princeton Monographs on Art and Archaeology 29. Princeton: Princeton University Press.

Rudenko, Sergei I. 1958. The Mythological Eagle, the Gryphon, and Winged Lion, and the Wolf in the Art of the Northern Nomads. *Artibus Asiae* 21(2): 101–122.

    1970. *Frozen Tombs of Siberia: The Pazyryk Burials of Iron Age Horsemen*. Trans. M. W. Thompson. Berekeley: University of California Press.

Saunders, Rachel. 2005. Pursuits of Power: Falconry in Edo Period Japan. *Orientations*, March, pp. 82–92.

Shelach, Gideon. 1999. *Leadership Strategies, Economic Activity, and Interregional Interaction: Social Complexity in Northeast China*. New York: Kluwer Academic/ Plenum Publishers.

Shelach, Gideon, and Yuri Pines. 2006. Secondary State Formation and the Development of Local Identity: Change and Continuity in the State of Qin (770–221 BC). In Miriam T. Stark (ed.), *Archaeology of Asia*. Malden: Blackwell, pp. 202–230.

Shennan, Stephen. 1994. *Archaeological Approaches to Cultural Identity*. London: Routledge.

So, Jenny F. 1995. Bronze Weapons, Harness and Personal Ornaments: Signs of Qin's Contacts with the Northwest. *Orientations*, November, pp. 227–234.

So, Jenny F., and Emma C. Bunker. 1995. *Traders and Raiders on China's Northern Frontier*. Washington, DC, and Seattle: Arthur M. Sackler Gallery and Washington University Press.

Stronach, David. 2005. The Arjan Tomb: Innovation and Acculturation in the last days of Elam. *Iranica Antiqua* 60: 180–196.

Tian, Guangjin, and Suxin Guo (eds.). 1986. *E'erduosi shi qingtong qi*. Beijing: Wenwu Press.

Tchlenova, N. L. 1963. Le Cerf scythe. *Artibus Asiae* 26(1): 27–70.

———. 1967. *Proiskhozhdenie i ranniata istoriia plemen tagarskoi kul'tury*. Moscow: Nauka.

Watson, Burton (trans.). 1961. *Records of the Grand Historian*. Vols. 1 and 2. Translated from the *Shiji* of Sima Qian. New York: Columbia University Press.

Weber, George W. 1973. *The Ornaments of Late Chou Bronzes*. New Brunswick: Rutgers University Press.

Wittfogel, Karl A., and Chia-sheng Feng. 1949. *History of Chinese Society: Liao, 907–1125*. Philadelphia: American Philosophical Society.

Wu, Juiman. 2004. The Late Neolithic Cemetery at Dadianzi, Inner Mongolia Autonomous Region. In Katheryn Linduff and Yan Sun (eds.), *Gender and Chinese Archaeology*. Walnut Creek: AltaMira Press, pp. 47–94.

Wu, Xiaolong. 2004. Female and Male Status Displayed at the Maoqinggou Cemetery. In Katheryn Linduff and Yan Sun (eds.), *Gender and Chinese Archaeology*. Walnut Creek: AltaMira Press, pp. 203–236.

Yu, Ying-Shih. 1967. *Trade and Expansion in Han China*. Berkeley: University of California Press.

CHAPTER 16

Blurring the Boundaries

Foragers and Pastoralists in the Volga-Urals Region

Laura M. S. Popova

TRADITIONAL MODELS of pastoral complexity in Eurasia have been primarily concerned with the interaction between nomadic populations and settled states, particularly in the Near East (Bates and Lee 1977; Rowton 1973), Central Asia (Vinogradova and Kuz'mina 1996; Hiebert 2002), and East Asia (Barfield 1993, 2001; Di Cosmo 1994). Researchers looked to the borders of Eurasia, because they assumed that the center was relatively homogeneous (politically, culturally, and economically) in the Bronze Age. Political activities were visible, supposedly, only when groups with diametrically opposed political institutions clashed. This border-focused research strategy, however, is based upon several linked assumptions. First, the type of political organization is associated with subsistence strategy, especially in the case of pastoralists. Second, it is assumed that certain types of subsistence strategies lead to more-advanced forms of politics (i.e., farmers have states, but pastoralists do not). Third, it is argued that politics cannot be understood without history, meaning that researchers who are serious about understanding Bronze Age political complexity need the textual sources from the early states on the fringe of Eurasia.

Politics, however, is just one aspect of social organization, which is determined by the economic, cultural, and ecological underpinnings of each society. What we study, as archaeologists, is the unique way power plays out on the ground. Who controlled key resources? How were inequalities in the social structure maintained? Where did people live? Who did people trade with and why? In order to answer these questions, we need to create a cohesive picture of political, economic, and ecological orders through time, focusing on changes in settlement, burial, and land-use strategies; productive economies; social relationships; and political institutions. Given this research goal, the question of political complexity seems to be outdated. Instead, we should focus on political diversity.

Studying political diversity in Eurasia, however, requires a revision of the way we look at the past in the region. First, pastoralism is a flexible subsistence strategy that is often combined with farming and foraging. Indeed, the combination of strategies used from year to year often changes. This fact makes it both exciting and frustrating for archaeologists to study pastoralists. Second, pastoral groups utilize a variety of unique political strategies that change over time, and thus the type of subsistence strategy cannot easily be linked to specific models of political complexity. Third, analyzing the interaction between pastoralists and farmers on the southern border of Eurasia does not help archaeologists understand what is happening in the middle of that world.

This chapter addresses the question of political diversity in Eurasia by switching the focus of research to another implied border of Eurasia – the forests of the north. Seen as the dividing line between pastoralists in the south and foragers in the north, the forested region of Russia provides a new location to analyze the relationship between politics and subsistence strategies. This analysis is accomplished by reconstructing the Bronze Age (3300 to 1300 BCE) landscape history of the Middle Volga region. By tracking changes in settlement, burial, and land-use strategies and their impact on the environment, it is possible to begin to uncover the complex ways in which different groups interacted with each other and negotiated over land and resources, demonstrating that complexity is not only manifested in urban structures or the remains of the military elite but also embedded in the complex relationships between people and places.

Political Complexity and Pastoralism

THIS FIRST assumption that must be confronted is the supposed uniformity of steppe groups, and the way in which this homogeny impacts politics. Chernykh (1992) has argued that during the Late Bronze Age the people who inhabited the steppe regions of Eurasia were united by the monotony of their culture. To some extent, this monotony is to be expected when one is relying on archaeological cultures, because they are based on similarities in the material culture, not differences. In its most simplified form, the term "archaeological culture" represents a suite of materials (ceramic, metal, lithic) and features (architectural and mortuary) that co-occur over various sites. However, archaeologists of the Eurasian steppe have added much more meaning to this term. The notion of an archaeological culture now carries with it major assumptions about the economic base, political organization, genetic lines of descent, and even the language family of the group (e.g., Telegin 2002; Bunyatyan 2003). Thus, the term "culture" tends to smooth out the differences that might be significant within a group of people with broadly similar sets of material remains. For this reason, archaeologists tend to look to the borders in order to answer questions of political complexity, because by definition the archaeological cultures are internally similar.

Moreover, this conception of homogeneity of steppe cultures has been linked to their inability to create cohesive political confederacies. For example, Barfield (2001: 14) wrote: "In any event, there was little economic diversity, unless perhaps one wanted to trade cows for camels. Everybody raised the same animals and produced the same products. This was excellent for subsistence but provided a weak internal economic base for the state." Thus, pastoralists were barred from establishing true states because of their form of economic subsistence. Larger pastoral confederacies formed instead only when highly mobile pastoralists raided and tormented their sedentary neighbors. It has been suggested that without these agrarian empires to pillage for exotic items, nomadic empires would quickly crumble. Barfield (2001) has called these nomadic confederacies "shadow empires" for this reason. Khazanov (1994) has argued this point further by suggesting that true nomads can be found only in association with "the outside world" or the agrarian societies.

A different picture of politics emerges, however, if we examine the pastoralists of Eurasia outside of the culture categories they have been constrained by. Pastoralism was relatively new in Eurasia during the Bronze Age, and thus it stands to reason that the strategies used were various and creative. Such diversity fits poorly under a blanket social evolutionary term like "tribe" or "chiefdom." Moreover, subsistence strategies such as pastoralism are part of larger spheres of socio-political, economic, and ecological control. By not taking the nuances of pastoral production for granted, we can gain a better perspective of the organization of society in general. Part of focusing on how pastoralism is practiced is a willingness to realize that people often use multiple strategies in order to meet subsistence needs. Pastoralists can be farmers and foragers as well, and thus this possibility should not be ignored when constructing research goals.

Foragers and Pastoralists

IT IS unusual to find a text that refers to the foragers and pastoralists of the Middle Volga region in the same section in Russian sources, because any group that subsists on foraging is categorized as a "Neolithic" or "forest" culture and is thus relegated to a different period, though in fact these groups were contemporaries of the pastoralists associated with the Bronze Age in the Eurasian steppe.[1] To some extent the categorization of foragers and pastoralists into separate periods perpetuates a social evolutionary logic that sees pastoralism as more advanced than hunting and gathering. It also explains a continued desire to see pastoralists of the steppe as agro-pastoralists or as nomads interacting with sedentary agricultural neighbors, because either pattern would indicate a move to the next level in the social evolutionary ladder. It seems that researchers are less comfortable considering the ways in which pastoralists of the steppe interacted with, and were influenced by, foragers from the forest zones.

In most cases, the line between forager and pastoralist should not be drawn so neatly. We still have a very poor sense of what plant remains were used on a regular basis by both groups, especially wild plants. In addition, we still must deal with the general paucity of faunal data available for most sites. Usually, there is simply a notation that the group practiced pastoralism with a list ranking the importance of the domesticated animals at the site based on ubiquity. In other cases, the archaeologist comments only that the remains of both domesticated and wild animals were uncovered.[2] Without more detailed faunal reports that discuss preparation, distribution, and consumption patterns, it is difficult to assess the difference in subsistence strategies between groups.

In the next section, I reconstruct a landscape history of the Middle Volga region during the Early, Middle, and Late Bronze Ages that incorporates the steppe, forest-steppe, and forest zones. I am focusing on these periods in particular, because data indicate there were major changes in the way in which pastoralism was practiced and the land was utilized and managed during this time period. The main region of focus is the land between the Chagra and Lower Kama rivers. To a certain extent, this history is not as detailed as I would like. There are very few radiocarbon

Table 16.1. The Bronze Age in the Middle Volga region divided by periods with associated horizon styles

Years BCE	Periodization	Horizon Style
1300	Late Bronze Age II	Srubnaya
1400		
1500	Late Bronze Age I	Pokrovka (or Early Srubnaya)
1600		
1700		
1800	Middle Bronze Age III	Potapovka and Abashevo
1900		
2000	Middle Bronze Age II	Abashevo, Poltavka
2100		
2200		
2300	Middle Bronze Age I	Poltavka, Vol'sko-Lbishchenskaya
2400		
2500		
2600		
2700		
2800	Early Bronze Age I	Yamnaya
2900		
3000		
3100		
3200		

dates for Bronze Age sites in the forest zone (Kazakov 2001: 204), and as a result it is difficult to anchor these places into the local chronology. Nevertheless, it is possible to delineate six distinct periods of interaction between pastoralists and foragers: Early Bronze Age I, Middle Bronze Age I, Middle Bronze Age II, Middle Bronze Age III, Late Bronze Age I, and Late Bronze Age II (see Table 16.1).

Early Bronze Age I (3300 to 2800 BCE)

FOR THE first 300 years of the Early Bronze Age in the Middle Volga region, the climate was relatively cool and moist, encouraging the growth of forests in the area (Kremenetski et al. 1999; Popova 2007). The Buzuluk forest stretching north from the bank of the Samara River thrived during this time period. Further to the west along the Samara River, broad-leaved trees, especially birch (*Betula*) but also hornbeam (*Carpinus betulus*) and hazelnut (*Corylus*), created lush riparian forests evincing this more humid climatic phase (Popova 2007). Over the course of the Early Bronze Age, however, the climate became increasing dry and warm, resulting in the expansion of steppe vegetation in the region (Kremenetski 1997; Kremenetski et al. 1997; Velichko et al. 1997; Popova 2007).

The shift from the Eneolithic (roughly 4500 to 3500 BCE) to the Early Bronze Age is defined archaeologically by a shift in burial traditions. During the Eneolithic, flat burial cemeteries were common, usually with all sub-sets of society represented (Vasil'ev and Matveeva 1986). By the Early Bronze Age, burial mounds were constructed in the region, clearly reserved as the final resting place for particular members of society, usually adult males (Vasil'ev et al. 2000).[3] Most of these early kurgans, associated with the Yamnaya horizon, consisted of one rectangular pit for a burial chamber located in the center of a simple earthen mound. In the Middle Volga region, some burials during this time period also contain burned wooden logs or thick organic mats placed above or below the interred. Usually there were no grave goods, though the deceased was almost always heavily covered with red ochre. When there were grave goods, they usually consisted of ceramic vessels, bone, flint, or some metal goods or jewelry, and/or stone implements (Turetskii et al. 2007).

It is primarily on the basis of distinction between flat burials and kurgan burials that archaeologists argue that new people migrated into the Middle Volga region from the Lower Volga region (e.g., Vasil'ev et al. 2000).[4] However, it should be noted that flat burials with Yamnaya-type ceramics continued to be constructed in the Middle Volga region, for example, at Ekaterinovka 5, suggesting that the tradition of burying people in flat burials was not immediately replaced. Moreover, there are some distinct differences between Yamnaya-type burials from the Lower Volga region and the Middle Volga region. Kurgans were built for one person or for a couple of people interred at the same time in the Middle Volga region. They were never reused. In addition, these kurgans tended to be small. For example, the kurgan cemetery at Lopatino 1 consisted of 40–50 little kurgans. In the Lower Volga region, on the other hand, the same kurgans were utilized over the centuries. As burials were added to a kurgan, the shape of the mound changed, resulting in much larger kurgans (Vasil'ev et al. 2000).

Here, I am primarily interested not in where the "kurgan culture" came from but rather how changes in burial practices reflect changing socio-political orders in the region. During the Eneolithic, flat burial cemeteries, such as the one found at Khvalynsk, consisted of shallow burials that contained representatives of all sub-sets of society buried over the course of several centuries (Vasil'ev and Matveeva 1986). The fact that this cemetery was used for such a long time demonstrates that the inhabitants of the region had a long-standing attachment to this particular place. Different graves contained varying levels of grave goods, and if one marks status on the basis of this criterion alone, it seems as if at least some members of society had more material wealth than others. During the Early Bronze Age, however, only a few people per generation were being buried, and it seems that those who were chosen fulfilled some sort of leadership role. When grave goods were found in the burials, they were often stunning objects of authority, such as the heavy bronze scepter found at Kutuluk (Turetskii et al. 2007). Objects such as this seem to have been without parallel in the region and were acquired only through complicated trade relations. The burial mounds themselves, visible on the more or less flat floodplains, were a constant reminder to the living of the deeds of the dead. In addition, they would have required considerably more effort to build than simple flat burials, suggesting that the dead leader (or his or her descendants) had enough power to bring together the whole community to participate in the funerary rite. Thus, this new type of burial would have served both to commemorate a leader and to define the community.

Kurgan burials are linked to *mobile* pastoralism in the archaeological literature (Vasil'ev and Matveeva 1986; Vasil'ev et al. 2000). Archaeologists determined that groups during the Early Bronze Age were mobile, mostly because they have been unable thus far to uncover settlement sites associated with the Yamnaya horizon in the Middle Volga region. They suggest that the economy of these groups was based on pastoralism, because faunal remains uncovered in kurgan burials tend to be from domesticated species such as horse, sheep or goat, and cattle, although wild species are occasionally represented as well (Vasil'ev et al. 2000). Without settlement data, however, it is very difficult to say for certain that the people who built these early kurgans in this region were primarily pastoralists. There is a tradition in the region of combining both foraging and pastoralism already in the Eneolithic. For example, in the Khvalynsk cemetery, faunal remains from both wild (boar, bear, wolf, deer, and others) and domesticated species (horse, cattle, and sheep) were common (Vasil'ev 1981). In addition, the presence of harpoons and hooks also suggest that fishing was an important part of the subsistence strategies of people living in the Middle Volga region during the Eneolithic. It is likely that, at least at first, Early Bronze Age groups practiced a complex form of pastoralism mixed with foraging and fishing. The fact, however, that the remains of domesticated animals became increasingly more common than those of wild animals in the kurgan burials suggests that people were beginning to value domesticated species more highly.

Early kurgan burials were established at considerable distances from one another, usually on the first terrace near major rivers in the region, including the Volga, Samara, Sok, Chagra, Kutuluk, and M. Kinel' rivers (Fig. 16.1). They are found in both the steppe and forest-steppe zones, though not farther north than the lower Sok River valley. Archaeologists have argued that each cluster of Yamnaya-type burials defines a distinct territorial unit (Vasil'ev et al. 2000).

There were many other groups in the region, however, that had a distinct influence on how these early pastoralists lived. For example, groups associated with the Volosovo horizon dwelled farther north in the forest-steppe and forest zones of the Middle Volga region. These people relied on foraging, fishing, and hunting for their livelihood and built long-term settlements, sometimes inhabited for more than a century (Krainov 1987). Highly skilled in the creation of flint weapons, tools, and figurines, these groups also had access to one of the richest flint sources in the region near the present-day city of Magnetigorsk in the Ural Mountains. It seems likely that the pastoralists of the steppe were interested in trading for flint because flint artifacts have often been found in Bronze Age contexts, becoming especially important during the late Middle Bronze Age (Vasil'ev et al. 2000). In addition, there are some similarities over time between Yamnaya- and Volosovo-type ceramics (Vasil'ev et al. 2000), which suggests that these two seemly separate groups shared technological and stylistic knowledge about ceramic production.

Middle Bronze Age I (2800 to 2300 BCE)

THROUGH THE end of the Early Bronze Age and into the Middle Bronze Age, the climate became increasingly continental with cold winters and hot and dry summers, and over time the extent of certain gallery forests decreased. As a result, the forest-steppe region would have become increasingly open, with larger tracts of grassland exposed for

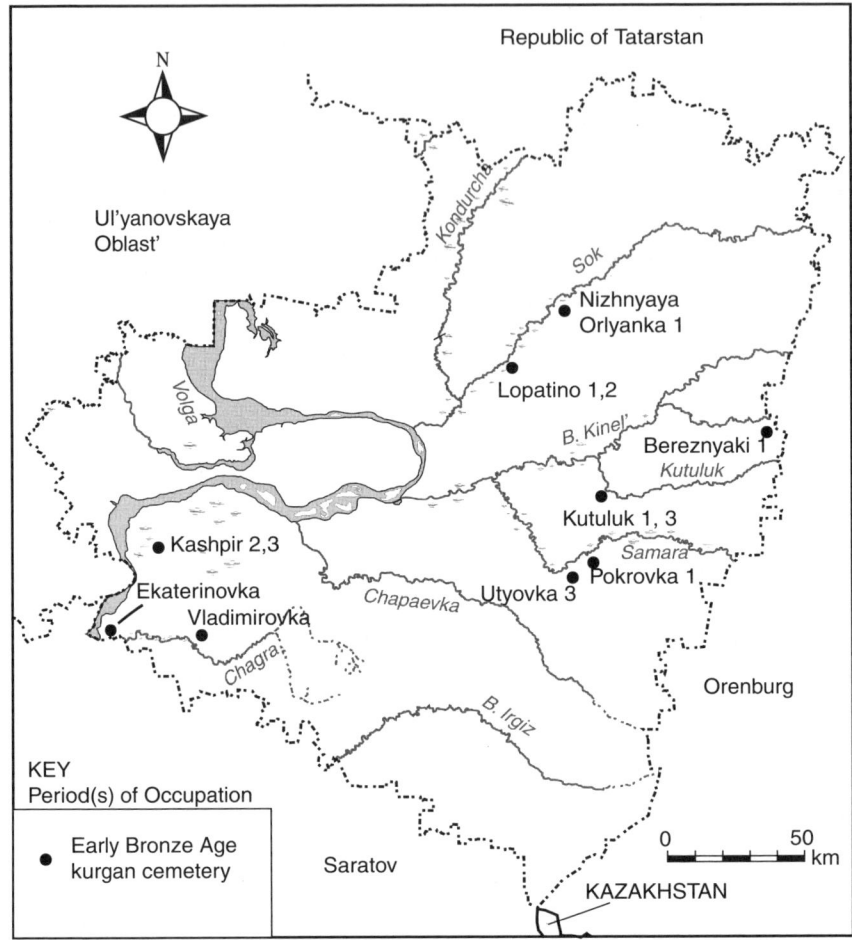

Figure 16.1. The location of Early Bronze Age kurgan cemeteries associated with Yamnaya-type ceramics in the Samara Oblast' (adapted from Kolev et al. 2000).

the expanding domesticated herds of the region (Popova 2007). At the same time, it would have become increasingly difficult to find appropriate summer pasture in the unbearably hot steppes to the south. Thus, it is likely that over time pastoralists changed their seasonal migration patterns in order to deal with these slowly changing climatic and vegetation patterns.

During the beginning of the Middle Bronze Age, the steppe of the Middle Volga region was the domain of pastoralists associated with the Poltavka horizon.[5] Although they constructed relatively few burial mounds, the kurgans they did build were striking in their complexity. The most common burial type in the Middle Volga region consisted of deep chambers with a wide step cut into the clay. Above the burial pit, a temporary roof of wood or mats would often be erected with posts, or the dead would be placed on thick organic mats. Red ochre, in differing amounts, and burnt offerings were common in most burials. In addition, faithful followers filled the burials with many fine objects like bronze axes, awls, adzes, jewelry, and ceramic vessels. In some instances, a round ditch was excavated in the subsoil around the perimeter of the kurgan only to be covered in the end by the monumental mound constructed to cap the burial (Vasil'ev et al. 2000). Kurgans during this period were still primarily built for a single individual, usually an adult male, though females were more common in this period than in the previous (Vasil'ev et al. 2000). It still seems likely that kurgan burials were reserved for the ruling elite.

During the Middle Bronze Age I period, kurgan burials were established throughout the steppe and forest-steppe zones of the Middle Volga region, particularly on the first terrace of the Samara River valley (Fig. 16.2). This region would have been highly valued during this time period, because it allowed pastoralists to use both the forests to the north and the abundant grasslands to the south without moving camp. Still, it seems that at this time pastoralists favored staying close to the meandering rivers of the region with their lush floodplains and marshes. As the pollen analysis of the Sharlyk marsh core demonstrates, the lower Samara River valley was actively used as a source of pasture throughout the Middle Bronze Age (Popova 2007). The pastoralists of this time period were considered to be highly mobile; archaeologists have been unable to uncover traces of long-term settlements (Vasil'ev et al. 2000), although they have found ceramic scatters associated with the Poltavka horizon.

There were other people, however, with a different way of life who lived in the Middle Volga region at this time.[6] Although the Samara River is considered to be the boundary between the forest-steppe and steppe, this border does not hold true for the region just west of the Volga River. With the water winding around the ancient Zhiguli Mountains, dense forests extended in a band along the river deeper into the steppe region. Here there was abundant wildlife more typical of the forest regions farther north. In addition, the Zhiguli region was home to many plant resources that were indigenous only to this particular

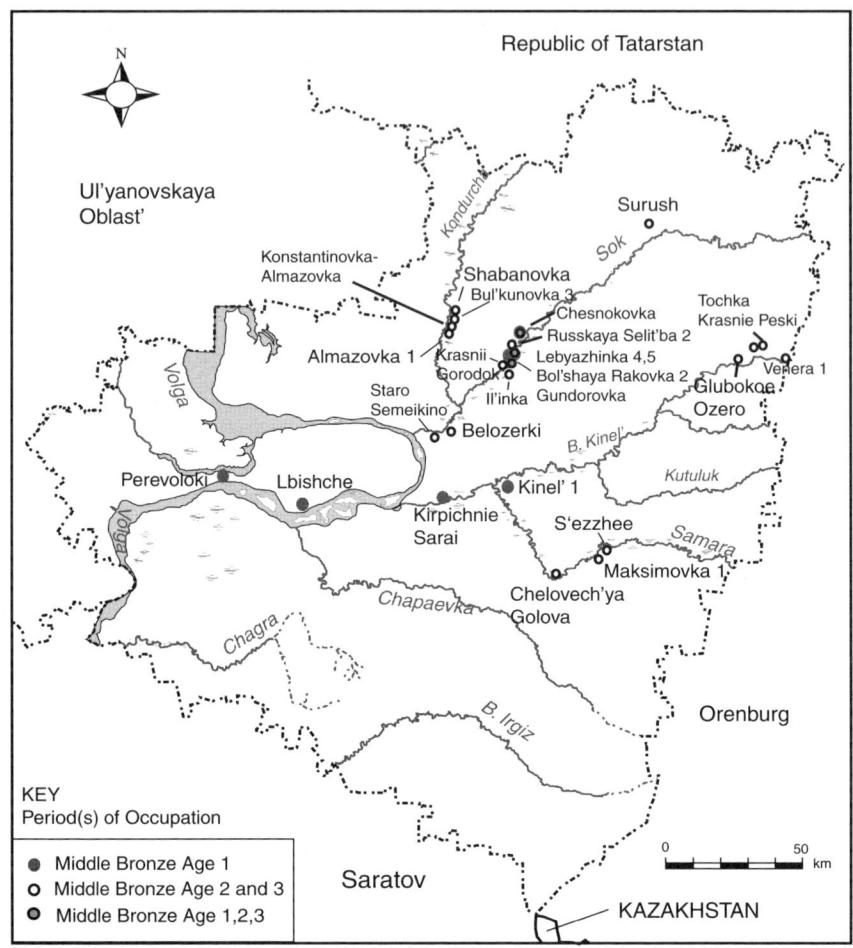

Figure 16.2. The distribution of places that have been interpreted as Middle Bronze Age settlements and campsites in the Samara Oblast' (adapted from Kolev et al. 2000).

ecosystem. In this location, nestled on the right bank of the Volga River, we find the settlements of people who were primarily foragers associated with the Vol'sko-Lbishchenskaya horizon.

These people built long-term settlements only on the right bank of the Volga in easily defended locations. The villages usually consisted of a few pit structures, household pits, and sometimes flat graves. Traditionally, the inhabitants of these villages have been described as foragers and pastoralists, because archaeologists have found the remains of both domesticated and wild animals at these settlements (Vasil'ev and Matveeva 1986). Still, the villages were located in areas that had abundant natural resources but were short on natural pasture, except perhaps on the bank of the Volga River. It would have been difficult to raise more than a few domesticated animals at such locations year-round. On the other hand, these settlements were close to both prime fishing spots on the Volga River and excellent hunting in the forests of the Zhiguli Mountains. Thus, it seems likely that foraging, fishing, and hunting were the primary subsistence strategies used by the people who lived in these particular villages.

There were no settlements associated with the Vol'sko-Lbishchenskaya horizon on the less forested left bank of the Volga River. However, there are several sites, perhaps ephemeral campsites, in which both Vol'sko-Lbishchenskaya and Poltavka materials are found together. All of these places are located in the forest-steppe regions north of the Samara River. It seems, then, that there was contact between the pastoralists of the south and foragers of the Volga River region.

Although the foragers lived in settlements that appeared to be defensive in nature, for the most part they interacted peacefully with the pastoralists. In fact, the relationship was mutually beneficial. The pastoralists could provide a steady supply of domesticated animals, wool yarn and rugs, metal ingots, and finished metal goods. Meanwhile, the foragers could have provided furs, flint, and ceramic vessels – for both their beauty and their contents. Food items could have also been an important component of trade, including smoked fish, wild grains, and products of the forest like berries, mushrooms and honey.

Middle Bronze Age II (2300 to 2000 BCE)

DURING THE second period of the Middle Bronze Age, the forest-steppe increasingly became the site of interaction in the Middle Volga region. To a certain extent, this switch makes intuitive sense, because the steppe regions would have become increasingly dry during the second half of the Middle Bronze Age, and it would have become precarious to use the steppe as a site for consistent summer pasture. On the other hand, the gallery forests were decreasing in size in the forest-steppe, opening up more areas for pasture, making this region more attractive to herders. However, as the pastoralists pushed farther north, they increasingly came in contact with the sedentary foragers in the northern forest zones, and the interaction between these groups was not always peaceful.

Although people in the forest-steppe in the Volga-Don and south-ern Urals regions started to live in larger permanent settlements dur-ing the second half of the Middle Bronze Age, most groups remained fairly mobile in the Middle Volga region.[7] Indeed, the settlements that we know of during this time period in the Volga-Urals forest-steppe are primarily small campsites (Fig. 16.2) that were repeatedly visited dur-ing a certain season every year (O. Kuz'mina 2000). These people were pastoralists relying on cattle and sheep or goat. In addition, the faunal assemblages from these sites contained a relatively high percentage of pig remains (Kosintsev 2001). In one case, at Krasnii Gorodok, bones of a beaver were uncovered as well (Kosintsev and Roslyakova 2002). Thus, it seems likely that hunting and gathering were important aspects of life at this time.[8]

Pastoralists associated with the Abashevo horizon expanded into the forest zone of the Middle Volga region, reaching all the way to the mid-dle Kama River valley. Here they interacted with foragers of the Kama River region (associated with the Volosovo horizon) and pastoralist-foragers (associated with the Balanovo horizon), moving into the region west of the Volga River (Koryakova and Epimakhov 2007). One can see the result of this push north as a result of changes in the way people lived. For example, the foragers of the Kama River region, who previ-ously had preferred to live on dunes and close to water sources, moved their settlements to higher, more defendable ground (Krainov 1987). In addition, there are some signs that, as these three different groups sought to utilize the resources of the same region, there was occasional conflict. On the right bank of the Volga in the forest zone, there is a kurgan (Pepkino) associated with the Abashevo horizon that contained more than two dozen burials. Archaeologists have determined that most of the people who were buried here met with a violent end (Bol'shov 1995). As Koryakova and Epimakhov (2007: 62–63) describe, "Twenty-eight young men were buried in a pit over 11 m long.... Traces of injuries – broken bones and skulls pierced with metal axes and stone arrowheads of the Balanovo type ... were detected on the bones of a large number of these skeletons. The bodies of the skeletons had been dismembered."

There is evidence of more mundane transactions between the groups as well. Abashevo-type ceramics and metal jewelry have been uncov-ered at many settlements in the forest zone of the Middle Volga region. Indeed, the distinct metal weapons and jewelry of the Abashevo horizon were spread over a wide territory of eastern Europe and the southern Urals (O. Kuz'mina 2000). The trading and influence were certainly not one way. Foragers of the Volosovo horizon were skilled hunters and were knowledgeable in the preparation of highly sought after furs, primarily beaver and sable. Indeed, there are some settlements where sable (*Martes zibellina*) bones make up 80% of the total faunal assemblage (Krainov 1987). A coat lined with sable would have been a considerable asset during the long winters in the forests and forest-steppes of Eurasia.

In addition, the pastoralists of the Middle Volga region may have had a considerable impact on the social orders and landscapes of the foragers to the north. By the end of the Middle Bronze Age, kurgan burials had been incorporated as the predominant burial ritual in the forest zone, although they were located closer to settlements than was normal in the forest-steppe. Similarly, the remains of domesticated animals started to become more common at the settlements of foragers. They perhaps began to raise the animals themselves, or they received animals as payment for other goods. Finally, there is some evidence that they began to utilize the local copper resources of the Kama River region.

Much of the contact and conflict that occurred in the Lower Kama region at the end of the Middle Bronze Age was in some way associated with the resources of the area. As noted, pastoralists associated with the Abashevo horizon established long-term settlements in both the Don-Volga region and the southern Urals. Both of these areas have major copper ore deposits. The Lower Kama region is also copper rich; however, large villages were never established in this region during the Middle Bronze Age.[9] It seems that competition by local forest groups for resources in the same area as the copper mines made this particular source more difficult to secure.

Middle Bronze Age III (2000 to 1800 BCE)

At the same time, toward the end of the Middle Bronze Age, new groups of pastoral leaders were consolidating their power in the Samara River valley and the lower Sok River valley. Staying close to the ecotone between the steppe and the forest-steppe, these groups must have utilized both regions. The location of burials associated with the Potapovka horizon along these major conduits through the Middle Volga region is significant. The Samara River, running roughly east-west, offers easy passage from the Ural River region to the Volga River. The Sok River, on the other hand, cuts diagonally across the forest-steppe from the Ural foothills farther north to the Volga River. Both river valleys provide a corridor of prime pasture from the more mountainous metallurgical centers in the east to the Volga River in the west.

It seems likely that this was a time when trade routes and resources were cautiously guarded. All over the Volga-Urals and southern Urals regions we see the appearance of pastoral rulers who were clearly honored for their military prowess and ability as successful herders. More than this, however, the nature of the burial remains suggests that it was not only the individual warrior but his or her entire family that was associated with a new role in society. Some archaeologists have argued that this demonstrates the creation of a warrior class that both actively hoarded metal objects and guarded the knowledge of how to create such objects (E. Kuz'mina 2000).

The burials at this time, at least in the Middle Volga region, are very different from kurgan burials of the previous period. During this time, a great deal of effort, wealth, and time was expended to bury particular individuals or families. First, a wide and deep rectangular pit with steps was dug into the ground. At the base of the pit, a shallow channel was excavated around the base perimeter. The base of the burial pit was often smeared with a layer of ashy clay. Finally, the burial was covered with thin planks of wood and eventually with an earthen mound (Kuznetsov and Semenova 2000). In many cases, several burials were added to the burial plot before the kurgan was added.

Feasts were associated with the funeral rites during this time, with either the whole animal or the head and feet of sacrificed animals being offered to the dead. In central burials, archaeologists found the remains of horse, bull, sheep, goat, and dog. In peripheral burials, it was more common to find the head and leg bones of bulls and sheep (Kuznetsov and Semenova 2000). Certain animals seemed to receive special treatment. In one case, at Utyovka 6, a whole horse was interred in a separate burial within the kurgan. In other burials, elaborately etched bone cheekpieces have been discovered. Both these elements point to an added emphasis on the importance of horses during this period.

Animal sacrifices were not the only offerings and gifts for the dead. Burials associated with the Potapovka horizon are some of the richest in the region. In many of the burials, bronze knives are placed near the head of the deceased. Archaeologists have also found bronze hooks, earrings, bracelets, needles, adzes, and sickles and one example of a hefty bronze spearhead from Utyovka 4 (burial 4, kurgan 6). Occasionally, jewelry made from gold and silver was also part of the burial offerings. Other items found in burials include bone combs, imported agate and faience beads, and many ceramic vessels, some of them quite diminutive (Kuznetsov and Semenova 2000). One of the most unusual items uncovered, however, and something that was unique to this period, was a complete set of flint projectile points. Flint objects were not very common in kurgan burials up to this time. Anthony (Chapter 4 in this volume) has suggested that the two types of flint projectile points found in burials associated with the Potapovka and Sintashta horizons were made specifically for arrows and javelins.

Good-quality flint is rare in Eurasia. There is poor-quality flint in the Samara region (Lastovskii 2000), but this type of stone was not used for the projectile points found in the Potapovka-type burials. Finer grades of flint are found only in the forested regions of Russia. For the Middle Volga region, one of the closest sources was near the present-day town of Magnetigorsk in the Ural Mountains. It is likely that the foragers who inhabited the forested regions of Eurasia had established vast trade networks that extended across this broad landmass early on. In a way, this trade network can be viewed as a northern Silk Road.[10] By the end of the Middle Bronze Age, when warfare was intensifying between pastoral groups in the steppe regions, not only flint but also large quantities of bronze objects were traded over this forest trade network. Most of these metal objects came from Siberia and are connected to the Seima-Turbino horizon.[11] Hoards of these metal weapons (spearheads, daggers, socketed axes) were found in Turbino on the Kama River, Seima and Reshnoe on the Oka River, and Sokolovka near the confluence of the Volga and Kama (Chernykh 1992). Although there is still much debate about what these deposits mean, Koryakova and Epimakhov (2007: 106–110) argue that these objects should be dated to around 1800 BCE, that they were more technologically sophisticated, and that they indicate the long-distance trade and/or the movement of highly skilled metallurgists through the forest-steppe regions of Russia. They claim that what is most interesting about these deposits is that they indicate that two competing technological systems co-existed and yet remained distinct during the end of the Middle Bronze Age.

Late Bronze Age I (1800 to 1600 BCE)

THUS, AT the end of the Middle Bronze Age, though the Lower Kama region was a hotbed of social and political interaction, mobile pastoralists held the lower forest-steppe region relatively unchallenged. As noted, burial sites associated with the Potapovka horizon were clustered in the Samara and Sok river valleys. By the beginning of the Late Bronze Age, this territorial control became even more apparent as pastoralists of the region began to build structures at traditional pastures in the same regions.

At the beginning of the Late Bronze Age, the climate became moister and cooler, similar to the climate of today (Popova 2007). This change resulted in the expansion of forests in the forest-steppe and perhaps the expansion of forests into regions south of the Samara River valley. However, this expansion of forests was met with an increased desire for forest products particularly for the construction of both dwellings and burial chambers. Thus, in the lower Samara River valley, there was no significant increase in tree cover overall.

These structures, partially dug into the ground and constructed of wood, were probably used on a seasonal basis by a single household or a kin group. Most of these structures were constructed near the Sok River near its confluence with the Volga River (Fig. 16.3). There were also a few settlements along the Volga and Samara rivers. Sachkovo, Yakovka, and Grigorievka on the Volga River are particularly interesting because they are located near the largest concentration of Late Bronze Age kurgan cemeteries. They are also strategically situated on the fringe of some of the largest marshes in the Middle Volga region (Kolev 2002; Sedova 2000).

These seasonal settlements were not evenly spaced throughout the Middle Volga region during the early Late Bronze Age. Instead, they were located primarily in three locations: the Samara, Sok, and Volga river valleys, usually near kurgan cemeteries. There were some exceptions to this rule, however. For example, Mikhailo-Ovsyanka, a unique mining site, was located deep in the hot and dry steppe zone of the Middle Volga region. The spatial organization of these sites seems to suggest that migration routes during this period became more routinized, especially as people visited specific pastures year after year. These seasonal settlement clusters were spaced far enough apart to allow a wide region to be used as pasture by each group. The settlements in each region, however, were located in relatively close proximity to each other, demonstrating that pastoralists would often move the location of settlements slightly in order to allow the previous settlement sites some time to recover from the trampling of people and the grazing of livestock.

The burials during the early Late Bronze Age were similar to burials from the previous period. Kurgans were usually built for only one or

Figure 16.3. The distribution of Late Bronze Age settlements in the Samara Oblast'. Other periods during the Bronze Age in which the same location was occupied are also noted (adapted from Kolev et al. 2000 and Kolev 2002). *Note:* Only Late Bronze Age settlements that have been sufficiently excavated (in most cases, more than 200 square meters) have been included on this map.

two very particular individuals. Animal sacrifices continued to be an important part of the burial rite. The most common grave offerings were bronze knives, flint arrowheads, bone and metal ornaments, and an occasional stone axe. Ceramic vessels were almost always included as well, and it seems that these vessels were made specially to be used in the burial rite. Certain kurgan cemeteries are defined by the particular style of decoration found on the ceramic vessels in each burial (Semenova 2000).

Late Bronze Age II (1600 to 1300 BCE)

ONE MIGHT expect that, as the forests slowly reclaimed the steppe, pastoralists would have stayed close to the border between the steppe and forest-steppe. Instead, during the second half of the Late Bronze Age, there was considerable expansion of pastoralists into the steppe, forest-steppe, and forested zones of the Middle Volga region (Fig. 16.3). The Sok and Samara river valleys continued to be utilized, and recent excavations suggest these villages were inhabited year-round (Anthony et al. 2005; Popova 2006). In addition, the marshy steppe between the Volga, Chapaevka, and Chagra rivers continued to be a popular location for both settlements and kurgan cemeteries. All of these river valleys were important in the previous period; however, in the second half of the Late Bronze Age they became pivotal centers of social, economic, and political activity in the region.

Nevertheless, there are many other regions that became the locations of new embedded seasonal encampments. For example, pastoralists moved to the western bank of the Volga River and established settlements that extend into the Zhiguli Mountain foothills. In addition, there were more seasonal settlements located on the smaller tributaries of larger rivers, especially farther north in the forest-steppe extending into the Ural Mountain foothills, including Malo-Mikushkino, Kibit, and Soplyaki (Kolev 2002, Sedova 2000). Finally, there were a few seasonal settlements, most likely winter encampments, that were situated farther south in the steppe zone along more ephemeral rivers such as the Chapaevka and B. Irgiz.

The distribution of these places throughout the Middle Volga region indicates that the land was being divided into more segments of territorial control, with certain powerful pastoralists controlling the movement of people throughout the region. The continued clustering of settlements in particular areas, such as the headwaters of the Bolshoi Kinel' River, seems to show that the same group was returning to the same region and that they were managing the land in very specific ways. On the other hand, the strict control of pastures and the movement of people throughout the region made it difficult for new herders to fit into the system. As a result, we see that increasingly over time pastoralists were exploring new areas for pasture. One of these new regions was the northern forest-steppe and forest region up to the Lower Kama River.

In the Lower Kama region, archaeologists know of approximately 500 settlements and 26 burials associated with the Srubnaya horizon (Kazakov 2001). This seems to indicate that over the course of the second half of the Late Bronze Age there was a steady stream of settlers who managed to make pastoralism work, at least seasonally, in the forest zone. According to the more predictable permanent presence of pastoralists in the region, the few foragers left in the area moved north of the Lower Kama River valley.

It is clear that, throughout the Middle and Late Bronze Age, pastoralism was the main productive economy for most groups, although wild grains formed an important supplement to their diets.[12] During the Late Bronze Age, however, there was a subtle shift in preference from maximizing sheep, goats, and horses to concentrating on cattle (Zhuravlev 1989; Kosintsev 2001). Cattle are more difficult to keep alive than sheep and goats, because they are more finicky eaters. In addition, it is difficult to recuperate from herd losses with cattle, because they reproduce more slowly than sheep and goats. Living in one place long term, it would have been realistic to keep only a handful of cows. If the herd of cattle, however, was split up and lent to other herders to be taken out into the pastures of the southern steppes during the harsh winter months, then the potential of having numerous animals by the spring season was higher. Thus, during the Late Bronze Age there were both sedentary pastoralists and mobile herders.

At a time when powerful pastoralists were establishing large settlements, less emphasis was placed on building large elaborate burial mounds to commemorate the leaders of society. Indeed, the second half of the Late Bronze Age was defined by relatively modest kurgans, often with dozens of interments from all sub-sets of society. The largest kurgan from this time period at Kaibel' contained 99 graves (Semenova 2000). Some archaeologists (Vasil'ev and Matveeva 1986) have interpreted this change in purpose of kurgan burials to indicate a switch to a more egalitarian political organization. Nevertheless, on the basis of evidence from the long-term Late Bronze Age settlement at Krasnosamarskoe, it seems that leaders maintained a sense of community by periodically holding elaborate feasts. Rather than bringing people together only upon the death of a leader, Late Bronze Age groups reaffirmed political ties several times a year. It seems that these regular meetings were needed to maintain the complicated migration schedules that were established during this time period.

Conclusion

TOO OFTEN discussions of social, economic, and political complexity in archaeology are reduced to questions of typology. Instead of talking about the ways in which a particular society functioned, archaeologists who still work within the culture-history framework work to fit groups into particular categories (e.g., chiefdom or state, nomad or farmer). In actuality, social complexity, including political and economic organization, is embedded in how people do things in the processes of daily life. By trying to fit people into awkward categories, it is easy to lose the very details that made that group interesting and innovative in the first place.

As I have demonstrated, it is necessary to look at the interaction between different societies in the Eurasian steppe and forest-steppe in order to break away from some of these overarching categories. In this chapter, I have focused on the landscape history of one particular area, the Middle Volga region, so as to show the myriad ways groups with differing social, economic, and political orders influenced one another. Moreover, I have tried to bring together two groups that are usually discussed separately, foragers and pastoralists, in order to show how their histories are linked. The result is a history that does not privilege any particular economic strategy or style of political organization and yet still demonstrates that the Eurasian steppe and forest-steppe were the locus of unique early complex societies and political diversity.

Notes

1. The definition of Neolithic is different in Western literature, usually implying that the groups practiced farming. The term Neolithic in the Russian archaeological literature tends to denote a group that still relies primarily on stone tools and weapons.
2. There are some fine articles on faunal data comparing certain periods or archaeological culture types (see, e.g., Kosintsev 2001 and Kosintsev and Roslyakova 2002). However, the level of detail is still too cursory to do a detailed re-analysis of any one site.
3. In the Middle Volga region, there are also examples of double burials with either a man and a woman or a child and an adult.

4. This argument relies heavily on the notion of a cohesive Yamnaya culture. Archaeologists have charted Yamnaya-type remains throughout the steppe and forest-steppe regions from the Ukraine to the southern Urals. In each case, kurgans are assumed to appear in a region only if Yamnaya *people* moved to the region. As I have tried to show elsewhere (Popova 2006), only a few burial traits hold together what is commonly referred to as the Yamnaya culture. There is still much debate about the origins of the Yamnaya culture that is beyond the scope of this chapter. For a brief discussion on the matter, see Turetskii et al. 2007: 70.

5. I am using the term horizon and not the traditional term culture or archaeological culture. By using the term horizon or horizon style, I wish to point out that there are indeed sets of similar materials (ceramic, metal, lithic) and remains (architectural, mortuary) that co-occur over various sites but that these remains are not automatically linked to a particular archaeological subject.

6. The dating of this particular archaeological culture is problematic. However, Turetskii et al. (2007: 120) argue that it is most likely associated with the early Middle Bronze Age.

7. I am referring here to the relatively broadly defined Abashevo horizon (Middle Volga variant).

8. There is some discussion of groups associated with the Abashevo horizon engaging in farming. So far in the Middle Volga region, there is very little convincing evidence that this was the case.

9. The exception to this rule is the Balanbash settlement on the Belaya River, which is a tributary of the Kama.

10. It seems that this could be connected in some way with "The Great Nephrite Road of the Bronze Age" referred to by Koryakova and Epimakhov (2007: 109).

11. Chernykh (1992) considers the Seima-Turbino to be a culture (or transcultural phenomenon) of metallurgists and warrior-horsemen. He believes that this small population of people spread Seima-Turbino–type artifacts (especially weapons) throughout the Eurasian forest and forest-steppe. See Koryakova and Epimakhov (2007) for an extended discussion of this phenomenon.

12. On the basis of the macrobotanical analysis conducted for samples from three Late Bronze Age places in the Middle Volga region (Krasnosamarskoe, Peschanyi Dol, and Kibit), groups in the region were not farmers. However, at each site, relatively large quantities of *Chenopodium* were discovered, suggesting that pastoralists did harvest local wild grains (Popova 2006).

References

Anthony, David. W., D. Brown, E. Brown, A. Goodman, A. Khokhlov, P. Kuznetsov, P. Kosintsev, O. Mochalov, E. Murphy, A. Pike-Tay, A. Rosen, N. Russell, D. Peterson, L. Popova, and A. Weisskopf. 2005. The Samara Valley Project: Late Bronze Age Economy and Ritual in the Russian Steppes. *Eurasia Antiqua* 11: 395–417.

Barfield, T. 1993. *The Nomadic Alternative*. Englewood Cliffs: Prentice-Hall.

⸻ 2001. The Shadow Empires: Imperial State-Formation along the Chinese-Nomad Frontier. In S. Alcock, T. D'Altroy, K. Morrison, and C. Sinopoli (eds.), *Empires: Perspectives from Archaeology and History*. Cambridge: Cambridge University Press, pp. 10–41.

Bates, D. G., and S. H. Lee. 1977. Role of Exchange in Productive Specialization. *American Anthropologist* 79(4): 834–841.

Bol'shov, S. V. 1995. Problemi kul'turogeneza v lesnoi polose Srednego Povolzh'ya v Abashevskoe Vremya. In I. B. Vasil'ev (ed.), *Drevnie Indoiranskie Kul'turi Volgo-Ural'ya*. Samara: SamGPU, pp. 124–140.

Bunyatyan, K. P. 2003. Correlations between Agriculture and Pastoralism in the Northern Pontic Steppe Area during the Bronze Age. In M. Levine, C. Renfrew, and K. Boyle (eds.), *Prehistoric Steppe Adaptation and the Horse*. Cambridge: McDonald Institute, pp. 269–286.

Chernykh, E. N. 1992. *Ancient Metallurgy in the USSR*. Trans. Sarah Wright. Cambridge: Cambridge University Press.

Di Cosmo, N. 1994. Ancient Inner Asian Nomads: The Economic Basis and Its Significance in Chinese History. *Journal of Asian Studies* 53(4): 1092–1126.

Hiebert, F. T. 2002. Bronze Age Interaction between the Eurasian Steppe and Central Asia. In K. Boyle, C. Renfrew, and M. Levine (eds.), *Ancient Interactions: East and West in Eurasia*. Cambridge: McDonald Institute, pp. 237–248.

Kazakov, E. P. 2001. Ob otnositel'noi datirovke kul'tur epokhi bronzi XVI–XIV vv. do n.e. v Nizhnem Prikam'e. In Yu. I. Kolev (ed.), *Bronzovii Vek Vostochnoi Evropi: kharakteristika kul'tur, Khronologiya i periodizatsiya*. Samara: NTTs, pp. 204–207.

Khazanov, A. M. 1994. *Nomads and the Outside World*. 2nd ed. Madison: University of Wisconsin Press.

Kolev, Yu. I. 2002. Kompleksi pozdnego bronzogo veka poseleniya Grigor'evka I v Samarskoi Oblasti. In A. A. Vibornov and V. N. Mishkin (eds.), *Voprosi Arkheologii Povolzh'ya*, Vipusk 2. Samara: Samara State Pedagogical University.

Kolev, Yu. I., A. E. Mamonov, and M. A. Turestskii (eds.). 2000. *Istoria Samarskogo Povolzh'ye s drevneishikh vremen do nashikh dnei: bronzovii vek*. Samara: Tsentr Integratsiia.

Koryakova, L., and A. V. Epimakhov. 2007. *The Urals and Western Siberia in the Bronze and Iron Ages*. Cambridge: Cambridge University Press.

Kosintsev, P. A. 2001. Kompleks Kostnikh Ostatkov Domashnikh Zhivotnikh iz Poselenii i Mogil'nikov Epokhi Bronzi Volgo-Ural'ya i Zaural'ya. In Yu. I. Kolev (ed.), *Bronzovii Vek Vostochnoi Evropi: kharakteristika kul'tur, khronologiya i periodizatsiya*. Samara: NTTs, pp. 363–367.

Kosintsev, P. A., and N. V. Roslyakova. 2002. Materiali po istorii zhivotnovodstva y naseleniya Samarskogo Povolzh'ya v bronzovom veke. In A. A. Vibornov (ed.), *Voprosi Arkheologii Povolzh'ya*. Vipusk 2. Samara: SamGPU, pp. 145–150.

Krainov, D. A. 1987. Eneolit tsentra Russkoi ravnini i Priural'ya: Volosovskaya Kul'tura. In O. H. Bader, D. A. Krainov, and M. F. Kosarev (eds.), *Epohki bronzi lesnoi polosi SSSR*. Moscow: Nauka.

Kremenetski, C. V. 1997. The Late Holocene Environment and Climate Shift in Russia and Surrounding Lands. In H. N. Dalfes, G. Kukla, and H. Weiss (eds.), *Climate Change in the Third Millennium BC*. Berlin: Springer, pp. 351–370.

Kremenetski, C. V., T. Bittger, F. W. Junge, and A. G. Tarasov. 1999. Late and Postglacial Environment of the Buzuluk Area, Middle Volga Region, Russia. *Quaternary Science Reviews* 18: 1185–1203.

Kremenetski, C. V., P. E. Tarasov, and A. E. Cherkinsky. 1997. Postglacial Development of Kazakhstan Pine Forests. *Geographie Physique et Quaternaire* 51: 391–404.

Kuz'mina, Elena. 2000. The Eurasian Steppes: The Transition from Early Urbanism to Nomadism. In J. Davis-Kimball, E. M. Murphy, L. Koryakova, and L. T. Yablonsky (eds.), *Kurgans, Ritual Sites, and Settlements: Eurasian Bronze and Iron Age*. BAR International Series 890. Oxford: Archaeopress, pp. 118–125.

Kuz'mina, Olga B. 2000. Abashevskaya Kul'tura v Samarskom Povolzh'e. In Yu. I. Kolev, A. E. Mamonov, and M. A. Turetskii (eds.), *Istoria Samarskovo Povolzh'e s Drevneishikh Vremen do Nashikh dnei: bronzovii vek*. Samara: Tsentr "Integratsiia," pp. 85–121.

Kuznetsov, P. F., and A. P. Semenova. 2000. Pamyatniki Potapovskogo Tipa. In Yu. I. Kolev, A. E. Mamonov, and M. A. Turetskii (eds.), *Istoria Samarskovo Povolzh'e s Drevneishikh Vremen do Nashikh dnei: bronzovii vek*. Samara: Tsentr "Integratsiia," pp. 121–151.

Lastovskii, A. A. 2000. Mezolit. In A. A. Vibornov, Yu. I. Kolev, and A. E. Mamonov (eds.), *Istoriya Samarskogo Povolzh'ya Drevneishikh Vremen do Nashikh dnei: kamennyi vek*. Samara: Tsentr "Integratsiia," pp. 81–141.

Popova, L. M. 2006. Political Pastures: Navigating the Steppe in the Middle Volga Region (Russia) during the Bronze Age. Ph.D. dissertation, University of Chicago.

———. 2007. A New Historical Legend: A Long-Term Vegetation History of the Samara River Valley. In L. M. Popova, C. W. Hartley, and A. T. Smith (eds.), *Social Orders and Social Landscapes*. Newcastle: Cambridge Scholars Publishing.

Rowton, M. B. 1973. Urban Autonomy in a Nomadic Environment. *Journal of Near Eastern Studies* 32: 201–58.

Sedova, M. S. 2000. Poselenia srubnoi kul'turi. In Yu. I. Kolev, A. E. Mamonov, and M. A. Turetskii (eds.), *Istoria Samarskovo Povolzh'e s Drevneishikh Vremen do Nashikh dnei: bronzovii vek*. Samara: Tsentr "Integratsiia," pp. 209–255.

Semenova, A. P. 2000. Pogrebalnie pamyatniki srubnoi kul'turi. In Yu. I. Kolev, A. E. Mamonov, and M. A. Turetskii (eds.), *Istoria Samarskovo Povolzh'e s Drevneishikh Vremen do Nashikh dnei: bronzovii vek*. Samara: Tsentr "Integratsiia," pp. 152–208.

Telegin, D. Ya. 2002. A Discussion on Some of the Problems Arising from the Study of Neolithic and Eneolithic Cultures in the Azov Black Sea Region. In K. Boyle, C. Renfrew, and M. Levine (eds.), *Ancient Interactions: East and West in Eurasia*. Cambridge: McDonald Institute, pp. 25–47.

Turetskii, M. A., I. N. Vasil'eva, S. A. Agapov, V. V. Kamaev, and N. P. Salugina. 2007. *Drevnie Kul'turi i etnosi Samarskogo Povolzh'ya*. Samara: Samarskii Dom Pechati.

Vasil'ev, I. B. 1981. *Eneolit Povolzh'ya.* Kuibyshev: Kuibyshevskii gosudarstvenyi pedagogicheskii institute.

Vasil'ev, I. B., P. F. Kuznetsov, and M. A. Turetskii. 2000. Yamnaya i Poltavkinskaya Kul'turi. In Yu. I. Kolev, A. E. Mamonov, and M. A. Turetskii (eds.), *Istoria Samarskovo Povolzh'e s Drevneishikh Vremen do Nashikh dnei: bronzovii vek.* Samara: Tsentr "Integratsiia," pp. 6–65.

Vasil'ev, I. B., and G. I. Matveeva. 1986. *Istokov istorii Samarskogo Povolzh'ya.* Kuibyshev: Kuibyshevskoe Knizhnoe Izdatel'stvo.

Velichko, A. A., A. A. Andreev, and V. A. Klimanov. 1997. Climate and Vegetation Dynamics in the Tundra and Forest Zone during the Late Glacial and Holocene. *Quaternary International* 41(42): 71–96.

Vinogradova, N. M., and E. E. Kuz'mina. 1996. Contacts between the Steppe and Agricultural Tribes of Central Asia in the Bronze Age. *Anthropology and Archaeology of Eurasia* 34(4): 29–54.

Zhuravlev, O. P. 1989. Skotovodstvo y naseleniya donskoi lesostepnoi srubnoi kul'turi. In D. Pryakhin (ed.), *Poseleniya srubnoi obshnosti.* Voronezh: University of Voronezh.

PART FOUR

SOCIAL POWER, MONUMENTALITY, AND MOBILITY

CHAPTER 17

Introduction

FRANCIS ALLARD

T HE THREE contributions in this section shift the attention to
Eurasia's eastern steppes, specifically Mongolia's rich archaeological
landscape during the Bronze and Iron Ages (second to first millennium
BCE). Already by the late nineteenth century, commentators had noted
the region's many types of ancient stone-built structures, including slab
burials, delineated by flat standing stones; structurally complex – and
often massive – khirigsuurs; standing deer stones; and small mounded
slope burials. This early archaeological record owes its enduring scholarly appeal not only to the high visibility of its numerous monumental
sites, which stand out among Mongolia's treeless valley bottoms and
hillsides, but also to the materials and information recovered from the
sites themselves. Khirigsuurs and the other types of burials have yielded
human and animal remains, while deer stones display enigmatic carved
designs of uncertain symbolism. It is fair to say that research on this
period of Mongolia's prehistory, including the three chapters in this
part, has in some way or other remained tethered to these prominent
sites and features.

Other aspects of Mongolia's early archaeological landscape have also
generated interest among scholars. Some of the monumental structures
seem to make an appearance at approximately the same time at the end
of the second millennium BCE, signaling the possibility of a significant
cultural transformation. All three chapters address, in some way or
other, the issue of the emergence and development of this monumental landscape. In addition, all of these visible structures are believed
to have had a ritual and/or funerary role, while also serving important

social functions. Not surprisingly, the absence of prominent habitation remains associated with these sites has encouraged the view that they were the products of mobile pastoralists who gathered at regular intervals to collaborate in their construction and participate in communal activities. How the period's monumental structures were built by mobile populations living at low population densities and what role these sites played in Mongolia's early pastoralist societies are lines of investigation common to much recent research.

Although few would dispute the idea that Mongolia's early monumental constructions represent a significant cultural development associated with a pastoralist lifeway, the fact that these developments span no less than 500 years alerts us to the likelihood that the period itself witnessed important trends and fluctuations in monument building and subsistence activities. However, charting such trends remains a challenge, owing to the relatively small number of absolute dates available so far. One can even make the legitimate charge that the first order of business remains the clarification of the region's culture history. In fact, not only is the chronological context uncertain, but the confirmation of a cultural-historical framework also suffers from the almost exclusive focus to date on the recording and excavation of the region's visible monuments, at the expense of the more ephemeral settlement remains. One may suggest that such an important component of a mobile lifeway is ignored because of the methodological challenge it presents (campsites are difficult to locate); because such sites rarely yield artifacts of aesthetic value; or, one suspects, because of the mistaken assumption on the part of archaeologists that they do not in fact present a significant interpretive challenge, as such transient sites appear to fit perfectly well within the "mobile pastoralist model" accepted by most archaeologists.

Although hampered by serious uncertainties regarding the region's cultural-historical framework, studies of early Mongolia have nevertheless attempted to explain the available data from a range of theoretical perspectives. Russian and Mongolian archaeologists have proposed models of culture change that rely on an assumed equivalence of archaeological cultures to culturally bounded populations, an approach that long ago lost favor among most Western archaeologists. Thus, areas of Mongolia characterized by the presence of deer stones, khirigsuurs, and slab burials, such as the Khanuy Valley, are said to have witnessed the overlap of three migrating peoples, with each group contributing its own distinctive cultural product to the local landscape. More recently, as illustrated by the three chapters here, archaeological studies of Mongolia have begun tackling the interpretation of data from the perspective of Western-inspired models. A central and welcome element of such a renewed effort at modeling culture change and processes has been the identification of spatial patterns in site location, type, and size.

Drawing on spatial patterning data from the Egiin Gol Valley in northern Mongolia, Honeychurch, Wright, and Amartuvshin propose in chapter 18 a model for the development of social inequality in this area during the second and first millennia BCE. They suggest that the gradual change from a monumental landscape dominated by khirig-suurs to one complemented by large numbers of newly built slab burials was associated with such emergent inequality. In support of their model, the authors point to a number of features of the changing archaeological landscape. Along with evidence for the appearance of horse riding, these include information on the slab burials themselves, such as the small number of burials (in relation to the purported population), their preferential placement close to pre-existing khirigsuurs and to routes providing efficient access in and out of the valley, the use of non-local materials as grave goods, variation in the size of the burials, and the large size of a few sub-adult burials. In the scenario proposed by the authors, those local lineages that happened to be preferentially positioned in the landscape were able to translate their access to external information and relationships into hereditary power. The emergence and maintenance of such inequality was dependent on the continued participation of large numbers of people at building events within existing monumental landscapes, with the smaller scale of the slab burials effectively communicating to everyone the greater exclusivity of the associated rituals and the dominance of the individual lineages responsible for their construction.

In chapter 19, Jean-Luc Houle focuses on the rich archaeological landscape of the Khanuy Valley in north central Mongolia. Drawing attention to the valley's significant number of khirigsuurs of various sizes and to the smaller and contemporary slope burials, he points to the clustering of these Bronze Age monuments in the valley and suggests that such "central places" were associated with social units of some complexity. Interestingly, modern-day ethnographic data, combined with the still limited Bronze Age settlement evidence, suggest the possible restricted mobility of such social units and thus their attachment to these monumental clusters. Recognizing that there is little evidence for coercion or economic control at this time, Houle proposes that the khirigsuurs and their associated rituals played a central role in integrating dispersed

populations, with larger monumental complexes identifying the presence of more powerful local elites. He suggests that the corporate nature
of building and ritual activities would have provided a sufficient basis for
the selection of local leaders, whose burials may very well have been the
central mounds of khirigsuurs. Significantly, he proposes that the later
slab burials may represent a transition to a society characterized by a
more "individualistic" and hereditary leadership.

In chapter 20, William Fitzhugh tackles the issue of the temporal,
functional, and spatial relationship among Mongolia's deer stones,
khirigsuurs, and slab burials. His attempt to clarify the region's culture
history incorporates previous data, as well as the results of recent fieldwork by the Smithsonian-Mongolian Deer Stone Project. Well over one
dozen radiocarbon dates are now available for the Ulaan Tolgoi deer
stone–khirigsuur complex and khirigsuurs in other regions of Mongolia.
Combined with the evident spatial relationship linking deer stones to
khirigsuurs at many sites (e.g., the proximity of the two monument
types, as well as an association of deer stones with stone rings typically
found at khirigsuurs), the chronology points to the possibility that the
two structures were temporally and functionally linked at ceremonial
complexes that date to around 1000 BCE. On the basis of a small number
of early radiocarbon dates, Fitzhugh also presents the possibility that the
small and stylistically simpler deer stones found in northern Mongolia's
Darkhad Valley may have been the precursors of the more developed
deer stones further south. In noting the consistency witnessed in the
spatial arrangement of ritual and structural elements, including horse
heads and khirigsuurs that point east, horse head mounds that are preferentially located on the eastern side of khirigsuurs, and sites that cluster
on the eastern sides of hills, Fitzhugh proposes the existence of a uniform cosmological world linking ritual behavior at the site, local, and
regional levels. In this scenario, larger ritual complexes would have acted
as foci of population and power.

Many would agree that foreign archaeological projects in Mongolia
over the past decade and a half have made a positive impact on our understanding of early steppe archaeology in this region. Combined with
the recent increase in the number of available radiocarbon dates, the
growing reliance on systematic field projects and efforts to collect non-
monumental data are helping to refine the region's culture history and
redirect attention away from a focus on impressive monumental structures and mortuary contents. Importantly, as illustrated in the three
chapters, a commitment to spatially bounded surveys ranging from the

site, local, and regional levels is now generating spatial patterning data that permit the testing and building of models of culture change whose potential cross-cultural validity remains an exciting avenue of research. As mentioned earlier, however, much remains to be done. Indeed, the three chapters also underscore the chronological uncertainty resulting from too few absolute dates, with disagreements evident in whether khirigsuurs increased or decreased in size over time, as well as the extent of temporal overlap between the construction phases of khirigsuurs and slab burials. No less significant, issues of terminology have yet to be resolved, with some of the archaeologists lumping khirigsuurs and what some call slope burials into the same category. Whether in regard to chronology or terminology, these differences among archaeologists do more than simply underscore uncertainties in the cultural-historical framework; they also impact on, and potentially weaken, the very interpretations that they put forward.

Reflecting a widespread concern in archaeological research, all three chapters attempt to tackle the issue of power and inequality in Bronze and Iron Age Mongolia. This is not a simple task. To many, the labor implications of the region's monumental landscape conflict uneasily with the assumed presence of a dispersed mobile population whose burials, especially the earlier ones, display little evidence of wealth, coercion, or warfare. In the end, we are left with the supposition that the human remains found under the central mounds of khirigsuurs and in slab burials, which were all labor-intensive constructions, were those of local leaders. Similarly, deer stones are said to have been the cenotaphs of absent leaders. Having assumed the presence of local elites, the chapters by Honeychurch et al. and Houle go on to propose developmental trajectories that culminate in the emergence of more-stable social hierarchies. Although the two models do differ, they both postulate a movement toward a condition of institutionalized inequality.

It is worth noting that none of the authors relies on labels such as "tribe" or "chiefdom" to identify the societies that are the focus of their research, an approach that is in keeping with a general move in archaeology away from the identification of distinct societal types within evolutionary schemes and toward the investigation of processes associated with social change (Chapman 2003: 41–45). In reality, such reluctance to assign labels to Mongolia's Bronze and Iron Age societies probably also reflects recognition of the distinctiveness of mobile pastoralist societies, which played no significant role in the neo-evolutionary schemes first developed by ethnographers and archaeologists. Having said this, one cannot but note an inclination on the part of two of the authors to chart the development of inequality and present it as a long-term unidirectional process. Although the emergence of power remains a legitimate topic of investigation, it is suggested here that successful models of culture change in ancient Mongolia should not ignore a central feature of social trajectories in that region, namely, the fact that periods characterized by demographic aggregation – as evidenced, for example, by public works or military campaigns – are invariably followed by periods that appear to be more decentralized. The features of this process of decentralization, encompassing the occasional loss of compliance on the part of mobile pastoralists, cannot be ignored in any model of culture change proposed to account for changes in the monumental landscape of ancient Mongolia.

The three chapters present challenging models of cultural interaction and change whose application can be extended to other mobile pastoralist populations of ancient Eurasia. Moving beyond a traditional focus on the contents of ritual and funerary structures, they incorporate a wide range of data and approaches that are already having an impact on methodologies and interpretations. Thus, along with a larger number of radiocarbon dates and a greater commitment to regional studies, there is now an increasing reliance on the systematic collection of settlement, spatial, and ethnographic data. The incorporation of paleoenvironmental data, of which a broader variety may be expected in the near future, may reveal some of the forces at play in the emergence, transformation, and dissipation of Mongolia's monumental landscape during the second and first millennia BCE. As a possible scenario, climatic amelioration may have resulted in more productive pastures, the intensification of pastoralist (and other?) activities, demographic "packing," and reduced mobility, all of which are in keeping with the greater labor demands and socio-political implications of monumental building and use at the local level.

Defining "egalitarianism" as a condition characterized by individual autonomy and an inability (by anyone) to control a society's wealth (as opposed to an absence of leadership or wealth differences), Salzman (2004: 29) relies on his cross-cultural analysis of past and present pastoralist societies to affirm that "acephalous, egalitarian, decentralized, nomadic [pastoralist] tribes are more likely to be found in remote regions far from centers of power, population, and trade." On the basis of what is presently known of the period before the appearance of slab burials, we would therefore conclude from Salzman's statement that those early pastoralist societies associated with the building of khirig-suurs were highly decentralized and acephalous. Possibly, as has been suggested, ritual played an important integrative role. If so, the appearance of inequality among Mongolia's khirigsuur societies may represent a parallel path to social complexity for which no recent examples exist (among pastoralists). Alternatively, local intensification and reduced mobility, if indeed this can be shown, may indicate the presence of social groups whose workings are more akin to those of settled and physically bounded communities. Putting these competing scenarios to the test represents an exciting avenue of research.

References

Chapman, R. 2003. *Archaeologies of Complexity*. London: Routledge.
Salzman, P. 2004. *Pastoralists: Equality, Hierarchy, and the State*. Boulder, CO: Westview Press.

CHAPTER 18

Re-writing Monumental Landscapes as
Inner Asian Political Process

WILLIAM HONEYCHURCH, JOSHUA WRIGHT,
AND CHUNAG AMARTUVSHIN

THE POTENTIAL for monumental structures to convert, transform, and communicate has been explored extensively in the archaeological literature (Trigger 1990; Sherratt 1990; Bradley 1998) and, most recently, with regard to social memory and its potential for re-invention through monumental re-use (Williams 2003; Van Dyke and Alcock 2003). The re-use of monumental sites in the eastern steppe region of Mongolia, Inner Mongolia, southern Siberia, Xinjiang, and eastern Kazakhstan (referred to here as Inner Asia), very likely dates back more than 3,000 years and persists in certain forms even today. While creating spatial and stylistic associations with former monuments of grandeur is a common method for bolstering political legitimacy in complex societies, especially in states and empires (Sinopoli 2003), understanding the role such practices play in the formation of initial socio-political complexity is a more subtle task.

In this chapter, we examine the remains of monumental activities in a northern Mongolian river valley over a period of 1,000 years during the first and second millennia BCE. Our study focuses on the question of what stone mounds and prominent burials might tell us about the character of political change in this local area and, by extension, across the eastern steppe. In the long process of building lineage-based institutions of inequality, local valley groups utilized monuments from their own past while simultaneously creating new monumental forms and associated activities. How might such practices of monumentalism and re-use have been different before the existence of "rulers and ruled" and how could these behaviors have contributed to the emergence and perpetuation of hereditary leadership?

We propose answers to these questions in the form of a preliminary model describing the transition to enduring inequality on the eastern steppe as an outcome of participation in rapidly expanding extra-local networks of contact and the manipulation of local relationships via monuments and mortuary activities. We argue that unequal access to network-based systems of value emerged rapidly among local lineage groups not because of the efforts of a few ambitious individuals but, quite unintentionally, owing to the intrinsic dynamics of expanding relational networks. Once the advantage of an exclusive position relative to other local groups was recognized, lineage members acted to secure, bolster, and institutionalize their new social position, in part through collective activities drawing on both old and new forms of monumentality.

A Sequence of Steppe Monuments

MORE THAN 3,000 years ago, steppe peoples constructed the first large mounds of stone, known today as *khirigsuur*, along low hill slopes, at major mountain passes, and in the broad valleys of the Inner Asian steppe lands. How these monumental stone piles and the collective labor they embodied created aspects of community or links between communities is the subject of debate and recent fieldwork. These massive stone markers were succeeded by monumental burials, ornate standing stone stelae, and impressive panels of rock art, all part of the cultural expression of the Late Bronze and Early Iron Age period (mid-second to late first millennium BCE). We study two types of monuments from this period, khirigsuur mounds and slab burials. The transition between these two monumental types was associated with dramatic changes in local social relations, technologies, lifeways, and transformation in the broader socio-political setting (Erdenebaatar 2002; Volkov 1967).

Khirigsuur monuments (Fig. 18.1) consist of a central stone mound with an internal cist, often surrounded by a circular or rectangular enclosure of stone and having small, circular rock heaps as satellites. When excavated, satellites are sometimes found to cover a shallow pit containing horse crania and vertebrae, occasionally accompanied by ceramics, beads, or bronze items (e.g., Erdenebaatar 2002: 211–213). Mean sizes for khirigsuur monuments, including outlying features, vary from region to region though most extend across a space of 25–40 meters. Some individual monuments, however, are substantially larger (Erdenebaatar 2004). Khirigsuurs occur as isolated features and in groups forming small and large complexes, sometimes numbering as

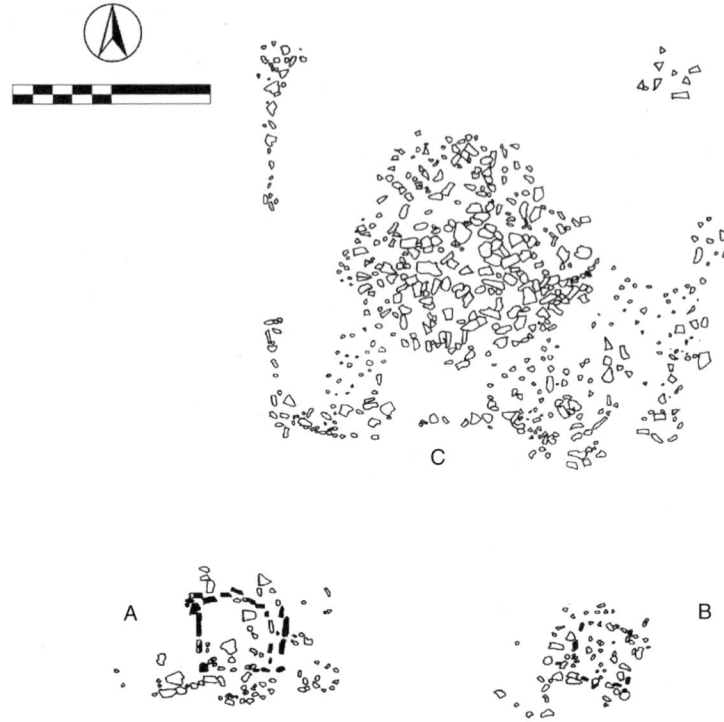

Figure 18.1. Site plan of a slab burial (A, B) and khirigsuur (C) complex from the Egiin Gol Valley (site EGS 153).

many as 50 monuments. Most researchers consider khirigsuurs to be ceremonial sites with evidence for mortuary activities in 40% of those so far excavated (Tsybiktarov 2003: 83). However, the most recent excavations in western Mongolia have consistently discovered human interments (Frohlich et al. 2008) suggesting geographic variability in the use of these monuments (Honeychurch and Amartuvshin 2009). Publications on khirigsuur chronology argue for construction periods dating from the mid-second to early first millennium BCE on the basis of stratigraphic and artifactual evidence (Tsybiktarov 1998: 141). Recent radiocarbon dates on human bone, horse teeth, and charcoal samples excavated from khirigsuur contexts from three regions across northern Mongolia provide support for this chronology (Frohlich et al 2008; Frohlich 2006; Allard and Erdenebaatar 2005; Torbat et al. 2003: 136).

Slab burials (Fig. 18.1) consist of large, upright stone slabs arranged in a rectangular formation, usually oriented east-west, and covering a burial pit up to 1.5 meters deep (Navaan 1975). The lengths of these burials are usually in the range of 1.5 to 3 meters, though some are in excess of 10 meters long with slabs standing up to 2 meters above the surface. These mortuary features appear individually or in small to medium-sized groups of 3 to 20 monuments. In addition to evidence for human interment, slab burials contain animal remains (horse bones being among the most common), and a wide range of artifacts including bronze, stone, ceramic, and bone items, which are usually found in disrupted contexts (Dikov 1958: 57). Slab burial assemblages have been stylistically dated from the terminal second to the mid-first millennium BCE, and this proposed range has been confirmed by several recently published radiocarbon dates (Tsybiktarov 1998: 103–104; Honeychurch 2004: 111–112).[1] Our model contextualizes these monuments, both socially and politically. We follow a brief theory section with expectations for an analysis of khirigsuur and slab burial distributions in the lower Egiin Gol Valley of northern Mongolia.

Social Change and Local Asymmetries

LOCAL SOCIAL groups or "communities" participate in broader external communities for biological and social maintenance. Relationships to neighboring groups, and those farther afield, support marriage arrangements, resource access and sharing, exchange opportunities, alliances during conflict, conflict resolution, and rights of movement across the landscape. Methods for creating and managing external relationships involved gift giving and exchange, ceremonial participation, feasting, fictive kinship, and perhaps gatherings and collective monument building, among others. During much of the Inner Asian Bronze Age, these relationships were limited to a region of direct social interaction and supplemented by contacts mediated by secondary and tertiary partners. By the end of the Bronze Age, the functional reasons for establishing external relationships beyond the local community remained unchanged. However, the means for creating those networks of contacts transformed radically and began to meet local needs at expanded scales, with farther reach, and having unforeseen consequences for both extra-regional and local interactions.

Our model places emphasis on local group perceptions, decision making, and opportunities within this rapidly expanding social information environment. Such an environment was brought about, in part, by the widespread adoption of horse riding across the central Mongolian steppe. Horse riding was probably an innovated or borrowed technique to assist and expand hunting or herding practices on a local basis (Allard and Erdenebaatar 2005), though it also made possible face-to-face interactions at much greater distances at higher frequencies than previously possible (Anthony 1998). Such compression of social space achieved through efficient transport produced higher levels of contact similar to what might be expected under higher population densities (Roscoe 1993). We view the impact of this technological change as creating a radically different social setting and an enlarged scale of social participation for local groups.

By virtue of the relatively rapid growth of relations over increasingly large areas and involving more groups, the social information environment became both more complex and more important to track. Where decision making and relationship maintenance became contingent on information about distant groups, social information itself became a valuable and manipulable commodity. Groups that did not participate actively in this information environment invited neighboring groups to distribute and control information about them relative to distant partners. For these reasons, both the benefits of participation and the costs of not participating provided a strong incentive for the growth of complex horizontal social networks across the central Mongolian steppe. These social dynamics have been studied in more detail on the central Eurasia steppe where horse riding probably arose at an early period (Anthony 1986: 303; Sherratt 1981). Horse riding as a novel "technology" likely facilitated social inequality through the creation of new forms of horse-based wealth; larger-scale and more reliable pastoral production; new trade and acquisition possibilities; and, in some regions, increased raiding and warfare (cf. Hamalainen 2003).

In conjunction with these factors, we argue that social inequality also emerged from differential access to and participation in the expansion of external relationships. Both recent network theory and ethnohistoric research have described the conjunction of rapidly expanding social networks and the emergence of "nodal" actors or sub-groups. A social network is a set of relationships, with past interactive histories, that enable or inhibit transactions of a wide variety on the basis of a conjunction of needs, settings, and opportunities at a given time (Emirbayer and Goodwin 1994: 1444–1445). In this sense, a network is more akin to a "field of potential" or a "probability space" rather than a formalized system of concrete exchanges. Social networks grow by way of individual and group decision making within the context of existing sets of relationships and the cultural and historical frameworks that give rise to them (Emirbayer and Goodwin 1994: 1438). In pursuing alliances, people or households already engaged in viable contacts with distant partners are often attractive to those parties seeking to establish new relationships. Choices about the formation of new relationships on the part of actors, therefore, have potential to favor other actors embedded in prior network relationships. This preferential attachment to existing "nodes" in an expanding network is an example of the relational logic by which "the rich get richer," otherwise known as scale-free network growth (e.g., Sindbaek 2007: 119–120; Bentley 2003; Barabasi and Albert 1999).

A similar dynamic lies behind the ability of some groups or individuals to work within expanding networks better than others by virtue of an existing investment in the broader social information environment. Dillehay (1990: 237–238) and Wiessner (2002: 244) both describe settings in Chile and Papua New Guinea, respectively, where positions of emerging or increasing inequality were enhanced by the rapid growth of social networks. These networks are such that only a few groups are positioned relationally to be capable of arranging extra-local ceremonial events because of the complexity of scheduling, deploying information, and coordinating between participants. These groups and individuals are characterized by Wiessner (2002: 252) as having "a strong information advantage, and ... much greater 'wealth in social ties' than others."

The concept of "wealth" in the form of a position embedded in an expansive network of relationships also suggests a way in which such wealth could have been transferred across generations. Unlike the accumulation of subsistence surplus or prestige goods or the control of a novel technology or important resources, a set of formalized relationships is created around and dependent upon a specific individual or group. Relationships are a more readily heritable form of wealth than material goods because lineal relatives can become involved in network activities as partners, assistants, or stand-ins and gain direct knowledge of and introductions to distant parties. As examples, transfers of long-distance relationships from father to son were evident in Tee cycle networks of the Enga (Wiessner 2002: 247), in trading partners among Andean herders (Flores 1979), and among later Inner Asian steppe elite whose children replaced their parents in existing alliance relationships (Cleaves 1982: 33).

We propose that by the late second or initial first millennium BCE, early adopters of riding technology in several regions of the Mongolian steppe had accrued unequal access to partners and knowledge of external networks and their function. By this time, some of these early positions may have also been transferred within an extended family unit on the basis of heredity. These sub-groups increasingly exercised their influence in negotiating the external affairs of their local communities, which brought them prestige and social indebtedness but also the costs of external reciprocation. How did these asymmetries in external ties and associated benefits and costs lead to qualitatively different interactions within the local community? We outline ways that activity sets surrounding local monument construction provided opportunities for transforming "wealth in social ties" into institutionalized forms of inequality.

"Re-writing" Monumental Sites, Identity, and Inequality

OUR IDEAS for monumental activities in north central Mongolia draw upon a conceptual framework for inequality developed in the sociological research of Charles Tilly. Tilly approaches inequality from a relational and interactive perspective (see Emirbayer 1997) and offers a helpful set of concepts with which to begin. He writes that "inequality emerges from asymmetrical social interactions in which advantages accumulate on one side or the other, fortified by construction of social categories that justify and sustain unequal advantage" (Tilly 2001: 362). While social categories denote grouping and distinction, these categories and the boundaries that define them do not necessarily imply inequality, only difference. However, when multiple transactions occur across such a boundary in which benefits unevenly accrue on one side and the parties obtaining those benefits re-invest some portion toward emphasizing the boundary itself, a durable form of interactive inequality comes into play. Such social categories can emerge from acts of social negotiation or imposition or by transfer from another setting (Tilly 2003: 34).

The process described by Tilly results in the formation of exclusive and asymmetric social categories that have been elsewhere referred to as "elite identities" (Schortman et al. 2001). Where monuments are a feature of local landscapes and traditions, monumental activities can become a salient part of this multi-sided process, specifically in the context of monument re-use. Monument building activities are often ceremonial, social and community-based, interactive, and traditional, and therefore they combine both a community-wide and a behavioral context in which new interactions might be embedded within older traditions. As studies of early complex societies show, the embedding of novel relationships within existing or older institutions supports the acceptance of new social roles. Monumental site re-use is just such a combination of continuity and discontinuity, which is often at the crux of social transformation (Wiessner 2002: 250; Johansen et al. 2004: 52–53; Porter 2002: 27–28; Dillehay 1990: 235–238).

Collective enactment of new monuments at an older monumental site presents an opportunity to "inscribe" a revised set of activities into the understanding of that setting on behalf of those promoting an event (Barrett 1990). These events draw upon the significance of the monumental site and the memory of its associated activities, while modifying specific actions, the criteria for performing certain roles, the nature of materials involved, and modes of participation. We propose that such monumental "re-writing" was one means for the social negotiation and acceptance of exclusive sub-group identities within the broader community (Hayden 2001: 262). Those sub-groups that acted as extra-local intermediaries and sought to reinforce and protect the benefits they accrued from networking with distant partners sponsored mortuary-based re-use events. Whether intended or not, the behavioral effect of these events was to sustain a factional identity by emphasizing intra-group distinctions through altering the balance of direct and indirect participation in ceremonial activities (e.g., Barrett 1990: 182; Sherratt 1997: 360); by incorporating significant use and display of non-local goods, styles, and forms as central to re-use ceremonies (Schortman et al. 2001: 314); and by gathering a dispersed local community for an activity in which attendance and participation acknowledges the social position of sponsors relative to participants (Rappaport 1979: 193–194).

Evaluating the Model in Northern Mongolia

WE PROPOSE an explicitly social and contextualized account of the rise of inequality among early steppe peoples and one process by which some groups capitalized on that inequality in order to institutionalize its supporting practices. Using the regional database available from a pedestrian survey of the lower Egiin Gol Valley, north central Mongolia (Honeychurch 2004; Wright 2006; Fig. 18.2), we develop a set of expectations to evaluate these ideas. In order to demonstrate a pattern of monument re-use, we expect to see clear spatial and structural relationships

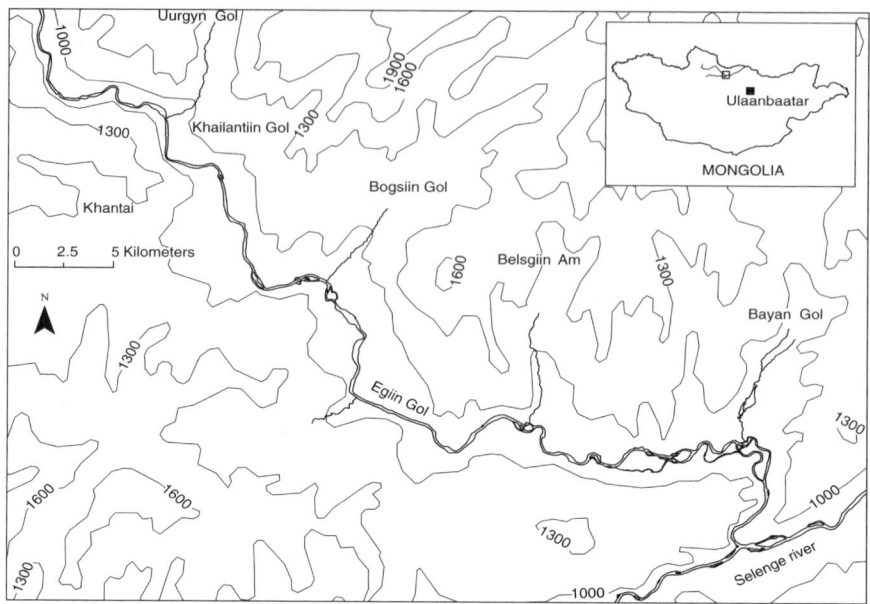

Figure 18.2. The Egiin Gol survey area and tributary valleys in north central Mongolia.

between monumental types. In particular, we expect to see a conjunction of two lines of evidence, the first involving relationships between slab burial placement and khirigsuur sites and the second being coincident changes in monumental practices and local lifeways.

In the first case, we suggest that slab burial events re-inscribed khirigsuurs as a way of facilitating new social relationships. This argues for a very precise association between the two site types at a socio-spatial scale that would have been most effective for such collective activities – at the scale of the site itself. In contrast to these continuities at the micro-scale, we expect to see discontinuities at socio-spatial scales above that of the site. Higher socio-spatial scales should relate less to the set of choices structuring relationships within the valley and more to the extra-local dynamics of the region beyond. We have hypothesized increases in the social scale of external interaction and increasing volume and complexity of the social information guiding that interaction. These changes influence how a group of people organize both themselves in relation to external groups and the landscape in response to novel functional and infrastructural priorities. We therefore do not expect evidence for a close "mapping" of slab burials onto khirigsuur distributions at these greater scales of practice.

The second set of evidence involves transformation in practices associated with the emergence of slab burial monuments. In contrast to older khirigsuur activities, we expect slab burial ceremonies to encourage subgroup exclusivity in the context of community-wide participation and to emphasize the use of non-local materials and ideas. Because changes in transport technology based on horseback riding are critical to our understanding of this period, the first evidence for equestrian activities should appear at Egiin Gol along with or slightly before the importation of slab burial monuments. We expect the practices associated with the new monuments to have left evidence for external network involvement, patterns of social differentiation, and some indication of lineal transfers of status.

Scales of Continuity and Discontinuity

THE MICRO-GEOGRAPHY OF MONUMENTS

Our first question for the Egiin Gol dataset is to what degree are slab burial monuments spatially associated with khirigsuur sites. We calculated a simple test measuring the distance between each slab burial (n = 86) and the first nearest khirigsuur neighbor (N1 mean distance = 196 meters). Over the same region, we generated random points equal to the number of khirigsuur features (n = 383) and repeated the nearest-neighbor measurement between slab burial sites and the randomly distributed points (N1 mean distance = 503 meters) (Wheatley and Gillings 2002: 136–139). We used the Wilcoxon rank sums test to determine the probability that these two point distributions could have been drawn from the same random population, and our results demonstrate that there is very little chance of that being the case (p < 0.0001). Therefore, slab burials are non-randomly distributed with respect to khirigsuur sites, and the majority of slab burials (90%) are in fact less than 500 meters from a khirigsuur. Furthermore, if we calculate the mean distance over which slab burials are situated from five nearest khirigsuur neighbors, that distance is a mere 620 meters.

This result provides evidence for slab burials having been intentionally located very close to khirigsuurs over much of the valley. Additional evidence from other sites includes intentional placement of slab burial–style ceramic vessels on top of khirigsuur mounds (Takahama 2005) and the use of khirigsuur enclosure stones as structural supports for slab burials (Tsybiktarov 1995; Batsaikhan 1996). How, then, were khirig-suur locations selected that were suitable for the construction of slab burials at Egiin Gol? According to the five-nearest-neighbors analysis, khirigsuur sites that display a degree of spatial clustering were favored, as were other monumental structural elements. We tested size correlations between the two monument types in order to discover whether larger khirigsuurs "attracted" the builders of larger slab burials to invest their activities in those areas. By creating a buffer of 200 meters around each slab burial and selecting from within that area the khirigsuur with the largest diameter to represent that locale, we tested the relationship between slab burial length and maximum khirigsuur diameter by calculating Spearman's rho. A moderate strength relationship exists ($r_s = 0.44$, $p = 0.0009$),[2] which suggests that larger monumental sites were re-used by groups emphasizing monumental size as an important variable, perhaps for collective display purposes.

THE SUB-REGIONAL GEOGRAPHY OF MONUMENTS

These tests demonstrate a conscious structuring of slab burial ceremonies with reference to existing sites and with attention paid to creating a degree of structural parity between older and newer monuments. Continuity with khirigsuurs, in this sense, was pursued by the builders of slab burial monuments at a spatial scale appropriate for interaction, performance, and participation. When we move to larger scales of organization, however, the spatial association of the two monuments changes in important ways. At what might be called the sub-regional level, consisting of the entire lower valley, there is a consolidation of slab burial investment in tributary valleys to the east (Bayan Gol, Belsegiin Gol, and Bogsiin Gol) and a virtual abandonment of the western portion of the khirigsuur distribution (Fig. 18.3; Table 18.1).

If we compare size calculations based on diameter and volume of khirigsuurs across the eastern and western areas of the valley, it is clear that those areas having slab burials are also those where khirigsuurs are more numerous and where average khirigsuur size declines (Table 18.1). Greater numbers of features in the eastern side valleys and structural differences between khirigsuurs in areas with and without slab burials may indicate a time when slab burials and khirigsuur activities overlapped, as the absolute chronology suggests. Allard and Erdenebaatar (2005) associate smaller khirigsuurs with later periods of construction at the Khanuy Gol site, south of Egiin Gol. This observation suggests that the western portions of Egiin Gol were intentionally neglected in terms

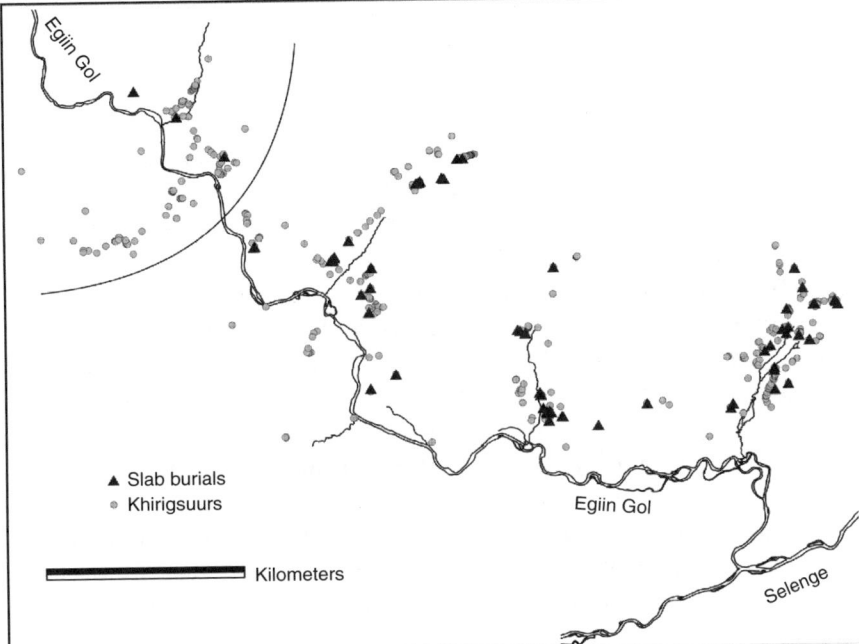

Figure 18.3. Egiin Gol Valley site distributions for khirigsuur and slab burial monuments. The arc segment marks the drop-off area for slab burial construction.

Table 18.1. Comparison of monument distributions in eastern and western Egiin Gol side valleys

Side valley area[a]	Khirigsuur count	Slab burial count	Average khirigsuur volume (cu. m)
Bayan Gol	112	26	92.5
Belsegiin Am	65	19	50.7
Bogsiin Gol	81	29	82.5
Khailantyn Gol	33	1	217.7
Uurgyn Am	31	1	104.5
Khantai	41	0	127.2

Note: Some sites cannot be clearly assigned to a side valley and so have not been included.
[a] Listed east to west in descending order.

of later monumental investment, despite the presence there of several khirigsuur monuments that are among the largest in the valley.

We favor an explanation that connects this sharply discontinuous pattern not with abandonment of the far western valley but instead with emphasis placed on the central and eastern valley specifically because they comprise the main regions of travel between two sections of the Selenge River to the south. Travel along the north side of the Selenge River basin is much easier if the lower Egiin Gol is used as a corridor. The flow of movement through the lower Egiin Gol exploits one pass in the upper Bayan Gol Valley to the east and another pass in the broad Khantai Valley accessible by way of a river ford still in use today. To test this idea, our survey collected ethnographic and ethnohistoric data on passes and routes used in the valley. We corroborated this information by way of both archaeological sites that mark passes and roadways, especially nineteenth-century Buddhist-inscribed stelae, and by using cost-distance analysis to determine efficient routes of movement according to topography (Wheatley and Gillings 2002: 151–158). By overlaying this model of Egiin Gol pathways with the locations of slab burial monuments, a fairly clear correspondence between the point and line distributions becomes evident (Fig. 18.4). If our hypothesis for the slab burial distribution following pathways of movement is correct, then it would seem that part of the longer-term process of monumental re-use involved initial control over places of significance on the landscape (khirigsuur sites), followed by a transformation in the nature of locational significance itself to reflect instead the importance of mobility and external routes of access.

Transforming Lifeways

CHANGES IN MONUMENTAL PRACTICES

The monumental sites we examine were not so much "built" as they were "enacted." They are the outcome of collective behaviors, often performance and ceremony oriented, which served to structure relationships between individuals and groups through an assignment of roles and prescribed interactions (Renfrew 1973: 555; Sherratt 1997: 353; Hanks 2002: 192). The re-use of prominent khirigsuur complexes by slab burial builders was a purposeful and structured association to invoke social memory of khirigsuurs in the production of such novel ceremonial relationships. A comparison of the construction and use of these two monuments provides evidence for changes that encouraged

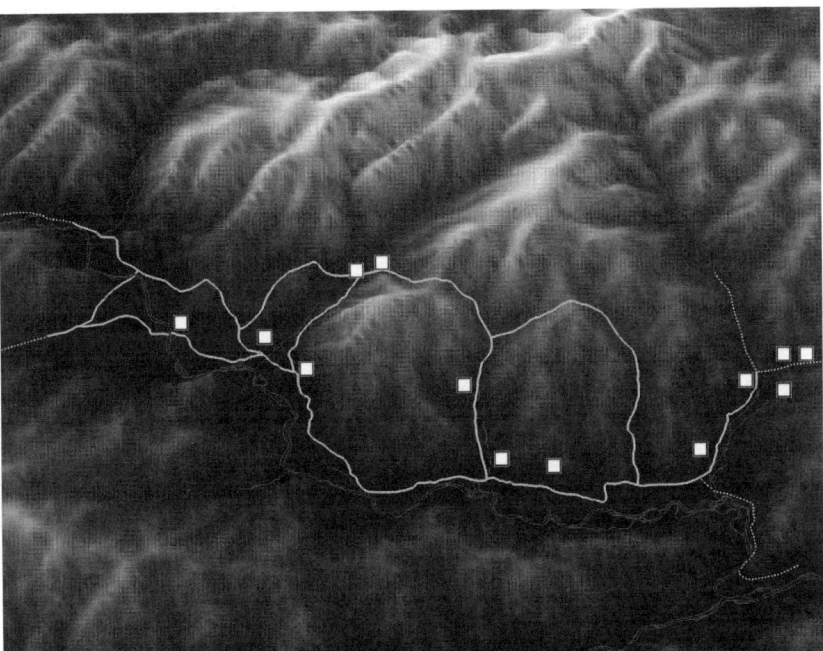

Figure 18.4. A three-dimensional rendering of Egiin Gol topography and probable pathways of movement. Major clusters of slab burial sites and large monuments are indicated by squares, whereas areas of valley entrance and exit are marked by broken lines.

sub-group exclusivity in the context of community involvement (see Table 18.2). Specifically, participation regimes were transformed by the use of a smaller feature requiring less labor, with fewer activity roles, and having decreased accessibility for community members. The smaller monumental structure of slab burials delimited the number of participants directly involved in creating the tomb, promoting a contrast to the larger-scale involvement at khirigsuur building events. Tomb evidence including large, soot-encrusted cooking pots and a range of large and small domestic animals represented by skull and leg portions suggests that ritual slaughter and possible feasting on meat may have occurred at slab burial events. If correct, food preparation and feasting may have increasingly become the main activity engaged in by otherwise "inactive participants" or "onlookers" removed from the center of performance.

Spatially, the new ceremony transfixed community focus on the central point of the burial feature and its subterranean chamber into

Table 18.2. Differences in the construction and use of Late Bronze and Early Iron Age monuments

Construction/ use category	Khirigsuurs (ca. 1400–800 BCE)	Slab burials (ca. 1100–400/300 BCE)	Interpretation
Attendance	Labor requirements suggest community-scale group involvement	Evidence for multi-horse consumption, cooking pots, and location among expansive field of khirigsuurs suggest large-scale group involvement	Continued emphasis on gathering local area group together (i.e., a dispersed local community)
Labor	Significant labor investment	Moderate labor investment	Reduction in the number of active individuals
Participation	Direct involvement in most or all construction activities on the basis of labor requirements	Direct participation limited by reduced size of monument and reduced construction activities	Direct participation reduced while indirect participation is encouraged
Activity spaces	Three distinct spatial areas: central mound, enclosed sector, and external sector with satellite features and khirigsuurs	Two distinct spatial areas: immediate burial feature, and expansive khirigsuur field around burial feature	Ceremonial space is compressed into two categories defined by proximity to a central point instead of distributed over a broader field

Construction/ use category	Khirigsuurs (ca. 1400–800 BCE)	Slab burials (ca. 1100–400/300 BCE)	Interpretation
Construction emphasis	Built primarily from the surface upward; occasional interments placed on soil surface or in shallow pits (0.15–0.20 meter)	Emphasis on interring beneath the surface and less so on building upward, mean depth of burial feature is 0.88 meter	Focus of attention changes from an open and widely visible edifice to a narrow context of restricted visibility
Material deposition	Few materials deposited relative to feature size, low diversity of item types, and mostly of local origin (e.g., ceramics, horse heads)	Many materials deposited relative to feature size, high diversity of items, emphasis on non-local origins (e.g., bronze and white bronze items, turquoise, cowries, helmets, ceramic tripods, "animal style" decorative repertoire)	Display and expenditure of non-local items increases their perceived value, creates new criteria for staging group ceremonies, and distinguishes between those with access and those without access to such materials

Note: Date ranges are based on radiocarbon results from contexts in the Egiin Gol, Bulgan province, Khanui Gol, Arkhangai province, and Delgermoron-Uushigiin Ovor, Khovsgol province (Frohlich et al. 2008; Frohlich 2006; Allard and Erdenebaatar 2005; Torbat et al. 2003: 136).

which items of non-local wealth were added. Some of these items bore the motifs of an imported "animal style" artistic tradition, while the burial form itself would have been understood as a structure introduced from outside. The act of creating and using a non-local monument form, interment of non-local goods, and the exclusionary allotment of participation are aspects of using "wealth in social ties" to fortify local categories of social distinction. The emphasis was upon "in-groups" and "out-groups" distinguished explicitly by those who had comprehensive external ties and those who did not. Most interesting, however, is that so much of the slab burial performance drew upon the older and, for some time, contemporary khirigsuur ceremony, including the use of prominent locations, stone monumentalism, horse meat feasting, large group attendance, and manipulations of ceremonial space.

Among these, the long tradition of drawing together segments of a dispersed local community to participate in a collective event was perhaps the deciding factor in staging slab burial activities at khirigsuur sites. The chronological overlap of these two monuments, indicated by more than 20 radiocarbon dates from three northern Mongolian valleys (see Table 18.2), suggests that new khirigsuurs may have been created simultaneously with slab burials. In this case, the main draw for the community may have been the construction of a khirigsuur, which, when combined with the modified form of slab burial construction, created community participation and public acknowledgment of slab burial proceedings. Attendance on the part of individuals at such an event constituted an affirmation of its significance and a willingness to enter into a particular public relationship with those organizing and managing the event. In terms of Rappaport's (1979: 179) framework for understanding public ritual, participation validates the "indexical" social information inherent to the proceedings of the ceremony. The new slab burial venue provided an opportunity to re-invest the khirigsuur construction ceremony with exclusionary implications and to affirm emerging sub-group identities that were to become a more permanent form of local elite.

HORSE RIDING

Central to changes in the broader social environment, in systems of social value, and in monumental practices was the availability of horses and riding technologies. It is ironic that in a place such as Mongolia where horse culture imbues so many aspects of life, archaeologists still understand very little about the advent of horse use and riding. As in many parts of the steppe, riding has been associated with specialized pastoral nomadism and the spread of Scythian cultures at the beginning of the first millennium BCE (Gryaznov 1980; Rudenko 1970; Askarov et al. 1992). We agree with Renfrew that it is quite difficult to understand why such a lengthy process led up to horse riding in Central Eurasia when many of the preliminary pieces of the puzzle seem to have been in place at a fairly early stage (Renfrew 2002). The use of domesticated horses on the Mongolian steppe is still not evidenced by archaeological data before the mid-second millennium BCE. The khirigsuur satellite contexts in which those earliest horse remains are found require more study to determine whether they were indeed domestic horses and how they might have been used (Allard and Erdenebaatar 2005; Honeychurch and Amartuvshin 2006: 259). Most archaeologists are comfortable placing horseback riding at the beginning of the first millennium BCE on the basis of a range of evidence from saddlery items to stylized images of riders (Erdenebaatar 2002; Bokovenko 2000; Dikov 1958: 59).

Egiin Gol is an important case study for this issue because teams of Mongolian, French, and American archaeologists worked in the valley over a 10-year period conducting intensive excavation and survey. The best evidence available from Egiin Gol points to the initial first millennium BCE as the time that horse riding was adopted. Evidence for horse-riding techniques in association with the earliest slab burials is so far the best indication of the development of "steppe horse culture" in the valley and in other parts of Mongolia and southern Siberia (Dikov 1958: 59; Tsybiktarov 1998: 148).[3] Evidence for this process is of various kinds. For example, more than half of the excavated slab burial contexts at Egiin Gol having faunal remains contained skeletal parts of equids (Torbat et al. 2003: 43). At least one slab burial context, dated to 940–800 BCE. (2ó calibrated date range), contained multiple horse crania, cheekpieces, and possible bridle decorations (Torbat et al. 2003: 31–32, 48).

In addition, this particular slab burial (EGS-7j, Mukhdagiin Am 1) contained an individual male, age 15–18, who had a skeletal anomaly potentially associated with extensive horseback riding, known as Poirer's facet (an anterior extension of the articular surface of the femoral head onto the upper portion of the femoral neck) (Nelson and Naran 1999: 6–7; Palfi and Dutour 1996; Miller 1992).[4] Petroglyphs depicting horse riders and stylistically dated to the early first millennium BCE decorate rock outcrops of the tributary valleys at Egiin Gol and are spatially associated with slab burial features nearby (Torbat et al. 2003: 41, 172). Finally, the spacing of large concentrations of slab burial monuments at a regional scale employing distances of 60 kilometers or more along pathways of movement (i.e., not by linear distance) possibly suggests dispersal beyond the relative compactness of regional khirigsuur sites because of changes in transport technology.

One point is clear from the Egiin Gol record. Though warfare and horse riding are commonly associated in socio-political theory on steppe cultures (Tsybiktarov 2003: 87; Erdenebaatar 2004; Chang and Tourtellotte 1998), very little evidence has been produced by the extensive work at Egiin Gol to suggest any conflict during the first millennium BCE. The skeletal record from this period in the valley shows no indication of conflict-related trauma (Nelson and Naran 1999, 2000). Likewise, very few artifacts have been recovered that are unquestionably implements of warfare as opposed to hunting or prestige items. Clear evidence for warfare is seen in other regions of Mongolia during the first millennium BCE. For example, burials of the Chandman-Ulangom culture, excavated in the western Mongolian Altai Mountains, provide clear evidence for weapon sets (war hammers) and the skull injuries inflicted by their use (Tseveendorj 1980; Novgorodova 1982). Egiin Gol evidence so far argues for a different local setting, one in which horse-based mobility was introduced and increased in the context of extra-local contact and network building instead of warfare. This is not to say that competition or conflict was not occurring but that these social activities were not being expressed through acts of collective violence at Egiin Gol. Instead, they may have been mediated by the effective building of inter-personal and inter-group alliances.

EXTRA-LOCAL RELATIONSHIPS

Evidence from slab burial contexts for extra-local contact is increasing with the help of compositional studies on metallic and ceramic artifacts (Honeychurch 2004; Hall et al. 1999; Hall et al., n.d.). The clearest example of the presence of items procured over very long distance are the well-known bronze helmets discovered in the valley from two slab burial contexts and stylistically dated to the early first millennium BCE. These artifacts most likely originated in eastern Inner Mongolia (Erdenebaatar 2002: 72–73; Erdenebaatar and Khudiakov 2000). Another slab burial contained fragments of a "white" bronze mirror that, on the basis of both manufacturing technique and materials, probably originated in Inner Mongolia or China. White bronze is created through a process of slow cooling, weak acid leaching, and the polishing of a bronze alloy consisting of copper, tin, and lead such that the lead and tin segregate and provide a silver sheen. This technique was long established in Chinese mirror crafting traditions of the first millennium BCE (M. Hall personal communication).

A lead isotope study of bronze artifacts from Egiin Gol slab burial contexts using ICP-MS techniques has shown that the copper used to make those artifacts (n = 5, from three burials and one diagnostic surface find) was not from the nearest regional copper source known to have been exploited during the Bronze Age (Erdenet source, ca. 85 kilometers from Egiin Gol) (Hall et al., n.d.). This conclusion is preliminary because more copper samples are needed from the Erdenet source in order to determine the extent of variability of the Pb-isotope signature at that location. Finally, the presence in six slab burials of beads made of turquoise also suggests imported materials or artifacts because turquoise is not present locally and is considered to be a long-distance exchange item in slab burial contexts (Okladnikov 1959: 49; Dubin 1987: 155). In fact, the burgeoning presence of artifacts and materials from outside the Egiin Gol valley, at this early time, is quite impressive (see Table 18.2) and hints not only at participation in relationships further afield but at the importance of constructing local relationships through the incorporation of such material in burials, which themselves were a non-local form.

MORTUARY DIFFERENTIATION AND HEREDITARY STATUS

The Egiin Gol mortuary record for most periods is problematic with regard to recovery of information. Slab burial excavations at Egiin Gol confirm the claims of widespread pillaging of these mortuary contexts reported in slab burial research elsewhere in the region. Additional disruption from rodent activity at Egiin Gol is evident from mortuary contexts for many periods. Careful measurements of burial surface size, however, are more resistant to the effects of disruption, especially because the stone surface features of these burials tend to be formidable and, in most cases, left intact despite various kinds of disturbance. In addition to the problem of disturbed contexts, another problem is small sample size. Of the 28 slab burials excavated by the various international projects at Egiin Gol, 14 yielded skeletal assemblages, and of those, only 7 were subsequently analyzed for age, sex, stature, and skeletal pathology (Nelson and Naran 1999, 2000). Given these limitations of the dataset, we pose two basic questions for the slab burial record at Egiin Gol. First, because slab burial size was considered in the process of khirigsuur site re-use and was therefore considered significant, is there evidence for distinctive size categories of burial suggestive of greater or lesser investment in burial activities, and perhaps linked to status? Second, do mortuary variables, potentially linked to status differences, distribute across age categories in a way suggestive of hereditary status (Peebles and Kus 1977: 431)?

To address the first question, a distribution of the length measurements for all slab burial features for which reliable data could be obtained (n = 78 out of 86 total, 91%) suggests the presence of two major groups and a pair of far outliers in the 10-meter range (Fig. 18.5). The size range of the smaller group is from 2.1–4.0 meters, while the group comprising larger burials ranges from 4.4–7.3 meters. These differences in construction size were likely due to the status or influence of the interred individual and his or her lineage group, and the extent of activities and numbers of people directly involved in monument construction and provisioning. The relevance of size as an indicator of prominence and external relationships can be assessed by examining the three burials mentioned with artifacts provenienced to Inner Mongolia or China. As might be expected, these are among the top 10% of burials in size, when outliers are excluded (6.2, 6.5, and 7.0 meters length). Furthermore, given the total number of slab burials recorded (n = 86) and their chronological duration of 600 years at Egiin Gol, it is clear that one or more exclusive groups of people had the capability to provide their dead with this form of mortuary treatment. Most of the local population did not receive mortuary monuments, and we have not discovered any evidence for their treatment at death.

We proposed earlier that the kind of group pursuing opportunities to create such distinction was organized around lineages that practiced the

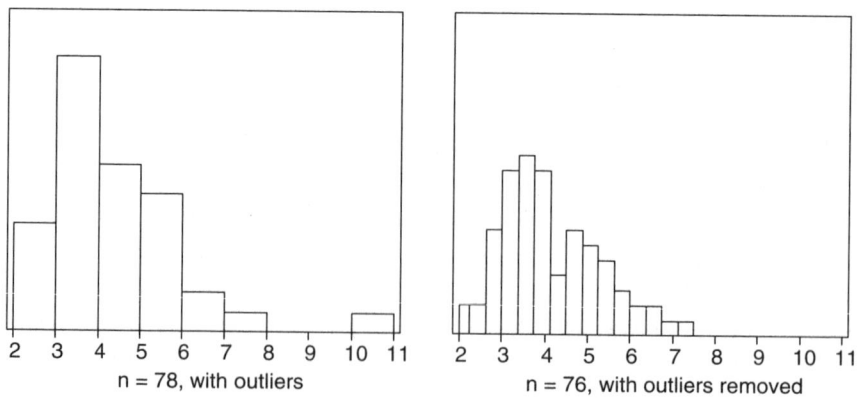

n = 78, with outliers   n = 76, with outliers removed

Figure 18.5. Egiin Gol slab burial length measurements in meters.

transfer of external alliance relationships between prominent deceased members and designated heirs. If we are correct, membership in such a lineage itself should be indicative of local prominence. One commonly cited indication of such hereditary-based status is the differential mortuary treatment given to potential heirs who, at the time of death, were too young to have achieved prestige of their own making. These sub-adults were prestigious because of their parentage and, for that reason, may have been the center of mortuary ceremonies greater in scale than many mature adult interments. Because we have argued that burial size is related to status differentiation, we provide a preliminary assessment of our proposition by comparing age at the time of death to size of slab burial construction (Fig. 18.6). There is indeed a preliminary pattern indicated among Egiin Gol slab burials in which sub-adults were provided with larger monumental structures and perhaps more elaborate mortuary contexts than some older individuals.

This is, of course, a minute mortuary sample accounting for a long period of time, and we agree that one or two large sub-adult burials do not necessarily signify membership in a society with hereditary inequality. While not terribly robust in isolation, these mortuary patterns become more meaningful when evaluated in conjunction with other

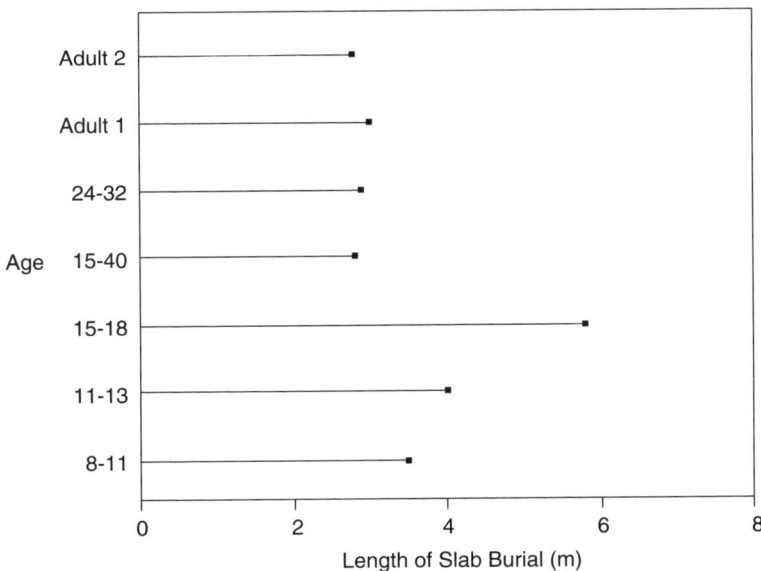

Figure 18.6. A comparison of slab burial size and the approximate age of the interred individual.

lines of evidence developed in the preceding discussion. A combination of evidence, including differential treatment of sub-adults, the monumental aspect of slab burials and their distinct size categories, the spatial and structural relationships observed with older monumental features, and the presence in some of the largest burials of very long-distance goods, better supports the argument for the emergence of hereditary inequality in the Egiin Gol valley during the first millennium BCE.

Conclusion

WE HAVE outlined here a detailed model for the emergence of social inequality among steppe nomadic pastoral groups. The changes we hypothesize, on the basis of the Egiin Gol record, were certainly part of a complicated array of transformations, including a variety of pathways for asymmetries to arise and persist. Our emphasis is upon inequality resulting from a change in transport technology and the creation of a novel external social environment in which information and relationships became forms of wealth. Those local lineages, having established effective external contacts, were in a position to accrue prestige and prominence locally. The re-use of monumental sites and modification of traditional monument construction ceremonies were ways of translating the privilege of external relationships into a local elite identity. Evidence from the Egiin Gol Valley for this argument includes spatial and structural associations between older and younger monuments; transformations in the activities surrounding monument construction, favoring greater exclusivity; contemporaneous indicators for initial horse riding; introduction to the valley of a suite of non-local materials, styles, and forms; and the beginning of local patterns of social differentiation and hereditary status.

Monuments and the tradition of activities that produced them were a promising context by which to negotiate new social positions and relationships. Most importantly, monument-building events were capable of bringing a community together in a structured and interactive setting. The re-use of older monumental sites to create new monuments with more exclusive proceedings set a precedent for intra-community divisions that were sanctioned by broad attendance and participation. Further, these asymmetric social divisions were reified through materialization in monumental form and embedded in social memory as the basis for a modified "tradition." These factors, we argue, are some of the relevant indicators that monuments, their activity sets, and processes of re-use were fundamental instruments involved in structuring and institutionalizing asymmetric relationships within steppe communities of the Late Bronze Age. This method of monumental re-use and co-option was so successful that it became one technique used to bring the expansive Xiongnu confederation to power across the steppe region hundreds of years later. The re-use of monumental sites was subsequently employed in Mongolia by leadership factions up to and including the nineteenth century, and even today this practice is still a venue for small-scale household ritual activities.

Notes

1. Tsybiktarov (1998) published some of the first radiocarbon analyses from systematically excavated slab burial contexts in Siberia. While our periodization is close to his, we take into account the large error ranges for dates obtained on some Siberian samples and have selected a more conservative start date for these features.
2. A Moran's I test on correlation residuals was used to confirm spatial autocorrelation in these observations. The resulting Spearman's rank coefficient was determined using a technique to adjust effective sample size accordingly (Dutilleul 1993; Dale and Fortin 2002: 164–166).
3. As of 2008, two unpublished studies on human skeletal pathologies possibly related to horse riding may push the date for riding evidence into the mid- to late second millennium BCE (Nelson et al. 2007: 42–47; Frohlich et al. 2008: 103). However, see the next note for the complexities of this effort.
4. Recent critical studies have shown that Poirer's facet is found among riding and non-riding populations. It may have potential as a suggestive indicator when taken as part of a suite of expected muscle- and bone-related pathologies, especially in a younger individual. However, significant comparative work is required on the problem of pathologies and horse riding, which is complex and inspires notable controversy among physical anthropologists (Garofalo 2004: 53, 70–73).

References

Allard, Francis, and Diimaajav Erdenebaatar. 2005. Khirigsuurs, Ritual, and Mobility in the Bronze Age of Mongolia. *Antiquity* 79(305): 547–563.

Anthony, D. 1986. The "Kurgan Culture." Indo-European Origins, and the Domestication of the Horse: A Reconsideration. *Current Anthropology* 27(4): 291–313.

——— 1998. The Opening of the Eurasian Steppe at 2000 BCE. In V. Mair (ed.), *The Bronze Age and Early Iron Age Peoples of Eastern Central Asia*. Philadelphia: University of Pennsylvania Museum Publications, pp. 94–113.

Askarov, A., V. Volkov, and N. Ser-Odjav. 1992. Pastoral and Nomadic Tribes at the Beginning of the First Millennium B.C. In A. H. Dani and V. M. Masson, (eds.), *History of Civilization of Central Asia: The Dawn of Civilization to 700 B.C.* Paris: Unesco, pp. 459–468.

Barabasi, A., and R. Albert. 1999. Emergence of Scaling in Random Networks. *Science* 286: 509–512.

Barrett, J. 1990. The Monumentality of Death: The Character of Early Bronze Age Mortuary Mounds in Southern Britain. *World Archaeology* 22(2): 179–189.

Batsaikhan, Z. 1996. *Bulgan aimagt ajillasan Selengiin ekspeditsiin tailan* (Report on Work in Bulgan Province by the Selenge Expedition). Archaeological field report. Ulaanbaatar: Mongolian Institute of Archaeology.

Bentley, A. 2003. An Introduction to Complex Systems. In R. A. Bentley and H. D. G. Maschner (eds.), *Complex Systems and Archaeology: Empirical and Theoretical Applications.* Foundations of Archaeological Inquiry. Salt Lake City: University of Utah Press, pp. 9–24.

Bokovenko, N. A. 2000. The Origins of Horse Riding and the Development of Ancient Central Asian Nomadic Riding Harnesses. In J. Davis-Kimball (ed.), *Kurgans, Ritual Sites, and Settlements: Eurasian Bronze and Iron Age.* Oxford: Archaeopress, pp. 304–310.

Bradley, R. 1998. *The Significance of Monuments: On the Shaping of Human Experience in Neolithic and Bronze Age Europe.* London: Routledge.

Chang, C., and P. Tourtellotte. 1998. The Role of Agro-Pastoralism in the Evolution of Steppe Culture in the Semirecheye Area of Southern Kazakhstan during the Saka/Wusun Period (600 BCE–400 CE). In V. H. Mair (ed.), *The Bronze Age and Early Iron Age Peoples of Eastern Central Asia*, vol. 1. Washington, DC: Institute for the Study of Man, pp. 264–279.

Cleaves, F. 1982. *The Secret History of the Mongols.* Cambridge, MA: Harvard University Press.

Dale, M., and M.-J. Fortin. 2002. Spatial Autocorrelation and Statistical Tests in Ecology. *Ecoscience* 9(2): 162–167.

Dikov, N. N. 1958. *Bronzovyi vek Zabaikal'ia.* Ulan-Ude: Siberian Division of the Academy of Sciences.

Dillehay, T. 1990. Mapuche Ceremonial Landscape, Social Recruitment and Resource Rights. *World Archaeology* 22(2): 223–241.

Dubin, L. 1987. *The History of Beads from 30,000 BC to the Present.* London: Thames and Hudson.

Dutilleul, P. 1993. Modifying the t Test for Assessing the Correlation between Two Spatial Processes. *Biometrics* 49: 305–314.

Emirbayer, M. 1997. Manifesto for a Relational Sociology. *American Journal of Sociology* 103(2): 281–317.

Emirbayer, M., and J. Goodwin. 1994. Network Analysis, Culture, and the Problem of Agency. *American Journal of Sociology* 99(6): 1411–1454.

Erdenebaatar, D. 2002. *Mongol nutgiin dorvoljin bulsh, khirigsuuriin soel.* Ulaanbaatar: Academy of Sciences.

———. 2004. Burial Materials Related to the History of the Bronze Age on the Territory of Mongolia. In K. M. Linduff (ed.), *Metallurgy in Ancient Eastern Eurasia from the Urals to the Yellow River.* Lewiston: Edwin Mellen Press, pp. 189–221.

Erdenebaatar, D., and Iu. S. Khudiakov. 2000. Nakhodki bronzovykh shlemov v plitochnykh mogilakh severnoi Mongolii. *Rossiiskaia Arkheologiia*, no. 2: 140–148.

Flores, J. 1979. *Pastoralists of the Andes*. Philadelphia: Institute for the Study of Human Issues.

Frohlich, B. 2006. Burial Mound Survey in Khovsgol Aimag, Mongolia. In W. Fitzhugh (ed.), *Mongolian Deer Stone Project Field Report, 2005*. Washington, DC: Arctic Studies Center, Smithsonian Institution, pp. 67–82.

Frohlich, B., Ts. Amgalantugs, and D. Hunt. 2008. Bronze Age Burial Mound Excavation in the Khovsgol Aimag, Northern Mongolia. In W. Fitzhugh and J. Bayarsaikhan (eds.), *The American-Mongolian Deer Stone Project: Field Report for 2007*. Washington, DC: Arctic Studies Center, Smithsonian Institution, pp. 101–103.

Garofalo, E. 2004. The Osteological Markers of Horseback Riding: An Examination of Two Medieval English Populations. M.Sc. dissertation, University of Bradford, UK.

Gryaznov, M. 1980. *Arzhan: tsarskii kurgan ranneskifskogo vremeni*. Leningrad: Nauka.

Hall, M., W. Honeychurch, J. Wright, Z. Batsaikhan, and L. Bilegt. 1999. Chemical Analysis of Prehistoric Mongolian Pottery. *Arctic Anthropology* 36(1–2): 133–150.

Hall, M., J. Yoshinaga, W. Honeychurch, Ch. Amartuvshin, U. Erdenebat, D. Erdenebaatar, and M. Yoneda. N.d. Lead Isotope Analyses of Bronzes from Northern Mongolia. Manuscript in preparation.

Hamalainen, P. 2003. The Rise and Fall of Plains Indian Horse Cultures. *Journal of American History* 90(3): 833–862.

Hanks, B. 2002. The Eurasian Steppe "Nomadic World" of the First Millennium BC: Inherent Problems within the Study of Iron Age Nomadic Groups. In K. Boyle, C. Renfrew, and M. Levine (eds.), *Ancient Interactions: East and West in Eurasia*. Cambridge: McDonald Institute, pp. 183–197.

Hayden, B. 2001. Richman, Poorman, Beggarman, Chief: The Dynamics of Social Inequality. In G. Feinman and T. Price (eds.), *Archaeology at the Millennium: A Sourcebook*. New York: Kluwer Academic/Plenum Publishers, pp. 231–272.

Honeychurch, W. 2004. Inner Asian Warriors and Khans: A Regional Spatial Analysis of Nomadic Political Organization and Interaction. Ph.D. dissertation, University of Michigan, Ann Arbor.

Honeychurch, W., and Ch. Amartuvshin. 2006. States on Horseback: The Rise of Inner Asian Confederations and Empires. In Miriam Stark (ed.), *Archaeology of Asia*. Malden, MA: Blackwell, pp. 255–278.

——— 2009. Timescapes from the Past: An Archaeogeography of Mongolia. In P. Sabloff and F. Hiebert (eds.), *Mapping Mongolia*. Philadelphia: University of Pennsylvania Museum Publications.

Johansen, K., S. Laursen, and M. Holst. 2004. Spatial Patterns of Social Organization in the Early Bronze Age of South Scandinavia. *Anthropological Archaeology* 23: 33–55.

Levine, M. 2004. Exploring the Criteria for Early Horse Domestication. In M. Jones (ed.), *Traces of Ancestry: Studies in Honour of Colin Renfrew*. Cambridge: McDonald Institute, pp. 115–126.

Miller, E. 1992. The Effects of Horseback Riding on the Human Skeleton. Paper presented at the meeting of the Palaeopathological Association, Las Vegas, March 31.

Navaan, D. 1975. *Dornod Mongolyn khurliin ue* (The Bronze Age of Eastern Mongolia). Ulaanbaatar: Academy of Sciences.

Nelson, A. R., M. Machicek, and J. Beach. 2007. *Joint Mongolian-American Expedition BGC 2007: Human Remains.* Ulaanbaatar: Academy of Sciences Institute of Archaeology.

Nelson, A. R., and B. Naran. 1999. *Bioarchaeology of the Joint Mongolian-American Egiin Gol Expedition Burial Sample: 1994–1999 Field Seasons. Physical Anthropology Analysis Report.* Ulaanbaatar: Mongolian Institute of History.

    2000. *Bioarchaeology of the Joint Mongolian-American Egiin Gol Expedition Burial Sample: 2000 Field Season. Physical Anthropology Analysis Report.* Ulaanbaatar: Mongolian Institute of History.

Novgorodova, E. 1982. *Ulangom: ein skythenzeitliches Graberfeld in der Mongolei.* Wiesbaden: Harrassowitz.

Okladnikov, A. P. 1959. *Ancient Populations of Siberia and Its Culture.* Russian Translation Series of the Peabody Museum of Archaeology and Ethnology 1:1. Cambridge, MA.

Palfi, G., and O. Dutour. 1996. Activity-Indiced Skeletal Markers in Historical Anthropological Material. *International Journal of Anthropology* 11(1): 41–55.

Peebles, C., and S. Kus. 1977. Some Archaeological Correlates of Ranked Societies. *American Antiquity* 42: 421–448.

Porter, A. 2002. The Dynamics of Death: Ancestors, Pastoralism, and the Origins of a Third-Millennium City in Syria. *Bulletin of the American School of Oriental Research* 325: 1–36.

Rappaport, R. 1979. *Ecology, Meaning, and Religion.* Richmond, CA: North Atlantic Books.

Renfrew, C. 1973. Monuments, Mobilization and Social Organization in Neolithic Wessex. In C. Renfrew (ed.), *The Explanation of Culture Change: Models in Prehistory.* London: Duckworth, pp. 539–558.

    2002. Pastoralism and Interaction: Some Introductory Questions. In K. Boyle, C. Renfrew, and M. Levine (eds.), *Ancient Interactions: East and West in Eurasia.* Cambridge: McDonald Institute, pp. 1–12.

Roscoe, P. 1993. Practice and Political Centralization: A New Approach to Political Evolution. *Current Anthropology* 34(2): 111–140.

Rudenko, S. I. 1970. *Frozen Tombs of Siberia: The Pazyryk Burials of Iron Age Horsemen.* Berkeley: University of California Press.

Schortman, E., P. Urban, and M. Ausec. 2001. Politics with Style: Identity Formation in Prehispanic Southeastern Mesoamerica. *American Anthropologist* 103(2): 312–330.

Sherratt, A. 1981. Plough and Pastoralism: Aspects of the Secondary Products Revolution. In I. Hodder, G. Isaac, and N. Hammond (eds.), *Pattern of the Past: Studies in Honour of David Clarke.* Cambridge: Cambridge University Press, pp. 261–305.

    1990. The Genesis of Megaliths: Monumentality, Ethnicity, and Social Complexity in Neolithic Northwest Europe. *World Archaeology* 22(2): 147–167.

1997. *Economy and Society in Prehistoric Europe: Changing Perspectives.* Edinburgh: Edinburgh University Press.

Sindbaek, S. 2007. Networks and Nodal Points: The Emergence of Towns in Early Viking Age Scandinavia. *Antiquity* 81: 119–132.

Sinopoli, C. 2003. Echoes of Empire: Vijayanagara and Historical Memory, Vijayanagara as Historical Memory. In R. Van Dyke and S. Alcock (eds.), *Archaeologies of Memory.* Malden, MA: Blackwell, pp. 17–33.

Takahama, Shu. 2005. Preliminary Report on Archaeological Investigations in Mongolia, 2004. Permanent Archaeological Joint Mongolian and Japanese Mission. *Newsletter on Steppe Archaeology* 15: 1–18.

Tilly, C. 2001. Relational Origins of Inequality. *Anthropological Theory* 1(3): 355–372.

2003. Changing Forms of Inequality. *Sociological Theory* 21(1): 31–36.

Torbat, T., C. Amartuvshin, and U. Erdenebat. 2003. *Egiin Golyn sav nutag dakh' arkheologiin dursgaluud* (Archaeological Monuments of Egiin Gol Valley). Ulaanbaatar: Mongolian Institute of Archaeology.

Trigger, B. 1990. Monumental Architecture: A Thermodynamic Explanation of Symbolic Behaviour. *World Archaeology* 22(2): 119–132.

Tseveendorj, D. 1980. *Chandmany Soel* (Chandman Culture). Arkheologiin sudlal Archaeological Research, 9. Ulaanbaatar: Academy of Sciences.

Tsybiktarov, A. D. 1995. Khereksury Buriatii, Severnoi i Tsentral'noi Mongolii (Khirigsuurs of Buriatiia and Northern and Central Mongolia). In P. B. Konovalov (ed.), *Kul'tury i pamiatniki bronzovogo i rannego zheleznogo vekov Zabaikal'ia i Mongolii.* Ulan-Ude: Nauka, pp. 38–47.

1998. *Kul'tura plitochnykh mogil Mongolii i Zabaikal'ia.* Ulan-Ude: Nauka.

Tsybiktarov, A. D. 2003. Central Asia in the Bronze and Early Iron Ages: Problems of Ethno-cultural History of Mongolia and the Southern Trans-Baikal Region in the Middle 2nd–Early 1st Millennia BC. *Archaeology, Ethnology and Anthropology of Eurasia* 13(1): 80–97.

Van Dyke, R., and S. Alcock. 2003. *Archaeologies of Memory.* Malden, MA: Blackwell.

Volkov, V. V. 1967. *Bronzovyi i rannii zheleznyi vek severnoi Mongolii.* Ulaanbaatar: Academy of Sciences.

Wheatley, D., and M. Gillings. 2002. *Spatial Technology and Archaeology: The Archaeological Applications of GIS.* London: Taylor & Francis.

Wiessner, Polly. 2002. The Vines of Complexity: Egalitarian Structures and the Institutionalization of Inequality among the Enga. *Current Anthropology* 43(2): 233–269.

Williams, H. 2003. *Archaeologies of Remembrance: Death and Memory in Past Societies.* New York: Kluwer Academic/Plenum Publishers.

Wright, J. 2006. The Adoption of Pastoralism in Northeast Asia, Monumental Transformations in the Egiin Gol Valley, Mongolia. Ph.D. dissertation, Harvard University.

CHAPTER 19

Socially Integrative Facilities and the Emergence of Societal Complexity on the Mongolian Steppe

Jean-Luc Houle

Nomadic Polities: The Problem

THE TENDENCY for pastoral groups to exploit marginal environments through high mobility and spatially extensive economies, resulting in very low population densities and unstable surplus production, has led many scholars to argue that nomadic pastoralism is conducive neither to political centralization nor to the emergence of institutionalized social hierarchy without regular interaction with already-existing agricultural state-level societies. This is often referred to as the "dependency" hypothesis (Sahlins 1968; Lattimore 1992; Khazanov 1978, 1994; Burnham 1979; Irons 1979; Krader 1979; Jagchid and Symons 1989; Barfield 1981, 1989; Kradin 2002). Accordingly, without such interaction with sedentary societies, pastoralists are expected to form at most "egalitarian" polities (Irons 1979; Burnham 1979; Salzman 1999, 2000, 2004).

The Mongolian case, however, is particularly perplexing in this regard, because impressive Late Bronze Age (mid-second to mid-first millennium BCE) ritual and funerary monuments suggest the development of early complex societal structures that exhibit some sort of formalized social differentiation at a time before regular interaction with large sedentary states in China existed. Significantly, while Mongolia is commonly considered as a "peripheral" area in early steppe socio-political dynamics, some of these monuments surpass in aboveground elaborateness anything else of this nature in the Bronze Age steppe. Furthermore, their appearance at the end of the second millennium

BCE is highly significant in that they precede the first large-scale Iron Age mortuary sites of Arzhan I and II in Tuva (ninth–eighth century BCE) and other so-called Scythian period royal burials in the Eurasian steppe. Nevertheless, there is currently very little other preserved material evidence, such as grave goods, that correlates specifically with social status or political authority. This poses the crucial question of how a complex society, apparently pastoralist, mobile, and dispersed to some degree, was organized and maintained in relative stability for several centuries. Although the purpose of this chapter is not to resolve this conundrum, it will serve to outline the spatial organization and spatial patterns of Mongolia's Late Bronze Age monuments at the local, sub-regional, and regional levels in order to examine the significance of socially integrative events such as mound building in the emergence and development of societal complexity among mobile pastoralists of central Mongolia. Furthermore, while the investigation of middle-range societies has been effectively investigated through the comparative analysis of sedentary, agricultural-based societies around the world, such complexity surrounding "pastoralists" has rarely been considered within broader comparative studies of trajectories of social complexity. In this regard, the Mongolian case offers a seemingly unique and significant case study for potential autonomous social development and corporate complexity that does not connect easily with such models as the "dependency" one. Such a case study appears to challenge traditional approaches and interpretations of causal factors and events that stimulate such social change and initiate such developments along certain pathways or trajectories.

Monuments of Central Mongolia and Societal Complexity

ALTHOUGH QUESTIONS still surround the temporal and functional relationship linking the various Late Bronze Age monumental constructions that dot the Mongolian landscape, there is no doubt that they represent the emergence and development of a distinctive cultural trend that appears to reflect a transition in social, economic, and political organization. Indeed, the monumental public works and mortuary complexes of the Late Bronze Age, notably the impressive khirigsuurs, along with deer stone stelae, "slope" burials, and "slab" burials, suggest the development of societal and political complexity in central Mongolia as early as 1200 BCE (Allard and Erdenebaatar 2005; Erdenebaatar 2002; Fitzhugh, Chapter 20 in this volume). Attention has focused especially on

Figure 19.1. Two khirigsuurs in Khanuy Valley, central Mongolia.

the khirigsuurs, a Mongolian version of the kurgans known from farther west, and consisting of massive central mounds of stones, surrounded by square or circular "fences" of surface stones, and satellite features with complex deposits of remains of horses and other domesticated animals (Figs. 19.1 and 19.6) (Allard and Erdenebaatar 2005). The conventional interpretation of Eurasian kurgans (Khazanov 1975; Grach 1980), given by a number of scholars, has suggested that khirigsuurs reflect the social place of the deceased as a member of a hereditary elite (e.g., Volkov 1967; Tsybiktarov 1995, 1998, 2003; Erdenebaatar 2002). Very few khirigsuurs have been excavated, and most were previously looted, but they usually contained central single human inhumations of both adults and sub-adults (Erdenebaatar 2002; Tsybiktarov 1998), suggesting hereditary ranking.

Slope burials, which are small graves without prominent tumuli or animal ritual deposits that occur in cemetery groupings along hill slopes, are taken to represent lower-ranking members of society (Fig. 19.2). This less monumental Late Bronze Age burial custom, unfortunately, is rarely considered in discussions concerning the social organization of Bronze Age Mongolia or is conflated into analyses of khirigsuurs proper because of their similar structure (cf. Houle forthcoming; also

Figure 19.2. Slope burials.

see Frohlich and Bazarsad 2005). A regional-scale roadside survey, however, suggests that these burials are often located in proximity to khirigsuur structures, while preliminary archaeological work in the Khanuy Valley suggests that they are built in direct association with contemporaneous settlements, probably winter campsites, if compared to local ethnographic patterns. As such, they thus suggest "household" or encampment burials. This two-tier burial tradition implies that social distinctions, at least in death, were reflected in spatial organization (Fig. 19.3).

The social function of deer stones remains an enigma, but the variable belt styles, chevron motifs, and tool kits depicted seemingly refer to a warrior or chief (Dikov 1958; Volkov 1981; Jacobson 1993; Magail 2003; Erdenebaatar 2004). The imagery and its style of presentation also parallels tattooed shamanistic elements or components found in shamans' ritual clothing (Bayarsaikhan 2005; Volkov 1981; Savinov and Chlenova 1978), and, in traditional Mongolia, a clan chief was sometimes both political leader and shaman (Jagchid and Hyer 1979: 171). By the very Late Bronze Age, although there is some evidence of chronological overlap (Honeychurch 2004; Tsybiktarov 1998: 103), slab burials are accompanied by animal remains (horse bones in particular), cowries and

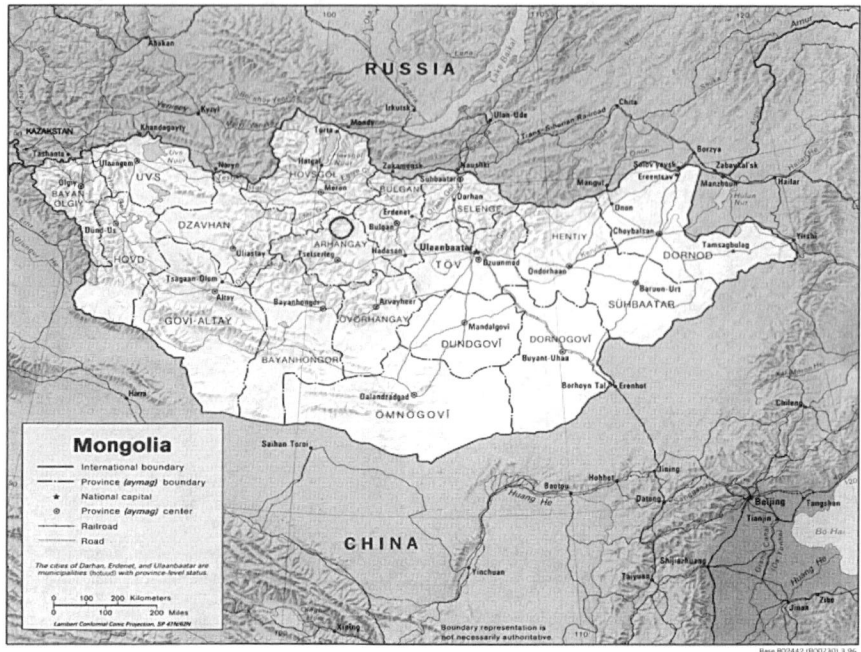

Figure 19.3. Research area showing location of khirigsuurs and slope burials.

mother-of-pearl (suggesting long-distance trade), bronze tools, hunting implements, weapons, helmets, ornaments, and horse-riding paraphernalia (Erdenebaatar 2002: 151–203, 239–252; 2004; Volkov 1995: 321; Ishjamts 1994: 151–152). Recent research in northern Mongolia has suggested that some sub-adults were provided with larger burials and more-elaborate offerings than some older individuals (Honeychurch 2004: 126), a further indication of hereditary ranking (Peebles and Kus 1977). Slab burials are frequently located in close proximity to khirigsuurs, often within the confines of these larger structures, thus suggesting either some type of connection to or co-option on the part of the peoples associated with these monuments. Khirigsuurs, while apparently emphasizing individuals (i.e., single inhumations), are nonetheless the only monuments to clearly exhibit important communal ritual activities.

Alternatively, it has been argued that khirigsuurs lack clear patterns of status differentiation when considering factors such as grave goods, spatial layout, and overall distribution (Allard 2006; Allard and Erdenebaatar 2005). Human remains are occasionally absent from the central cist and others seem to be cenotaphs (Erdenebaatar 2002; Takahama 2004; also see Ionesov 2002; Kroll 2000), which has led some to label them "ceremonial" rather than mortuary structures (Jacobson 1993; Honeychurch 2004; Honeychurch and Amartuvshin 2006; Wright 2006) and to see the societies that built them as acephalous segmentary groups of a corporate kind (Allard 2006).

This lack of consensus regarding the socio-political organization of Late Bronze Age societies of central Mongolia is in great part due to the lack of regional analysis (but see Honeychurch 2004; Honeychurch et al., Chapter 18 in this volume) and the fact that there are still few data on habitation sites and their regional distribution – all vital to discussions of socio-economic and socio-political systems among mobile herders (Casimir and Rao 1992; Irons 1979). Nevertheless, if it is viewed from a proxemic perspective, that is, how humans use and organize space at different scales (Hall 1966; Tringham 1973), the patterning of khirigsuurs and other Late Bronze Age monuments can help to *outline* the structure of the societal organization at the local, sub-regional, and regional levels. In other words, while we can hope that the debate regarding the Late Bronze Age socio-political organization will be resolved through more regional survey and more focused archaeological research that includes residential sites, the spatial arrangement of Late Bronze Age monuments may allow mapping of social groupings and spatial activity patterns that were emphasized at different scales during the Late Bronze Age of Central Mongolia. Preliminary results of an ongoing archaeological project in the Khanuy River valley (Arkhangai Aimag) will provide a case study for this discussion.

The Khanuy River Valley Research Area

THE KHANUY River valley, which is located to the north of the Khangai mountain range in Arkhangai Aimag (N48°05′/ E101°03′), is part of the extensive non-urbanized grasslands of present-day central Mongolia. It is a remote region far from the direct intersection with centers of power such as China (Fig. 19.4). In addition, the Khanuy Valley is located at the geographic meeting point of the three major forms of Late Bronze Age monument construction: khirigsuurs, deer stones, and slab burials (Novgorodova 1989: 256). Furthermore, although khirigsuur complexes and deer stones are found across a fairly large territory, they are mostly

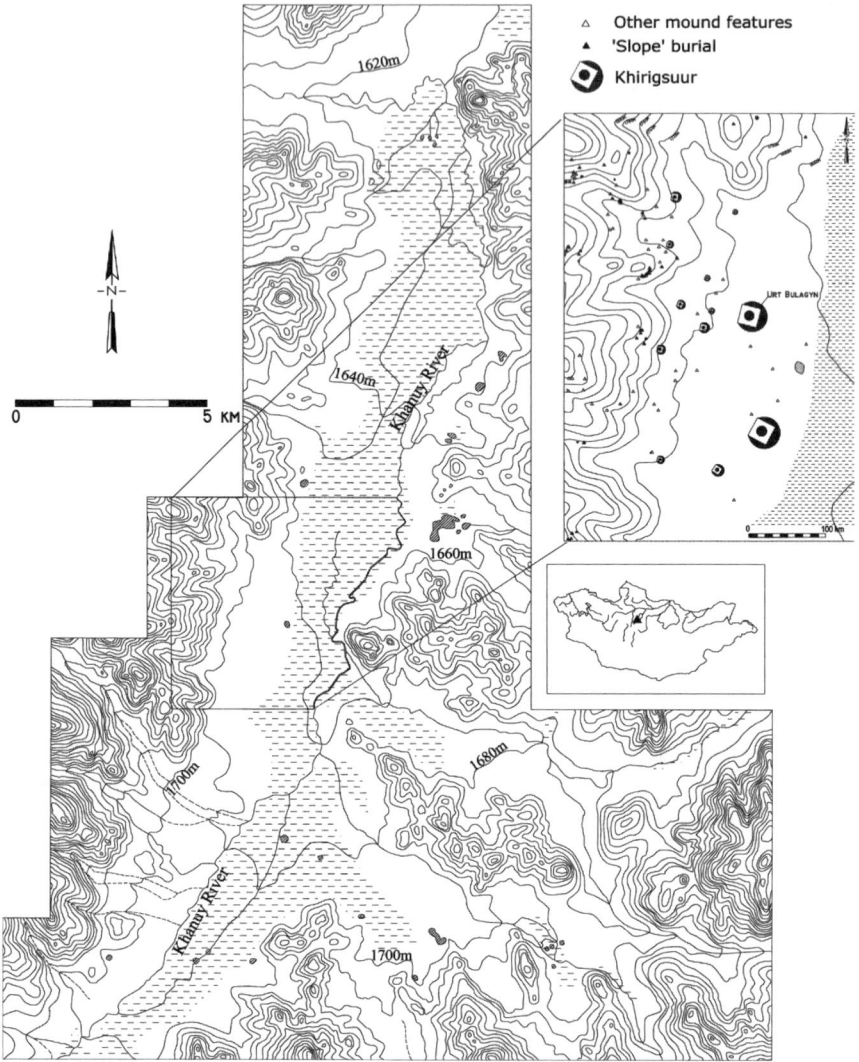

Figure 19.4. Location of the Khanuy River valley research area in Mongolia.

concentrated along major river valleys, such as the Khanuy Valley, located between the Khangai mountain range in central Mongolia and the regions of Gorno-Altai, Buryatia, and Tuva in southern Siberia (Volkov 1981: 123; Tseveendorj et al. 1999; Tsybiktarov 2003). This may not be a coincidence because there are many lines of evidence that suggest that the Late Bronze Age societies of central Mongolia may have had some type of connection with the contemporary and highly sophisticated metal-producing Karasuk culture of southern Siberia (Volkov 1967, 1995; Gryaznov 1969: 98; Askarov et al. 1992). Moreover, Mongolia may have acted as a pathway for the diffusion of Karasuk-type bronze artifacts toward China (Legrand 2004). Although this chapter does not propose any particular solution to the conundrum related to deer stones and social structure (see Fitzhugh, Chapter 20 in this volume), it is worth mentioning that their distribution conforms strongly to the distribution of khirigsuurs (Volkov 1981: 123).

On a local scale of analysis, even though khirigsuurs are distributed in a network-like pattern throughout these valleys, there are a number of areas that show particularly high densities of these monumental structures and suggest that they are places of higher centrality. In fact, a recent roadside survey in the Khanuy Valley region revealed that several large concentrations of these monuments are separated by "empty" or relatively vacant "buffer zones," thus emphasizing that these clusters may indeed indicate areas of greater spatial institutionalization of social organization or centrality (Fig. 19.5) (see Honeychurch 2004: 116–118 for a similar pattern in northern Mongolia). Significantly, the scale of many khirigsuurs, the construction of which must have involved organized activities of entire communities, and the elaborate seasonal ceremonial activities carried out at these complexes, including feasting (Houle et al. 2004), fit the archaeological description of central places – that is, the nexus of a larger web of social interaction. One of the two largest khirigsuurs (Urt Bulagyn [KYR1]) in the Khanuy River valley research area measures over 400 by 400 meters with a 5-meter-tall and 26-meter-diameter mound at its center and may consist of more than half a million stones (Fig. 19.6). Furthermore, while a number of "satellite" features containing animal deposits regularly accompany khirigsuurs, usually ranging from 12 to 40, and exceptionally as many as 150 (Erdenebaatar 2004), this khirigsuur has more than 1,700 small mounds. Each of these structures contains an east-facing horse skull and/or vertebrae or leg bones. In addition to these, more than 1,100 stone circles containing cremated animal bone fragments of various animal species are situated at the periphery (Allard et al. 2006; Allard and Erdenebaatar 2005).

Although these monuments may have been reused over time (Wright 2006), overlapping dates obtained from inner and outer satellite mounds at this khirigsuur (i.e., BP 2970–2780 and BP 2980–2770) (F. Allard personal communication) suggest a short building period. Furthermore,

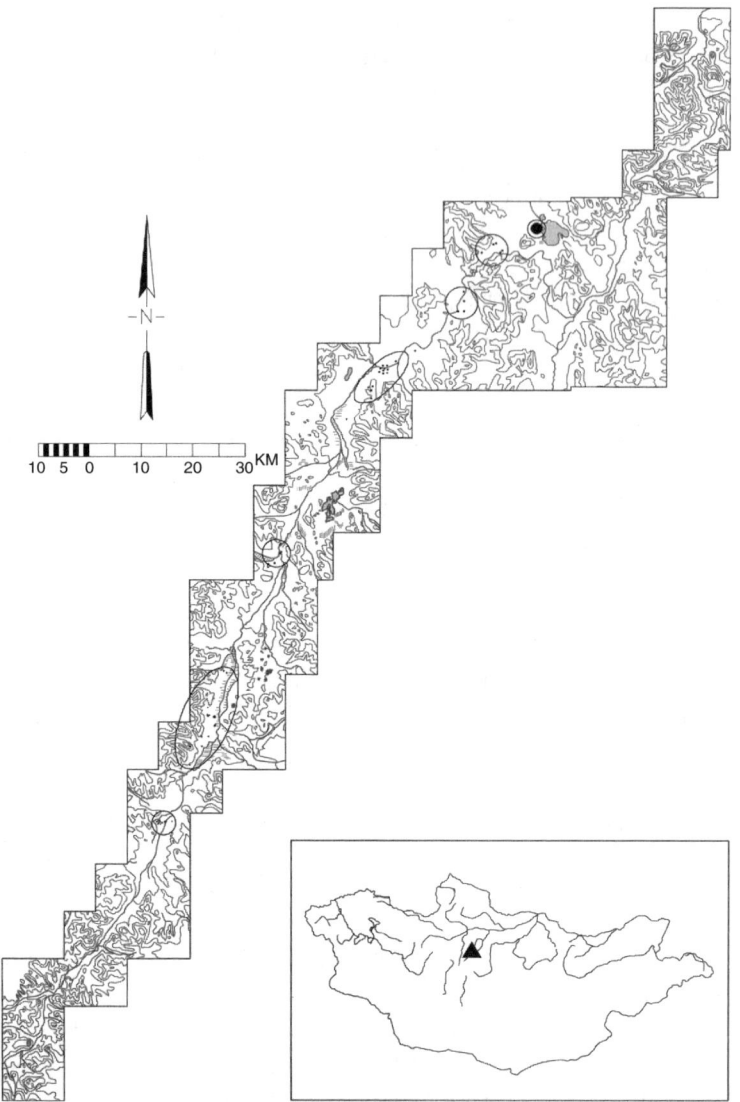

Figure 19.5. Khanuy River valley showing khirigsuur clusters and "buffer zones" (sites not to scale and not all represented in cluster areas).

two other samples obtained in the Khanuy River valley produced a date of 1390–910 cal. BCE for a small khirigsuur [KYR57], and a date of 930–785 cal. BCE for a larger one [KYR119] (Allard and Erdenebaatar 2005). Although many more dates are needed to test the hypothesis, these

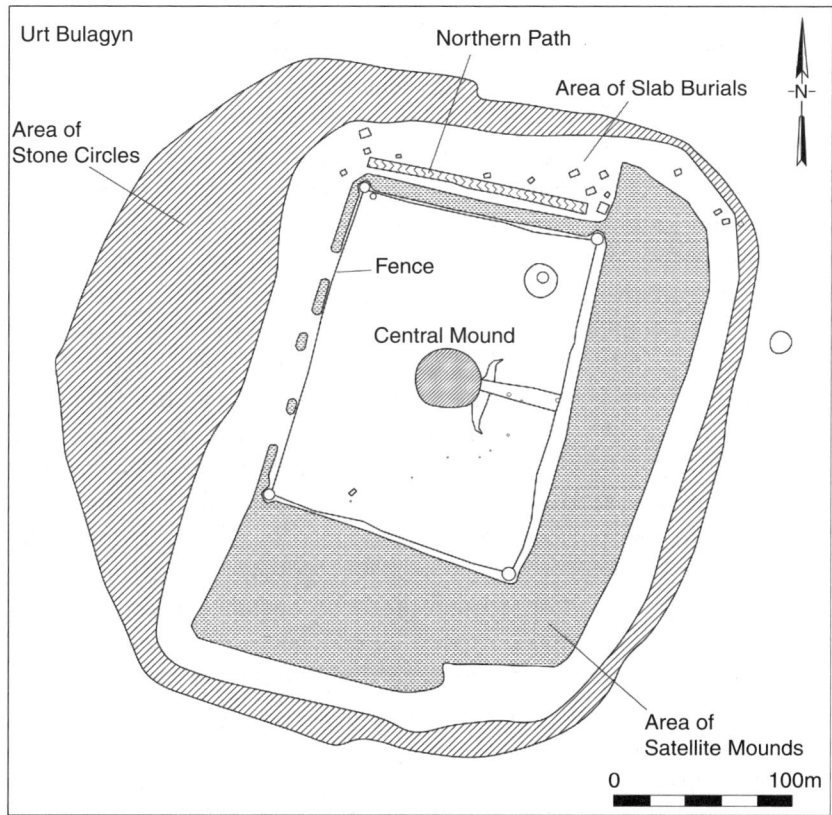

Figure 19.6. Khirigsuur Urt Bulagyn (Khanuy Valley) (from Allard and Erdenebaatar 2005).

preliminary results suggest that small khirigsuurs may predate larger ones. If this is true, then we may consider the idea that small khirigsuurs reflect relatively small corporate unit ritual or funerary activities, whereas larger khirigsuurs may represent the actions and achievement of an emerging elite that had succeeded in "integrating" larger groups into communal activities. The labor involved in the construction of the khirigsuur Urt Bulagyn, like many others in the research area, suggests the participation of numerous settlement units as its construction far exceeds the realistic contribution of an isolated social unit. The widespread regularity of ritual practice witnessed at all levels and space within and between these structures (see Allard and Erdenebaatar 2005), as well as the practice of depositing particular horse remains in a specific pattern for a period of more than 500 years, has suggested to some studying similar patterns for the contemporary metal-producing Karasuk culture (thirteenth–eighth century BCE) in southern Siberia that this could develop only in stock-rearing groups with a stable economic structure, who were relatively prosperous and who had advanced far beyond the relatively egalitarian groups that preceded them in terms of social development (Gryaznov 1969: 129; Legrand 2006).

Finally, with regard to metallurgy, diffusion may not have been the only source of bronze-production technologies. Test excavations at two distinct occupation areas in the Khanuy Valley revealed a small amount of slag and fragments of bronze-casting debris. Both these sites are located within a zone where there are numerous khirigsuur structures, including Urt Bulagyn. Interestingly, there is increasing evidence for independent metal production centers in Mongolia, including the central Khangai forest-steppe region, which embraces the Khanuy Valley research area (Erdenebaatar 2004: 218). Although research on the origins of bronze production in Mongolia is in its initial phase, metal goods found in Late Bronze Age slab burials suggest that metal-based craft industries may have developed in association with the evolution of mobile herding economies in this region (Erdenebaatar 2004). In any event, by the very Late Bronze Age, the groups that inhabited the central steppes of Mongolia had created complex economies based on metallurgy, livestock, and horses. Furthermore, while there is still no overt evidence of warfare and violence during this period, there is an increasing emphasis on the military nature of the burial of prominent people, an emphasis that is reinforced when one considers deer stones. The pictorial elements found on these stelae (knives, axes, daggers, swords, battle-picks), which are typologically similar to weapons found in slab burials, suggest the "representation of an armed individual and most likely one having some social distinction in local society" (Erdenebaatar 2004: 193).

A Proxemic Analysis

FROM A proxemic perspective (Hall 1966; Tringham 1973), and specifically through the study of the "built environment," what Edward T. Hall (1966: 103) called the "fixed-feature space" – that is, one of the basic ways of organizing the activities of individuals and groups – the spatial arrangement of khirigsuurs and other Late Bronze Age monuments within the Khanuy River valley offers us a way to evaluate social structures that may have been emphasized locally, sub-regionally, and regionally during the Late Bronze Age of Central Mongolia:

1. Although the khirigsuurs represent a pan-regional phenomenon with broad commonalities in "architectural vocabulary" across vast regions, they are mostly concentrated along major river valleys located between the Khangai mountain range in central Mongolia and the regions of Gorno-Altai, Buryatia, and particularly Tuva in southern Siberia (Volkov 1981: 123; Tseveendorj et al. 1999; Tsybiktarov 2003). This pattern suggests network-like structures that connect specific areas.

2. At the sub-regional level, though khirigsuurs are distributed throughout the Khanuy Valley, there are a number of distinct areas that show particularly high densities of these monumental structures, thus suggesting places of higher centrality. In many ways, these represent "central places" in that they reflect areas of spatial institutionalization of social organization.

3. Locally, once they are understood in their overall sub-regional context, khirigsuurs are almost always located in conjunction with the less monumental slope burials forming groupings that may reflect local societal structures. These groupings do not usually exceed 10–15 kilometers, a distance that is consistent with ethnographically recorded localized migratory circuits in the region. This in turn may have helped to reinforce social and political contacts within a defined territory. While a number of ecological-territorial zones can be distinguished in Mongolia on the basis of prevailing ecological conditions, natural geographic boundaries, and herding methods, ethnographic research by Bazargur (2005), Erdenebaatar (2000), and this author on mobility patterns in the Khangai range of central Mongolia has substantiated patterns of relatively localized seasonal migratory circuits that reflect a region of constant and high productivity. In fact, the Russian ethnographer Simukov, who carried out research in the 1930s on mobility patterns in Mongolia, identified a system of movement that he called "Khangai" in the region that includes the Khanuy Valley. He pointed out that owing to the constant and high productivity of the region, including the presence of different complementary types of pasture within a short distance, there was no need to make long migrations in response to drought (Simukov 1934). He estimated the diameter of the annual movement cycle in this region to be no more than 7–8

kilometers, a pattern still prevalent today (on this mobility pattern, see also Vainshtein 1980; Novgorodova 1989). Archaeologically, no aboveground structures related to habitation sites are visible in the Khanuy Valley, yet a high-resolution survey of two zones of 25 square kilometers revealed Late Bronze Age occupation areas and a settlement system that might be indicative of a restricted mobility pattern not unlike the one observed by Simukov and still prevalent today (Houle forthcoming). This micro-regional mobility pattern has profound implications in terms of territoriality and the nature of a given pastoral economy (Rosen 1992; Meadow 1992; Barth 1961; Koster 1977; Spooner 1973; Irons 1974; Tapper 1979; Cribb 1991). It also has important implications for issues of regional demography, resource mobilization, centralization, and the nature of societal complexity (Burnham 1979; Irons 1979; Salzman 1967). In the Iranian area, for example, there seems to be a correlation between lush and predictable pastures (leading to shorter migration routes), higher population densities, and strong chiefly control (Barth 1961: 128; Tapper 1979: 97).

Socially Integrative Facilities, Centralization, and the Development of Regional Networks

IN NON-STRATIFIED societies, order often depends more on integration and cooperation than on force; and rituals – especially above the household level – are often essential to social integration (Hegmon 1989). First, rituals reinforce social norms and promote social solidarity. Second, ritually communicated information is "sanctified" and can thus serve to promote the acceptance of important social decisions such as the choice of a leader. Finally, in some cases, rituals may help regulate aspects of the socio-cultural system. To the extent that rituals are conducted in a built environment, then architecture plays an important role in the ritual and thus in social integration. Substantial public works, ritual ones in particular, have been shown to serve such "integrative" functions (Adler 1989; Hegmon 1989). Because architecture used for ritual purposes is often built by the shared labor of those who will use them, architecture may help to define groups of individuals and contribute to the integration of these individuals into a social group or community. As suggested by Hegmon (1989:7, 9), architecture contributes to integration by defining boundaries and by symbolically reinforcing ideology and social norms. As such, the scale and labor required to build substantial public works that require unusual construction investments, such as khirigsuurs, can be indicative of the geographic scale and extent of community integration. On the basis of the scale of some khirigsuurs, as well as the required labor to build them and the number of animal deposits, social integration probably far exceeded the immediate community surrounding these monuments.

Although khirigsuurs are quite homogeneous with standardized patterns in architecture and ritual activities, there is some variability in design (Tsybiktarov 1995; Allard and Erdenebaatar 2005), suggesting that khirigsuurs were not built under the auspices of a regional-level authority. This is important for understanding these spatial arrangements, because khirigsuurs and their placement may be seen as products of local decision making within a common standardized pan-regional ideological system that included religious ideas. Consequently, these arranged monuments formed locales that would have been nonetheless immediately recognizable to others in the north central steppes of Mongolia, themselves occupying similar places.

Thus, within a mobile pastoralist system, this sort of purposeful planning at the local level suggests substantial efforts at maintaining an integrated community that went beyond the immediate locale and existed in the absence of everyday face-to-face interaction. The locales would have had a unifying effect, therefore providing, if not the shared experience of a central communal place, a familiar experience that both reflected cultural and social cohesion and reinforced it. This may well have been an important aspect of social integration. Mobile pastoralists are potentially highly segmentary, and social as well as political integration depends on the existence of social groupings that crosscut other social segments. The shared use of khirigsuurs may have contributed to social integration by discouraging social segmentation. Certainly, khirigsuurs likely served several purposes, some of which may have been more important for some members participating in the ceremonial events than for others. Feasting, aggrandizement, community, and lineage integration are all possible functions, perhaps all working at the same time through the events associated with their construction. The action of khirigsuur construction, especially in the form of adding to an already existing composition, may also have made statements of group power, ancestry, and alliances. Yet, along with these group-oriented integrative activities, there is also some visible emphasis on symbolism relating to individuals, such as single interments in the central mound of some khirigsuurs, "warrior-elite" slab burials, and deer stones. The presence of both group-oriented and individualistic symbolism, perhaps related to both achieved and ascribed status, has been documented in many "transegalitarian" societies and may reflect a transitory situation in which a system of hereditary status and a class of chiefs are not yet firmly established. Mortuary display, for instance, could be at its greatest when the concept of inheritance is accepted, but when there is still some uncertainty in the attribution of relative status positions within society (Cannon 1989; Randsborg 1982; Schulting 1995).

Thus, the critical peculiarity of these Late Bronze Age societies is that their monumental structures suggest organized labor investments, differential mortuary treatment for some elite individuals, hints of incipient hereditary principles, and supra-local centralized organization consistent with a hierarchical political structure. However, other formal indicators usually characteristic of "ranked" societies, such as increase in population density, socio-economic centralization, complex technologies, and increase in structural and functional specialization, are apparently missing (cf. Kradin 2002; see also Johnson and Earle 2000), although this needs to be empirically substantiated with more work on the domestic component. Nevertheless, this seemingly paradoxical situation is important because, in effect, it may signify the first stage in the emergence of political organization operating beyond the descent group (cf. Clark and Blake 1994; Hayden 1995). It may also reflect hierarchical social relations based on the control of non-material resources, or ritual-based polities, rather than hierarchical social relations based on economic variables (Potter 2000; McIntosh 1999). It may as well reflect a path to societal complexity that was entirely different from those followed in other parts of the world. As previously stated, the lack of settlement data significantly limits our ability to understand the nature and scale of these societies and to explain how these societies were organized at the local and regional levels. While the monumentality of khirigsuurs and other Late Bronze Age burials suggests organized labor and perhaps differential mortuary treatment for some elite individuals, there is still insufficient evidence regarding the socio-political organization of steppe groups during this period (Tsybiktarov 1998). We lack the more direct evidence of social status, political authority, and economic specialization that might come from the investigation of residential remains. Furthermore, this lack of data on habitation sites and their regional distribution also makes assessments of population size, subsistence practices, degrees of mobility, and territorial behavior highly speculative. Thus, while the study of ritual and monumental structures may help to define and direct the parameters of inquiry, it will not replace the need to investigate the domestic components if we hope to better understand the social organization of the groups that inhabited these regions. Fortunately, this concern is increasingly attracting the attention of archaeologists working in the Eurasian steppe.

Acknowledgments

I WISH to thank Bryan K. Hanks and Katheryn M. Linduff for their kind invitation to contribute to both this volume and the Eurasian Steppe Symposium held at the University of Pittsburgh in February 2006. Many thanks also go to Diimaajav Erdenebaatar, the co-director of the Khanuy Valley Archaeology Project, as well as to Francis Allard for his help in the initial survey work and mapping. Recognition must also go out to the Mongolian and foreign students who participated in the survey work, as well as to the people from the Khanuy Valley research area. Without their help and hard work, this project would not have been possible. Finally, I am grateful for the generous support from the Department of Anthropology, University of Pittsburgh.

References

Adler, M. A. 1989. Ritual Facilities and Social Integration in Nonranked Societies. In W. D. Lipe and M. Hegmon (eds.), *The Architecture of Social Integration in Prehistoric Pueblos*. Crow Cortez, CO: Canyon Archaeological Center, pp. 35–52.

Allard, F. 2006. Investigating the Bronze Age of Khanuy Valley, Central Mongolia. Unpublished paper, University of Pittsburgh, Pittsburgh, PA.

Allard, F., and D. Erdenebaatar. 2005. Khirigsuurs, Ritual and Mobility in the Bronze Age of Mongolia. *Antiquity* 79(305): 547–563.

Allard, F., D. Erdenebaatar, and J.-L. Houle. 2006. Recent Archaeological Research in the Khanuy River Valley, Central Mongolia. In David L. Peterson, Laura M. Popova, and Adam T. Smith (eds.), *Beyond the Steppe and the Sown: Proceedings of the 2002 University of Chicago Conference on Eurasian Archaeology*. Leiden: Brill.

Askarov, A., V. Volkov, and N. Ser-Odjav. 1992. Pastoral and Nomadic Tribes at the Beginning of the First Millennium B.C. In A. H. Dani and V. M. Masson (eds.), *History of Civilizations of Central Asia*, vol. 1: *The Dawn of Civilization: Earliest Times to 700 B.C.* Paris: Unesco Publishing, pp. 459–472.

Barfield, T. J. 1981. The Hsiung-nu Confederacy: Organization and Foreign Policy. *Journal of Asian Studies* 16(1): 45–61.

——— 1989. *The Perilous Frontier*. Oxford: Blackwell.

Barth, F. 1961. *Nomads of South Persia*. New York: Humanities Press.

Bayarsaikhan, J. 2005. Shamanistic Elements in Mongolian Deer Stone Art. In W. Fitzhugh (ed.), *The Deer Stone Project: Anthropological Studies in Mongolia, 2002–2004*. Washington, DC, and Ulaanbaatar: Arctic Studies Center, Smithsonian Institution, and National Museum of Mongolian History, pp. 41–45.

Bazargur, D. 2005. *Geography of Pastoral Animal Husbandry* [in Mongolian]. Ulaanbaatar: Mongolian Academy of Science, Institute of Geography.

Burnham, P. 1979. Spatial Mobility and Political Centralization in Pastoral Societies. In Équipe écologie et anthropologie des sociétés pastorales (ed.), *Pastoral Production and Society = Production pastorale et société: Proceedings of the International Meeting on Nomadic Pastoralism, Paris, 1–3 Dec. 1976*. Cambridge: Cambridge University Press, pp. 349–360.

Cannon, A. 1989. Historical Dimensions in Mortuary Expressions of Status and Sentiment. *Current Anthropology* 30(4): 437–458.

Casimir, J., and A. Rao. 1992. *Mobility and Territoriality: Social and Spatial Boundaries among Foragers, Fishers, Pastoralists and Peripatetics*. Oxford: Berg.

Clark, J. E., and M. Blake. 1994. The Power of Prestige: Competitive Generosity and the Emergence of Rank Societies in Lowland Mesoamerica. In E. M. Brumfiel and J. W. Fox (eds.), *Factional Competition and Political Development in the New World*. Cambridge: Cambridge University Press, pp. 17–30.

Cribb, R. 1991. *Nomads in Archaeology*. Cambridge: Cambridge University Press.

Dikov, N. N. 1958. *Bronzovyi vek Zabaikal'ia* (The Bronze Age of Zabaikal'e). Ulan-Ude: Siberian Division of the Academy of Sciences.

Erdenebaatar, D. 2000. *Bulgan Aimagiin Khutag-Ondor sumyn Khantai bagiin nutag Egiin Golyn khondiid yavuulsan etnografiin ekspeditsiin sudalgaany tailan* (Research Report of the Ethnographic Expedition to the Egiin Gol Valley of Khantai Baga, Khutag-Ondor Sum, Bulgan Province). *Field Report*. Department of Archaeology and Ethnology, Ulaanbaatar University.

2002. *The Four-Sided Grave and Khirigsuur Cultures of Mongolia* [in Mongolian]. Ulaanbaatar: Mongolian Academy of Sciences, Institute of History.

2004. Burial Materials Related to the History of the Bronze Age in the Territory of Mongolia. In K. M. Linduff (ed.), *Metallurgy in Ancient Eastern Eurasia from the Urals to the Yellow River*. Lewiston: Edwin Mellen Press, pp. 189–223.

Frolich, B., and N. Bazarsad. 2005. Burial Mounds in Hovsgol Aimag, Northern Mongolia: Preliminary Results from 2003–2004. In W. Fitzhugh, J. Baiyarsaikhan, and P. Marsh (eds.), *The Deer Stone Project: Anthropological Studies in Mongolia, 2002–2004*. National Museum of Natural History, Smithsonian Institution, and the National Museum of Mongolian History. Washington, DC, and Ulaanbaatar: Arctic Studies Center, pp. 57–88.

Grach, A. D. 1980. *Drevniye kochevniki v tsentre Asii*. Moscow: Nauka.

Gryaznov, M. 1969. *Southern Siberia*. Geneva: Nagel Publishers.

Hall, E. T. 1966. *The Hidden Dimension*. Garden City, NY: Doubleday.

Hayden, B. 1995. Pathways to Power: Principles for Creating Socioeconomic Inequalities. In T. D. Price and G. Feinman (eds.), *Foundation of Social Inequality*. New York: Plenum Press, pp. 15–85.

Hegmon, M. 1989. Social Integration and Architecture. In W. D. Lipe and M. Hegmon (eds.), *The Architecture of Social Integration in Prehistoric Pueblos*. Cortez, CO: Crow Canyon Archaeological Center, pp. 5–14.

Honeychurch, W. H. 2004. Inner Asian Warriors and Khans: A Regional Spatial Analysis of Nomadic Political Organization and Interaction. Ph.D. dissertation, University of Michigan.

Honeychurch, W., and Ch. Amartuvshin. 2006. States on Horseback: The Rise of Inner Asian Confederations and Empires. In M. T. Stark (ed.), *Archaeology of Asia*. Oxford: Blackwell, pp. 255–278.

Houle, J.-L. In press. Investigating Mobility, Territoriality and Complexity in the Late Bronze Age: A Perspective from Monuments and Settlements. In J. Bemmann, H. Parzinger, E. Pohl, and D. Tseveendorj (eds.), *Current Archaeological Research in Mongolia*. Papers from the first international conference on "Archaeological Research in Mongolia." Bonn Contributions on Asian Archaeology 4. Bonn, Germany.

Houle, J.-L., K. Taché, and F. Allard. 2004. "Feasts of Burden": Interpreting Bronze Age Khirigsuurs of Central Mongolia. Paper presented at the 69th SAA Annual Meeting. Montreal.

Ionesov, V. I. 2002. *The Struggle between Life and Death in Proto-Bactrian Culture: Ritual and Culture*. Lewiston: Edwin Mellen Press.

Irons, W. 1974. Nomadism as a Political Adaptation. *American Ethnologist* 1: 635–658.

   1979. Political Stratification among Pastoral Nomads. In Équipe écologie et anthropologie des sociétés pastorales (ed.), *Pastoral Production and Society = Production pastorale et société: Proceedings of the International Meeting on Nomadic Pastoralism, Paris, 1–3 Dec. 1976*. Cambridge: Cambridge University Press, pp. 361–373.

Ishjamts, N. 1994. Nomads in Eastern Central Asia. In J. Harmatta (ed.), *The Civilizations of Central Asia: The Development of Sedentary and Nomadic Civilizations*, vols. 2 and 3. Paris: Unesco, pp. 151–171.

Jacobson, E. 1993. *The Deer Goddess of Ancient Siberia*. Leiden: Brill.

Jagchid, S., and P. Hyer. 1979. *Mongolia's Culture and Society*. Boulder, CO: Westview Press.

Jagchid, S., and V. J. Symons. 1989. *Peace, War, and Trade along the Great Wall*. Bloomington: Indiana University Press.

Johnson, A. W., and T. Earle. 2000. *The Evolution of Human Societies: From Foraging Group to Agrarian State*. Stanford: Stanford University Press.

Khazanov, A. M. 1975. *Sotsial'naya istorija Skifov*. Moscow: Nauka.

   1978. Characteristic Features of Nomadic Communities in the Eurasian Steppes. In W. Weissleder (ed.), *The Nomadic Alternative: Modes and Models of Interaction in the African-Asian Deserts and Steppes*. The Hague: Mouton Publishers, pp. 119–126.

   1994. *Nomads and the Outside World*. 2nd ed. Madison: University of Wisconsin Press.

Koster, H. A. 1977. The Ecology of Pastoralism in Relation to Changing Patterns of Land Use in the Northeast Peloponnese. Ph.D. dissertation, University of Pennsylvania.

Krader, L. 1979. The Origin of the State among the Nomads of Asia. In Équipe écologie et anthropologie des sociétés pastorales (ed.), *Pastoral Production and Society = Production pastorale et société: Proceedings of the International Meeting on*

Nomadic Pastoralism, Paris, 1–3 Dec. 1976. Cambridge: Cambridge University Press, pp. 221–235.

Kradin, N. N. 2002. Nomadism, Evolution and World-Systems: Pastoral Societies in Theories of Historical Development. *Journal of World-Systems Research* 8(3): 368–388.

Kroll, A.-M. 2000. Looted Graves or Burials without Bodies? In J. Davis-Kimball, E. M. Murphy, L. Koryakova, and L. T. Yablonski (eds.), *Kurgans, Ritual Sites and Settlements: Eurasian Bronze and Iron Age.* BAR International Series 890. Oxford: Archaeopress, pp. 215–222.

Lattimore, O. 1992 [1940]. *Inner Asian Frontiers of China.* Oxford: Oxford University Press.

Legrand, S. 2006. The Emergence of the Karasuk Culture. *Antiquity* 80: 843–859.
   2004. Karasuk Metallurgy: Technological Development and Regional Influence. In K. M. Linduff (ed.), *Metallurgy in Ancient Eastern Eurasia from the Urals to the Yellow River.* Lewiston: Edwin Mellen Press, pp. 139–161.

Magail, J. 2003. Entre steppe et ciel. In J.-P. Desroches (ed.), *Mongolie, le premier empire des steppes.* Arles: Actes Sud/Mission archéologique française en Mongolie, pp. 182–208.

McIntosh, S. K. 1999. Pathways to Complexity: An African Perspective. In S. K. McIntosh (ed.), *Beyond Chiefdoms: Pathways to Complexity in Africa.* Cambridge: Cambridge University Press, pp. 1–30.

Meadow, R. 1992. Inconclusive Remarks on Pastoralism, Nomadism and Other Animal-Related Matters. In O. Bar-Yosef, and A. Khazanov (eds.), *Pastoralism in the Levant: Archaeological Materials in Anthropological Perspectives.* Madison: Prehistory Press, pp. 261–269.

Novgorodova, E. A. 1989. *Drevnyaya Mongoliya.* Moscow: Nauka.

Peebles, C. S., and S. M. Kus. 1977. Some Archaeological Correlates of Ranked Societies. *American Antiquity* 42: 421–448.

Potter, J. M. 2000. Ritual, Power, and Social Differentiation in Small-Scale Societies. In M. W. Diehl (ed.), *Hierarchies in Action: Cui Bono?* Center for Archaeological Investigations, Occasional Paper no. 27. Carbondale: Southern Illinois University.

Randsborg, K. 1989. Theoretical Approaches to Social Change: An Archaeological Viewpoint. In C. Renfrew, M. J. Rowlands, and B. A. Segraves (eds.), *Theory and Explanation in Archaeology.* New York: Academic Press, pp. 423–430.

Rosen, S. 1992. The Case for Seasonal Movement of Pastoral Nomads in the Late Byzantine/Early Arabic Period in the South Central Negev. In O. Bar-Yosef and A. Khazanov (eds.), *Pastoralism in the Levant: Archaeological Materials in Anthropological Perspectives.* Madison: Prehistory Press, pp. 153–164.

Sahlins, M. D. 1968. *Tribesmen.* Englewood Cliffs, NJ: Prentice-Hall.

Salzman, P. C. 1967. Political Organization among Nomadic Peoples. *Proceedings of the American Philosophical Society* 3(2): 115–131.
   1999. Is Inequality Universal? *Current Anthropology* 40(1): 31–61.
   2000. Hierarchical Image and Reality: The Construction of a Tribal Chiefship. *Comparative Studies in Society and History* 42(1): 49–66.
   2004. *Pastoralists: Equality, Hierarchy, and the State.* Boulder, CO: Westview Press.

Savinov, D. G., and H. L. Chlenova. 1978. Zapadnyye predely rasprostraneniya olennykh kamney i voprosy ikh kul'turno-etnicheskoy prinadlezhnosti. In A. P. Okladnikov (ed.), *Arkheologiya i Etnografiya Mongolii.* Novosibirsk: Nauka pp. 72–94.

Schulting, R. J. 1995. Creativity's Coffin. Innovation in the Burial Record of Mesolithic Europe. In S. Mithen (ed.), *Creativity in Human Evolution and Prehistory.* London: Routledge, pp. 203–226.

Simukov, A. D. 1934. Mongol'skie kochevki. *Sovremennaia Mongoliia* 4(7): 40–46.

Spooner, B. 1973. *The Cultural Ecology of Pastoral Nomads.* An Addison-Wesley Module in Anthropology 45. Reading, MA.

Takahama, S. 2004. Preliminary Report of the Archaeological Investigations in Mongolia. Edited by the Permanent Archaeological Joint Mongolian and Japanese Mission. http://web.kanazawa-u.ac.jp/~steppe/new15.pdf.

Tapper, R. 1979. *Pasture and Politics: Economics, Conflict and Ritual among the Shahsevan Nomads of Northwestern Iran.* London: Academic Press.

Tringham, R. (ed.). 1973. *Territoriality and Proxemics: Archaeological and Ethnographic Evidence for the Use and Organization of Space.* Warner Modular Publications, book 4. Andover, MA.

Tseveendorj, D., N. Urtnasan, A. Ochir, and G. Gongorjav (eds.). 1999. *Historical and Cultural Monuments of Mongolia* [in Mongolian]. Ulaanbaatar: Mongolian Academy of Humanities.

Tsybiktarov, A. 1995. Khereksury Buriatii, Severnoi i Tsentral'noi Mongolii. In P. B. Konovalov (ed.), *Kul'tury i pamiatniki bronzovogo i rannego zheleznogo vekov Zabaikal'ia i Mongolii.* Ulan-Ude: Nauka, 38–42.

1998. *Kul'tura plitochnykh mogil Mongolii i Zabaikal'ia.* Ulan-Ude: Nauka.

2003. Central Asia in the Bronze and Early Iron Ages: Problems of Ethno-cultural History of Mongolia and the Southern Trans-Baikal Region in the Middle 2nd–Early 1st Millennia BC. *Archeology, Ethnology and Anthropology of Eurasia* 13(1): 80–97.

Vainshtein, S. 1980. *Nomads of South Siberia: The Pastoral Economies of Tuva.* Trans. M. Colenso. Cambridge Studies in Social Anthropology, no. 25. Cambridge: Cambridge University Press.

Volkov, V. V. 1967. *Bronzovyi i ranii zheleznii vek Severnoi Mongolii.* Ulaanbaatar: Nauka.

1981. *Olennie Kamni Mongolii.* Ulaanbaatar: Nauka.

1995. Early Nomads of Mongolia. In J. Davis-Kimball, V. A. Bashilov, and L. T. Yablonski (eds.), *Nomads of the Eurasian Steppes in the Iron Age.* Berkeley: Zinat Press, pp. 319–333.

Wright, J. 2006. The Adoption of Pastoralism in Northeast Asia: Monumental Transformation in the Egiin Gol Valley, Mongolia. Ph.D. dissertation, Harvard University.

CHAPTER 20

Pre-Scythian Ceremonialism, Deer Stone Art, and Cultural Intensification in Northern Mongolia

William W. Fitzhugh

Deer stones and khirigsuur mounds are the most visible features of the archaeological landscape in the steppe region of northern Mongolia. Standing as slender stelae silhouetted against the sky and dark mounds against contoured landforms, deer stones and khirigsuurs evoke a past that contrasts with as much as it conforms to the light-drenched landscape of gers and herders today. One world we see and understand; the other is only faintly visible through the stony remains of spiritual and ceremonial life. This chapter explores a time when stone men walked and spirits plied the mountaintops; when shamans sang, and warriors rode deer spirits to heaven. Once considered a marginal late-Scythian derivative, Mongolia's deer stones and khirigsuurs have recently been dated several hundred years earlier than Arzhan, and their elaborate burial mound architecture and artistic stone monuments indicate a level of socio-political intensification and complexity previously unknown among Bronze Age societies of the eastern steppe.

The Deer Stone–Khirigsuur Complex

The archaeology of central and eastern Asia has seen dramatic change during the past two decades. Some areas like Kazakhstan, Inner Mongolia, and Mongolia, which have been long closed to Western scholars, have become accessible now, while others like Tibet are almost completely unexplored. Despite uneven research activity, the high plateaus, steppe, and semi-desert regions between the Central China Plain and the Siberian taiga forests have emerged as productive loci for

378

investigating archaeological issues related to pastoral nomadism first explored by Lattimore (1938, 1992), including its origins, development, and economic and organizational structure (Ingold 1980; Vasil'evski 1985; Khazanov 1994; Desroches 2003; Rogers et al. 2005) and the relationship of these societies to neighboring states and empires (Linduff 1995; Barfield 2001; Di Cosmo 1999, 2002; Honeychurch 2003; Honeychurch and Amartuvshin 2004, 2006a, 2006b, 2007; Honeychurch et al. 2007; Honeychurch and Wright 2008).

Unlike the hilly flanks of Mesopotamia or lowland regions of China and Southeast Asia, the cold high plateaus of Inner Asia did not originate the first plant domesticates or earliest examples of urbanism, centralized states, or empires. However, this climatically harsh region has witnessed surprising instances of political development, economic centralization, and military prowess. Although Inner Asia was not in the vanguard of social and political complexity in Eurasia as a whole, steppe polities are now seen as having been advanced players in processes that contributed to such developments in neighboring regions, particularly the rise of nomadic pastoralism in the second millennium BCE. In fact, Mongolia may have led in the introduction of complex social and religious organization, at least in the eastern steppe region. In addition, Mongolian nomadic polities occasionally achieved impressive levels of self-generated development and periodically dominated larger agricultural states during the Late Bronze Age, Xiongnu, Uighur, and Mongolian Empire periods through the use of their superior cavalry, urban siege tactics, and deft economic and political manipulation.

In contrast to Mongolia's first "quasi-state" society, the historically known Xiongnu (ca. 2200–1800 BP), its prehistoric Bronze Age and Neolithic past is obscure. Potanin (1881) and Radlov (1893) were the first to comment on khirigsuurs (kurgans) and deer stones. However, it was not until Okladnikov (1954) and Dikov (1958) began investigating deer stones and kurgans in Trans-Baikal, and Gryaznov (1950, 1980) and Rudenko (1970) opened frozen mounds at Pazyryk and the huge Arzhan complex, that khirigsuurs were explored actively in Mongolia. But because the Mongolian mounds and deer stones produced no human remains or artifacts, it was assumed they had some other ceremonial purpose, especially as they were often accompanied by square stone slab arrangements set vertically into the ground that contained human remains, artifacts, and animal bones. At first, it appeared the slab burials were the missing burials belonging to the deer stone–khirigsuur complex; but it soon became evident the slab burials were architecturally intrusive and deer stones were frequently found recycled unceremoniously as corner posts and retaining walls. Recent research, including radiocarbon dating, confirms that slab burials represent an intrusion of Karasuk-related culture and people from the northeast, which marks the termination of the deer stone–khirigsuur complex (Jacobson 1993: 147; Amartuvshin 2003; Takahama et al. 2006: 77). Unfortunately, this fascinating subject and its implications (see Tsybiktarov 1997, 2003) are beyond the scope of this chapter.

In the absence of grave goods at Mongolian sites, interest quickly turned to deer stone engravings. Okladnikov suggested the stones were anthropomorphic cenotaphs to dead warriors and dated them to the Karasuk culture, circa 2500 BP, on the basis of tool typology. Research soon followed on Mongolian deer stones (Dikov 1958; Chlenova 1962; Volkov and Novgorodova 1975; Khudyakov 1987; Novgovodva 1989) and on deer stones and kurgan complexes of the Gorny Altai (Kubarev 1979). In 1981 [2002] Volkov published his extensive descriptive inventory of deer stone sites in northern Mongolia. This work was recently updated by Savinov (1994) and has been summarized in English by Jacobson (1993: 142–158; Jacobson-Tepfer 2001) and Tsbiktarov (2002; 2003). By the 1990s interest in deer stone research in Mongolia slowed as a result of political changes, absence of radiometric dates, and insufficient archaeological data to resolve conflicting interpretations. Soviet archaeologists concentrated on new surveys and excavations of frozen sites of the Pazyryk culture, producing important new interpretations (Molodin 2000; Polosmak and Molodin 2000). Jacobson (1998; 2002; Jacobson et al. 2001) investigated western Mongolian rock art. Sanjmiatav (1995) and a few others have worked on deer stone iconography, cultural associations, and related Bronze Age subjects, but in the absence of excavation, questions concerning chronology, relationships between khirigsuurs and deer stones, Scytho-Siberian animal style art connections, and its place in Central Asian culture history remained unanswered.

A new phase of research began when Kanazawa University began work at the large Uushigiin Uver deer stone and khirigsuur site in 1999 (Takahama et al. 2006). Concurrently, an interdisciplinary Smithsonian program in Khovsgol Aimag began to study northern Mongolia's connections to Siberia and the circumpolar world, one focus of which related to Bronze Age ceremonialism and art. This phenomenon has important implications for Inner Asian cultural development but has been overlooked because of the paucity of grave goods and the nearly complete absence of settlement sites and domestic material culture. Even if some mounds date to later periods, the hundreds of deer stones and tens of thousands of khirigsuurs found throughout the northern Mongolia steppe zone exceed by quantum factors the archaeological traces of all previous periods of Mongolian prehistory. Furthermore, the deer stone–khirigsuur complex in central northern Mongolia, with its adherence to a single artistic genre, a single iconic deer motif, and production of architecturally complicated yet standardized public works at the core of large-scale ceremonial events, implies significant hierarchical social and political control. The deer stone–khirigsuur complex therefore marks a major advance in social and political evolution over a large region of Inner Asia where pastoral nomadism developed and came to dominate much of the Eurasian grassland core, setting in motion a long-standing conflict between settled agriculturalists and nomadic peoples that changed the course of world history.

Khirigsuur Mounds

Of the two monument types, mounds are far more numerous and widely distributed. Unlike simple mounds or rock pavements known from later periods, khirigsuur architecture signifies complex ceremony and extensive social investment, in both construction and use (Allard and Erdenebaatar 2005; Frohlich and Bazarsad 2005; Takahama et al. 2006). Physically, they have a concentric arrangement of components around a mound of boulders or a circular pavement of flat slabs or rocks (Fig. 20.1). Mound sizes range from a few meters to 40 meters in diameter and up to 4–8 meters in height. The central mound is almost always surrounded by a circular or square fence-like construction of rocks. The space between usually has a cobble or rubble pavement that is today frequently covered by wind-blown deposits. The size of the enclosure conforms to the size of the central mound and may be only a few tens or hundreds of square meters for a small mound, while large khirigsuurs like Urt Bulagyn in Arkhangai Aimag have enclosures thousands of square meters in area. Circular fences are generally perfectly round, whereas most square fences have trapezoid-like proportions (Allard and Erdenebaatar 2005: fig. 5), frequently have small rock mounds or pavements at their corners, and are often marked by a standing slab.

Two types of satellite features usually accompany round- or square-fenced khirigsuurs. Small rock mounds 2–3 meters in diameter are found concentrated outside the eastern fence, and after the most prestigious locations were utilized, mounds begin to be constructed along

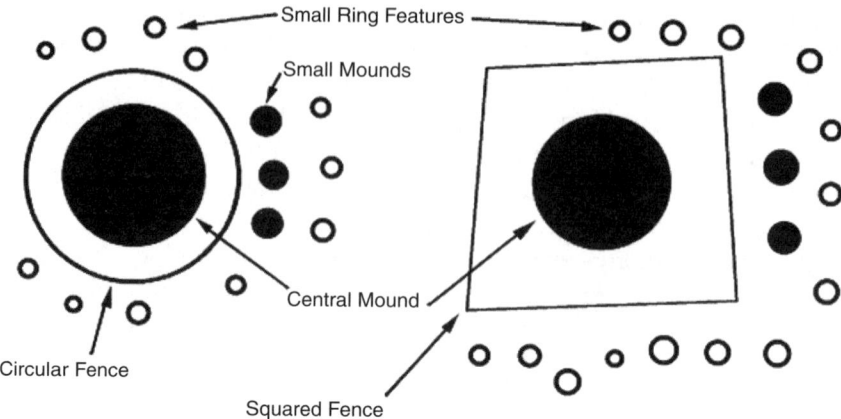

Figure 20.1. Plan of idealized square and round khirigsuur types (Frohlich et al. 2005).

the south and north sides of the fence. At Urt Bulagyn (ca. 800 BCE) satellite mounds are arranged in orderly geometric rows along its southeast and southwest fence walls (Allard and Erdenebaatar 2005: Fig. 20.2). On or just below the ground surface at the center of these mounds one almost invariably finds an east-facing horse skull with its maxilla and mandible articulated, accompanied by articulated cervical vertebrae and hooves. Frequently, cut marks indicate dismemberment or de-fleshing, and rarely a ceramic fragment or bones of other animals are present. It is widely supposed that these mounds represent horses sacrificed to honor the individual buried in the central khirigsuur mound, and one presumes, because the number of horse mounds correlates with the size of the khirigsuur, the number of horses sacrificed is a proxy for the status of the individual buried.

The fourth component, often but not always present, are small satellite boulder ovals or rings, 1–1.5 meters in diameter. These hearth features, which contain charcoal and cremated or burned bones of sheep, goats, and larger mammals, may be scattered without obvious geometric arrangement but also can occur in regular lines and ranks outside the horse mounds. While horse burials are rarely found on the west or north sides of the fence, ring features are frequently found here, and at Urt Bulagyn more than 1,000 oval rings are located in this position. Rings may also be present inside the fence together with other stone features, like a stone apron pavement or "horns" extending from the east side of the central mound toward the east fence (Takahama et al. 2006: pl. 2).

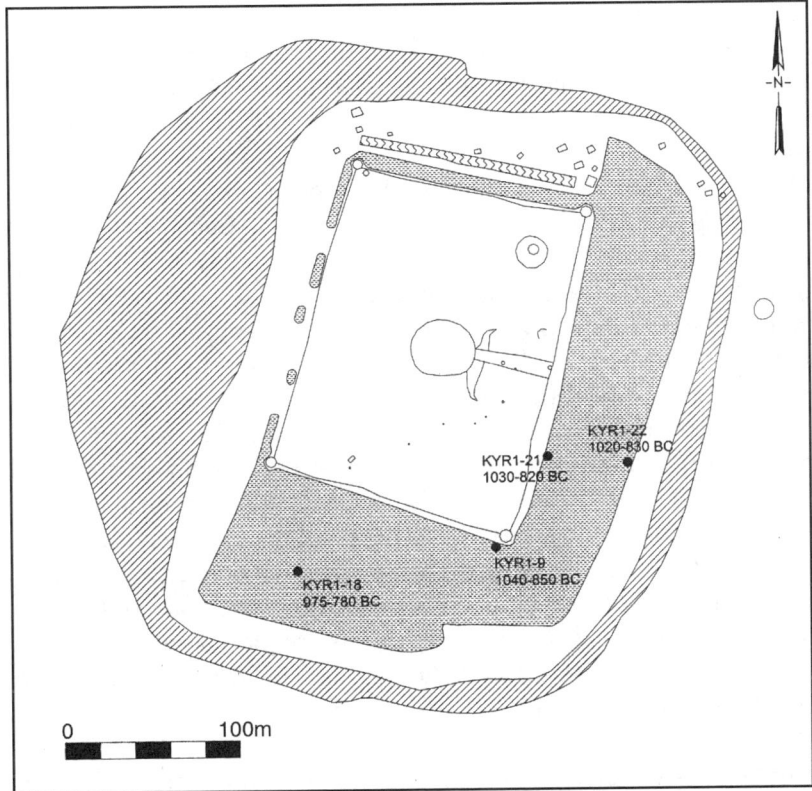

Figure 20.2. Urt Bulagyn KYR1 mound showing radiocarbon-dated horse head features KYR1–9, 18, 21, and 22 (dates for features 9 and 18 courtesy of Francis Allard; map modified from Allard and Erdenebaatar 2005: fig. 3).

Sometimes individual "key-stones" are found beyond the horse mounds and oval rings, perhaps to align the khirigsuur with a specific hill or mountain peak on the eastern horizon, if not a celestial position. These features and horse mounds give khirigsuurs a strong east- or southeast-facing orientation (Allard and Erdenebaatar 2005: 554, Fig. 3).

The foregoing description applies to khirigsuurs from Khovsgol and Arkhangai Aimag and neighboring areas of northern Mongolia and Trans-Baikal (Tsybiktarov 1997, 1998, 2003). Further to the south and west, mounds display greater architectural variation, and in Tuva khirigsuurs, known locally as kurgans (Kubarev 1979; Savinov 1994), are even more diverse, signifying different cultural traditions and perhaps different chronology. The classic northern Mongolian deer stones display a similar relaxation of their formal organizational pattern concurrent with increasing diversity of motifs and elements as one proceeds toward the north and west where Type II and III deer stone styles predominate.

Many Khovsgol khirigsuurs that have been excavated or observed after looting have a rectangular stone crypt at or slightly below ground level near the center of the mound. We have found both human and animal bones in these tombs, and all six mounds excavated in 2006 contained human burials (B. Frohlich personal communication 2007). However, some excavated mounds lack obvious burial features and human remains (e.g., Takahama et al. 2006: 67). Sometimes a few personal artifacts are found, such as small bronze buckles or harness ornaments, a bronze knife, or a few fragments of ceramics, but little else (Erdenebaatar 2002). Some mounds have flat tops, but most are domed or conical unless their stones have been robbed in later years. Stone robbing and absence of human remains has led some to suggest that mounds served as altars or ritual platforms rather than burials (Jacobson 1993: 146). However, because most burials are placed within the upper 1.5 meters of soil, or were simply laid on the surface of the ground and covered with boulders, absence of human remains may simply reflect poor organic preservation.

Until recently (Tsybiktarov 1997, 2003; Takahama et al. 2006), khirigsuurs were often not mapped systematically and frequently were not excavated carefully. Recent studies by Allard and Erdenebaatar (2005) explored celestial orientation of khirigsuurs and horse head burials in the Khanuy Valley (both are oriented to the east-southeast, with about 45 degrees variation), dimensions and shapes of "square" fences (wider on the east sides than on the west), and the distribution of mounds in the landscape. Two horse mounds from Urt Bulagyn produced dates of cal. 1040–850 BC and 975–680 BC (Allard and Erdenebaater 2005: 551), suggesting large khirigsuurs were used over hundreds of years. However, excavations conducted jointly with the Smithsonian in 2006 (Fig. 20.2), which were done to test the length of occupancy by dating a horse mound in the first rank outside the fence (presumably an early sacrifice) and a mound in the outermost rank (one of the last sacrifices) produced identical dates (KYR-1–21, cal. BC 1030 to 820, and KYR-1–22, cal. BC 1020 to 830). This suggests that the site's 1,700 horses could have been sacrificed and buried within a few years or decades, or possibly even in a single monumental ceremony.

A detailed regional mapping project being conducted by Bruno Frohlich employs exhaustive surveys and precision GPS mapping to investigate mounds in selected regions near Muren and in the Darkhad Valley (Frohlich et al. 2003, 2004, 2005; Frohlich and Bazarsad 2005; Wallace and Frohlich 2005). These data enable statistical quantification of settlement patterns, relationships of size and types of mounds with landforms, and orientations to landforms and also facilitate use of mound data as proxy indicators of local population size, gender, age, social structure, social hierarchy, and complexity.

Deer Stones

REPRESENTING HUMAN figures with bows, swords, and belts with hanging daggers, axes, and other articles of male materiality (Fig. 20.3), and with torsos displaying images of stylized cervids with giant flowing antlers, deer stones extend across Mongolia's northern steppe and southern Sayan Mountain foothills from the Altai of western Mongolia and nearby Russia nearly to Baikal. For more than 50 years, studies have established the anthropomorphic nature of these monuments (Okladnikov 1954; Dikov 1958; Gryaznov 1984; Novgorodova 1989; Jettmar 1994), whose general ceremonial nature has been inferred from a combination of Herodotus's descriptions of Pontic Scythian ritual and archaeological reconstructions of deer stone images. Few excavations have been conducted to determine archaeological context, relationship to khirigsuurs, or absolute age and stylistic development.

In the Khovsgol region, deer stones are often found singly or in small isolated groups that are usually, but not necessarily, associated with khirigsuurs. Some deer stones are embedded in (or even used as construction material for) khirigsuur mounds (Allard and Erdenebaater 2005: fig. 9; Hatakeyama 2002) in addition to their irreverent use as construction materials in slab burials (Amartuvshin 2003; Tsybiktarov 2003).

When found in groups of two or three, deer stones often are set without obvious geographic orientation, but when present in larger numbers, such as at Ulaan Tolgoi (5 stones) or Uushigiin Uver (15 stones), they tend to occur in north-south alignment (Volkov 2002: 78; Takahama et al. 2006: pl. 2). Deer stone shapes vary from broad, low stones a meter high to slender stones from 1 to 3.5 meters in height. As Jacobson-Tepfer (2001: 33) has noted, research on deer stones has been complicated by problems of classification and terminology, which has resulted

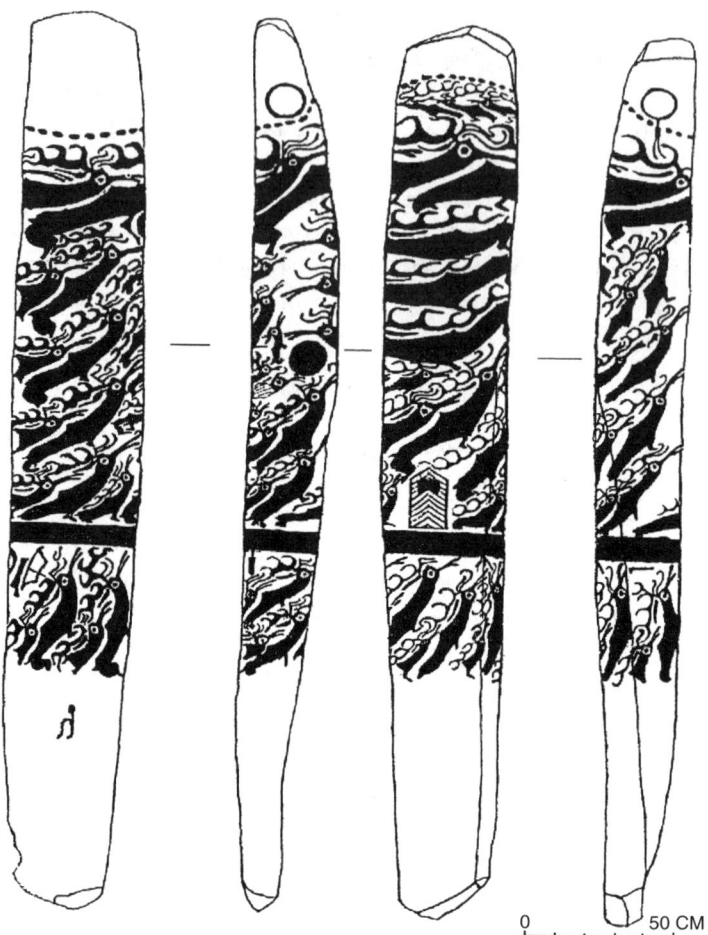

0 _____ 50 CM

Figure 20.3. Deer Stone 1, Undur Ulaan, Arkhangai Aimag (Volkov 2002: pl. 111).

in confusion in discussions of geographic distribution and cultural affiliation. For instance, it has been claimed that deer stones are found as far west as the Black Sea, Georgia, and even the Elbe (Chlenova 1962; Savinov and Chlenova 1978, cited in Jacobson 1993: 142). These stones have little resemblance to classic Mongolian deer stones, and their affiliation with the eastern stones is questionable. Even in the more restricted region of Mongolia, Tuva, and Altai, deer stones have various styles and exist in different architectural and cultural contexts, and some of these variants probably date to different chronological periods (Savinov 1994;

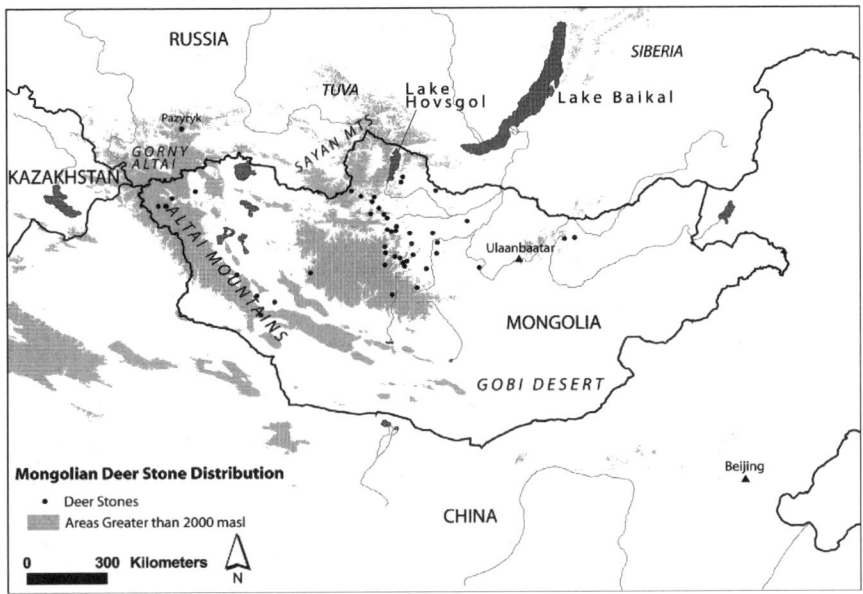

Figure 20.4. Deer stone sites in Mongolia (after Tseveendorj et al. 1999: 68).

Jacobson 1993; Jacobson-Tepfer 2001). Volkov identified 300 deer stones in Mongolia (Fig. 20.4) and noted that another 300 existed in neighboring Tuva, Altai, Kazakhstan, and China. However, Kubarev (1979) reported more than 500 in Mongolia, 30 in Tuva, and 50 in the Russian Altai. Extrapolating from new deer stones we have found in Khovsgol Aimag since 2002, more than 1,000 deer stones may exist in Mongolia alone.

While there are three or more regional deer stone style variants present in the core area of northern Mongolia, Sayan-Tuva, Gorno-Altai, and nearby regions of China, we are concerned here with the "classic" or "Mongolian" deer stones described by Volkov and Novgorodova (1975; Volkov 2002; Novgorodova 1989), called Type I by Savinov (1994), which are found predominantly in northern Mongolia. In its characteristic form, the Mongolian deer stone is a square or rectangular slab of hard rock – usually granite but sometimes of graywacke or diabase, usually having an angled top, with carvings on one or multiple sides with deer images wrapping entirely around all four sides. Sometimes the carvings wrap around the entire stone. These carvings have been the subject of intense scholarly interest with regard to semantics, meaning, and function. Most stones are set with their front side facing east.

Deer stones typically have three ornamented sections, each covering about a quarter of the stone's length, with the fourth undecorated section embedded in the ground (see Fig. 20.3). The top section often has a set of large round grooved rings with dangling ornaments carved into opposite sides of the stelae. When occurring with the carving of a human face on one of the other sides – known in only a few cases, most notably on Uushigiin Uver (UU) Deer Stone 14 (Volkov 2002: pl. 79) – these rings have sometimes been interpreted as mirrors with handles, but their context and form more certainly identify them as Late Bronze Age earring hoops with bifurcated pendant attachments. Much more common are stones without carved faces but whose earrings on opposite sides of the stone, with or without dangling ornaments, establish anatomical reference. However, adjacent smaller rings may also accompany these large rings, and in these cases a multi-symbolic function links human anatomy with images of the sun and moon (UU4; Takahama et al. 2006: pl. 14). More confident identifications of the sun or moon occur when these motifs are depicted as large and small discs engraved into the central panel area among images of deer (UU 2, 8; ibid. pl. 13, 17). Deer stone semantics also frequently have shamanic reference. The UU 14 stone face has a rounded mouth suggesting enactment of shamanic breath or singing, a common feature of circumpolar shaman ritual. This otherworldly and perhaps explicitly celestial upper part of the stone is often set off from the middle panel by a looping necklace-like line of engraved pits or cupmarks.

In contrast to this celestial or anatomical reference in the stone's upper domain, the lower part of the tableau invariably deals with worldly power and status. This is usually represented by a geometrically hatched or textured warrior's belt from which hang such items as dagger, axe, sword, fire starter, or other implements. Rather than being a standardized warrior's kit, the inventory and the form of objects vary from stone to stone, creating the impression that they represent the specific assemblages belonging to individual warriors. Tools and implements also sometimes float above the belt in the lower part of the middle panel.

The middle panel carries images of an abstract, stylized, elongated deer with a distinctive peaked withers and antlers that flow along its back in a series of wave-like curls. Okladnikov thought the two forward-pointing tines over the brow identified this animal as a reindeer, but the absence of the broad expanding reindeer brow tine and the elegant wave-like antlers more likely signifies the Asian roe deer, also known as elk or maral (*Cervus elaphus sibiricus*). The most prominent deer figures are positioned singly or stacked in tightly nested ranks and are usually shown in slanted, ascending attitude, with smaller deer added into the tableau to fill blank spaces, as though it was important to include as many of these images as possible on one stone. Legs are recumbent, folded as though running or flying, never extended straight, and a prominent round eye is present. While the antler form indicates the roe deer, the curiously elongated, bulbous snout, shown partly open as though calling or speaking, is distinctly un-deer-like and resembles the bill of a large water bird, suggesting the image may depict a cervid–water bird trans-formation spirit. Unlike belt accoutrements, whose implement types and forms vary from stone to stone, the deer spirit figure is always rendered in a single iconic form. Despite the standardization of this crucial motif, the number and arrangement of deer images vary from one stone to the other and, like the tool belt distinctions, give the impression of unique personalized representation.

In addition to the standard elements and tri-partite structure, other motifs are common. In addition to "sun" and "moon" discs, re-curved bows, quivers, and other implements, the central torso may include a chevron image that is often interpreted as a warrior's shield but which may be an emblem of shamanic power, perhaps representing an animal's skeletal or palate image. Like Savinov and Chlenova, and Vainshtein (cited in Jacobson 1993: 154), Bayarsaikhan (2005) has noted its simi-larity to the skeletal-like motif found on ethnographic shaman drum handles and drum beaters, a widespread theme in circumpolar shaman-ism and art and a probable indication of Siberian cultural affiliation for deer stone iconography.

The structure of deer stone art embodies a stylized anthropomorphic reference whose distinctive belt and tool varieties and the arrangement of deer motifs, chevron, and other forms suggest they represent a par-ticular warrior or leader. This seemingly personalized treatment is set against a standardized torso tableau dominated by stylized deer-bird, sun, and moon motifs whose essential features never change except in number and placement. That this exact iconic image also occurs on rock art in western Mongolia (Jacobson-Tepfer 2001: 50), and Hovsgol (see Fig. 20.10) suggests a deer cult central to the cosmology of this period and culture. Early on, Okladnikov and Dikov offered the opinion that deer stones commemorated and represented powerful warriors and chiefs and that the three panels reflected cosmology of heaven, earth, and underworld, and this idea has persisted with minor differences of inter-pretation (Chlenova 1962; Kovalev 1987; Dobzanski 1987; Volkov and

Novgorodova 1975; and Savinov 1994; Jacobson-Tepfer 2001: 38). Given the ancient practice of tattooing, known archaeologically from Pazyryk and similar sites and which can be inferred from Asian rock art, Jomon figurines, and ethnographic clothing from ancient to modern times, it seems likely that the deer image served as "armor" to protect warriors from injury or harm and that the art shown on the deer stones replicates images tattooed on the bodies of the leaders they represent (Gryaznov 1984; Jettmar 1994; Polosmak and Molodin 2000: 83; Polosmak 2000). The presence of shamanic elements and absence of human burial remains may indicate that deer stones were erected to honor specific fallen heroes who died elsewhere and could not be buried in the normal manner in khirigsuurs. We may further speculate on the basis of widespread ethnographic and archaeological evidence in eastern and northeastern Asia and North America, that the souls of these chiefly warriors nevertheless needed to be dispatched to the upper world with the aid of shamans and deer-bird spirits in ceremonies centered on the erection of personalized monuments with feasting and the sacrifice of horses.

Previous Research

This review of aboveground features reveals a welter of research problems related to deer stones, khirigsuurs, and Late Bronze Age ceremonial life and belief systems. What was the function of the mounds? Was it for human burials, altars for sacrifices or offerings, or some other purpose? As has been argued by others, are the elaborate mound architecture and satellite features based on celestial or solar models, chariot wheel images, or something else? Was the purpose of the fence to create an inner sacred space for the mound and its contents and ceremony apart from worldly matters of horses and feasting? What was the significance of the square versus round enclosures? Do they signify the gender of the deceased, or lineages, clans, or some other concept – for instance, the 3,000-year tradition of circular and square sacred mounds and temples expressed in Han and Ming concepts of *mingtang* (sky: square, earthly temple) and *biyong* (earth: circular, heavenly temple)? What were their dating ranges, and how did this relate to demography, regional variation, and settlement patterns? There are also a host of questions involving the relationship between khirigsuurs and deer stones. Even less information is available on the deer stones themselves. What did the stones commemorate? What was the meaning of their human form, symbols, and deer images? What was their origin, and how long were they in use?

What was their distribution and geographic variation? What is the age of slab burials, and how do they relate to the deer stone complex?

General consensus is based on Savinov's (1994; Jacobson-Tepfer 2001: figs. 3–5) analysis identifying deer stone "types," each having a slightly different geographic and cultural context but with areas of overlap. Type 1, the classic belted anthropomorphic northern Mongolian stone, is dominated by the stylized iconic image of a great deer with flowing antlers (Savinov 1994: figs 1–3). Type 2, centered in the Sayan Mountain in Tuva, carries freestanding images of animals such as deer, horses, and goats, including, sometimes, tigers, often shown standing with legs extended (Savinov 1994: figs. 4, 5). Type 3, found mostly in the Altai, has few animals and rare instances of the deer motif, but it has circle grooves, slash marks, and necklace-like pits near the top of the stone (Savinov 1994: fig. 6), linking this Altai type with Type 1 Mongolian stones more than the Sayan type. Until 2003 none of these types had been radiocarbon-dated, although various authors had suggested ages from circa 2800–2000 BP on the basis of weapon styles, association with square burials, and other criteria. We are concerned here with dating and context of the Type 1 "classic Mongolian" deer stone rather than a re-evaluation of style, internal analysis, or interpretation.

Jacobson's critical assessment of a century of deer stone studies as containing much "wishful mythologizing" (Jacobson-Tepfer 2001: 38) also applies as well to the study of khirigsuurs. The Smithsonian-Mongolian Deer Stone Project set out to establish a foundation for studying this complex. Our first objective was to establish a radiocarbon chronology, because until now deer stone dating has been based on typological comparisons only. A second goal was to explore a particular deer stone site, Ulaan Tolgoi, to determine spatial relationships, identify associated features, and learn about site organization and duration of use. A third goal was to examine the relationship between deer stones and khirigsuurs. We also hoped to learn about deer stone semantics and form, typological variety, and the role deer stones play in Mongolian Bronze Age life and cosmology.

Landscape, Ritual, and Cosmology

THE ULAAN Tolgoi site, named for a small and isolated 250-meter-high hill rising from the valley floor eight kilometers west of Erkhel Lake, contains hundreds of khirigsuurs and five granite deer stones (Figs. 20.5 and 20.6). Near the deer stones are several of the largest khirigsuurs

Figure 20.5. GPS mapping at Ulaan Tolgoi deer stone site, view south.

in the Erkhel Valley, and on the south and east side of the hill large numbers of smaller khirigsuurs extend up the rocky slope to the summit. All mounds are located on the eastern and southern slopes. The presence of human remains in mounds opened by looters and recent khirigsuur excavations conducted by Frohlich west of Muren suggest that most if not all are burials and that the absence of human remains in some mounds results from poor preservation.

Bruno Frohlich has conducted precision GPS-based surveys in the west Darkhad Valley and Muren region of Khovsgol Aimag since 2003. These surveys are investigating variation in mound form and size, identification of associated features, geographic relation to landscape and topography, clustering of mound types, and other questions. Preliminary results indicate that the pattern in the placement and types of khirigsuurs at Ulaan Tolgoi is replicated in the West Darkhad and Muren regions (Frohlich and Bazarsad 2005). On the basis of these data, it appears that mound density per given area reflects some sub-set of the population of a local tribe or clan and thus may provide a key for reconstructing population size. However, the distribution of mounds over the landscape is not even, as there is a tendency for clumping in relation to geographic

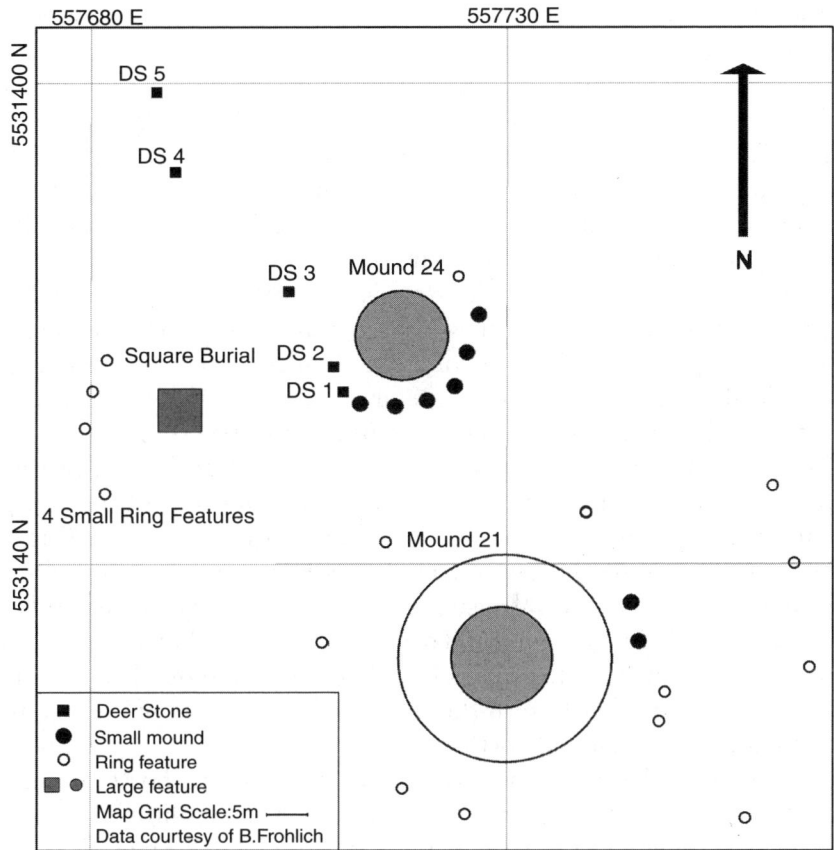

Figure 20.6. Ulaan Tolgoi khirigsuurs 21 and 24, and Square Burial in vicinity of Deer Stones 1–5.

features. Khirigsuur size also correlates with geography. Larger, more complex (Type 1) khirigsuurs are found in relatively few numbers in the valley lowlands; smaller and less complex (Type 2) structures are more common and occur along the valley margins; and still smaller (Type 3) mounds, often the most common, are found on higher east-facing hill slopes. The consistent construction of mounds in the southeastern sectors of hills, like the preference for placement of satellite mounds and of east-facing horse skulls, is generally believed to reflect orientation to the rising sun, an explanation often given by modern herders. Even today, the custom continues of placing the skulls of departed horses in eastern site orientations and at the tops of special hills. Finally, comparisons between the Muren, Erkhel, and Darkhad regions indicate north-south gradients in the numbers and sizes of khirigsuurs and deer stones, with the largest khirigsuurs and greatest numbers of deer stones around Muren and progressively fewer in the Darkhad Valley. These trends most likely have an environmental basis, as they conform to ecological gradients like water availability, elevation, temperature, and pasture quality and to economic indicators like herd size, human population levels, and proximity to population centers.

Distribution studies suggest a model of scalar replication and an eastern-orientation principle may explain individual, local, and regional social and spiritual obligations of a deer stone–khirigsuur cosmology. Specifically, mound and deer stone site density at the regional geographic scale appears to parallel the architectural structure of *individual* khirigsuurs and *individual* horse burial features. Ulaan Tolgoi, clearly a sacred eminence, is the local community-scale equivalent of the khirigsuur's central mound, and the east-facing khirigsuurs on Ulaan Tolgoi's slopes are the geographic equivalent of satellite mounds and ovals found around the east side of khirigsuurs, and of east-facing horse skulls at the scale of individual mound and deer stone features. Furthermore, the concentration of hillside khirigsuurs, like satellite mounds and ovals, diminishes in number toward the south and north sides of the hill and are absent to the west and northwest.

Replication of the eastern-orientation principle in horse features, satellite mounds, khirigsuur construction, and sacred hills indicates a unified cosmology at the core of the deer stone–khirigsuur relief system. This arrangement suggests a hierarchy of relationships in which an individual's relations with society – in this case, the spiritual society that exists after death – are structurally mapped out from the center of the khirigsuur as in a kinship chart: "ego" – the deceased – is at the center, with people having social or political obligations with that individual represented by horses offered in the satellite mounds, and families or other social groups consuming animal offerings represented in the oval feasting hearths that surround the khirigsuur. At the local geographic level, the central hill – in this case Ulaan Tolgoi and its universe of khirigsuurs – represents the spiritual death community of ancestors bound to the spirit of the hill itself, acting as the collective ancestral spirit of the entire local community. Higher levels of social integration through the celebration of death and renewal ritual may follow the pattern of political hierarchy above the local level at some regional scale. This ceremonial scale seems represented by larger ceremonial sites like those at

Uushigiin Uver (Takahama et al. 2006) or other locations where the size of khirigsuur monuments increases by significant factors, as clearly seen in the huge monument of Urt Bulagyn and others in the Khanuy Valley (Allard and Erdenebaatar 2005). In this way, the material features of the Bronze Age belief system seem to replicate a shared cosmological model at the full range of social and political scales – horse, individual, community, regional, and macro-social scales – all of which have separate geographic orientations to deities, spirits, or ancestors residing in appropriately scaled cosmological levels with equivalently structured sets of topographic eminences, from horse mounds to khirigsuur mounds to local hills to sacred mountains. Such a model would be consistent with the hierarchical social structure that seems to be a pervasive principle in Bronze Age society in general.

Ulaan Tolgoi

TURNING TO deer stones, we spent a week each season from 2002–2006 mapping and excavating at the Ulaan Tolgoi site (Fitzhugh 2003, 2004a, 2004b, 2005). This site has five deer stones positioned in north-south alignment in the midst of a jumbled field of rocks protruding from the grassy steppe. Four of the stones are approximately 1–1.5 meters tall and are boldly ornamented, while a fifth, DS 2, is approximately 3.5 meters high and is covered with exceptionally beautiful, lightly cut engravings (Fig. 20.7). This stone has a diagonal zone of differential weathering that indicates it had fallen and was partially buried in the soil at a 20–30 degree angle for many years. Local herders told us it was pulled upright by a tractor and re-set several decades ago. A few meters to the south, DS 1 had also fallen and was re-set in concrete by Russian archaeologists. A slab grave a few meters west of DS 2 also seems to have been excavated, but the northern deer stones appeared to be undisturbed. All of these monuments were probably quarried from the granite hill bordering the south shore of Erkhel Lake, where slabs of similar size and quality can be found today.

During our first season we mapped DS 5 and its surroundings and excavated a trench, 1 by 2 meters, 50 centimeters south of the deer stone. The purpose of the trench was to explore the deer stone setting for burials, artifacts, and ritual activity and to search for datable bone or charcoal (Fitzhugh 2005: 14–21). The excavation produced no human or animal bone, but a small charcoal sample recovered was dated (cal. 2150–1960 BP – see Table 20.1). The deer stone had been set in a narrow vertical hole just large enough for its base without any other ritual preparation.

Figure 20.7. Laser image of the top portion of Ulaan Tolgoi Deer Stone 2 (courtesy V. Karas and Harriet Beaubien).

We also excavated a small oval ring feature 50 meters east of DS 5. This feature was identical to oval features found in the outer tier of khirigsuur complexes, but in this case it was associated with deer stones. In its center we found charcoal stains and calcined bone fragments of small and large mammals, which were too fragmented to be identified.

In 2003 and 2004 we mapped the Ulaan Tolgoi deer stones and expanded the excavations at DS 4 and its surroundings (Fig. 20.8; Fitzhugh 2004, 2005b; Frohlich et al. 2004, 2005). Once again, nothing was found in direct contact with the deer stone, but we were surprised to discover a series of small circular or oval buried rock features encircling the base of the deer stone, each containing a horse head, vertebrae,

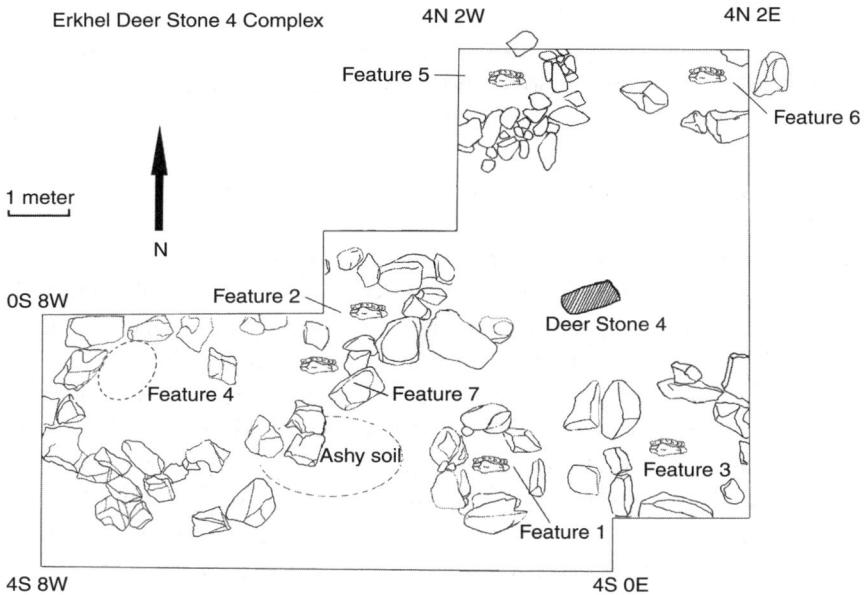

Figure 20.8. Excavation plan of Ulaan Tolgoi Deer Stone 4 showing circle of east-facing horse head burial features.

and hooves. Subsequent work through 2005 revealed seven horse head features. Six contained horse skulls and mandibles buried facing east with the cervical vertebrae column along the south side of the skull and hooves usually placed along the north side. The horse features did not include artifacts or other remains except for a few hand-sized pecking stones. In the upper levels of the soil, we found a few thick undecorated ceramics and a fragment of a stone bowl, which we determined to be part of a later ephemeral post–deer stone component linked to a circa 2000 BP date.

Excavations at the base of DS 4 failed to produce charcoal or other datable material, but a charcoal sample (S7, B-182959) associated with a pecking stone and a small piece of burned ceramic at the base of the cultural deposit west of DS 4 was found with charcoal dating cal. 3220–2950 BP. This pecking stone had been abraded around its entire surface. Several other pecking stones recovered around the base of the deer stone, mostly in and between F1 and F3, were made of hard "greenstone" and had battered working edges matching the rounded grooves of the stone carving. Finding pecking stones within the horse burial features, and associated with the cultural level, allows us to directly link the horse features with the production of deer stone art. This connection was later supported by a series of dates of horse bone and teeth from the cluster of horse head burials surrounding DS 4, which dated between cal. 3200 and 2800 BP (Table 20.1).

Averaging the six DS 4 features gives a date for the ceremonies connected with this deer stone of cal. 2995–2895 BP. An almost identical age date is indicated for the average of two horse head burial dates from DS 5, cal. 2925–2885 BP. Reliability is high because most are collagen AMS dates from dense horse tooth. A slightly earlier date has been obtained from a horse head burial at the Tsatstain Khoshuu deer stone site south of Tsaaganuur in the northern Darkhad Valley (cal. 3330–3060 BP). These dates are consistent with two dates from khirigsuur satellite horse mounds at Urt Bulagyn (1040–850 BCE and 975–680 BC; see Fig. 20.2) and only slightly earlier than other horse mounds dated in the Khanuy Valley (1390–910 BC and 930–785 BC, see Allard and Erdenebaatar 2005: 551–553). These ages are consistently 200–300 years earlier than dates from Arzhan and other early Scythian-related sites in southern Siberia and Central Asia (Sementsov et al. 1998). These dates and identical horse rituals begin to resolve questions raised by Jacobson (2001: 48–51) and others as to whether deer stones and khirigsuurs belong to a single culture, a fact accepted by most Russian and Mongolian specialists for many years, or whether khirigsuurs and their satellite features were created as short-term events or accumulated as palimpsests over decades or even hundreds or thousands of years (Allard and Erdenebaatar 2005: 561). It has also been debated as to whether deer stones and khirigsuurs were components of a single ceremonial system or were independent. Importantly, our dates now show that deer stones and khirigsuurs are ceremonial aspects of the same culture.

In order to investigate deer stone–khirigsuur relationships, we began excavations at Mound 1, a large square khirigsuur 100 meters south of the Ulaan Tolgoi deer stones. This is the largest khirigsuur at Ulaan Tolgoi region. The east side of its central mound has been robbed of stones in recent times, creating a deep depression but not diminishing the original height. Everywhere else, the building stone was simply dumped to create the mound but on the north side of the mound the top portion was laid up in nearly horizontal courses. An apron of small cobbles extends from the east side of the mound to the fence wall. In all there are 110 satellite mounds: 54 are lined up in four parallel ranks outside the east fence, 18 in two ranks along the north fence, and 36 in three

Table 20.1. Radiocarbon dates from deer stone and khirigsuur sites in Khovsgol and Arkhangai Aimags, 2002–2006

Site / Feature	Location/Year	Sample no.	Material	Uncorrected	Calib. (2-sig)
Ulaan Tolgoi DS 4 S-17	Erkhel / 2003	B-182958 AMS	Charcoal	2170 ± 40 BP	BP 2320–2050
Ulaan Tolgoi DS 4 S-7	Erkhel / 2003	B-182959 AMS	Charcoal	2930 ± 40 BP	BP 3220–2950
Ulaan Tolgoi DS 4 F1	Erkhel / 2004	B-193738 AMS	Bone collagen	2530 ± 40 BP	BP 2750–2470
Ulaan Tolgoi DS 4 F2	Erkhel / 2004	B-193739 AMS	Bone collagen	2950 ± 40 BP	BP 3240–2970
Ulaan Tolgoi DS 4 F3	Erkhel / 2004	B-193740 AMS	Bone collagen	2810 ± 40 BP	BP 2990–2800
Ulaan Tolgoi DS 4, F5	Erkhel / 2005	B-207205 RAD	Bone collagen	2790 ± 70 BP	BP 3220–2800
Ulaan Tolgoi DS 4, F6	Erkhel / 2005	B-207206 RAD	Bone collagen	2740 ± 70 BP	BP 3150–2780
Ulaan Tolgoi DS 5, T1	Erkhel / 2002	B-169296 AMS	Charcoal	2090 ± 40 BP	BP 2150–1960
Ulaan Tolgoi DS 5, F1	Erkhel / 2005	B-215694 AMS	Tooth collagen	2800 ± 40 BP	BP 2980–2790
Ulaan Tolgoi DS 5, F2	Erkhel / 2006	B-222535 AMS	Tooth collagen	2830 ± 40 BP	BP 3050–2850
Ulaan Tolgoi M1, F1	Erkhel / 2005	B-207209 AMS	Bone collagen	1880 ± 40 BP	BP 1900–1720
Ulaan Tolgoi M1, F2	Erkhel / 2005	B-215692 AMS	Tooth collagen	2860 ± 40 BP	BP 3080–2870
Ulaan Tolgoi M1, F2	Erkhel / 2005	B-215644 AMS	Charcoal	2980 ± 40 BP	BP 3310–3000
Ulaan Tolgoi M1, F3	Erkhel / 2005	B-215693 AMS	Tooth collagen	2950 ± 60 BP	BP 3320–2940
Nukhtiin Am DS 1/2, F1	Galt / 2006	B-222534 AMS	Tooth collagen	2830 ± 40 BP	BP 3050–2850
Evdt 2 DS 2 Circ. feat.	Evdt Valley	B-215643 AMS	Charcoal	3030 ± 40 BP	BP 3350–3090
Tsatstain Kh DS 1, F1	Tsaagan / 2005	B-207208 AMS	Tooth collagen	2920 ± 40 BP	BP 3160–2920
Tsatstain Kh DS 1, F2	Tsaagan / 2005	B-207207 AMS	Tooth collagen	3000 ± 40 BP	BP 3330–3060
Urt Bulagyn KYR1–21	Khanuy / 2006	B-222532 AMS	Tooth collagen	2780 ± 50 BP	BP 2980–2770
Urt Bulagyn KYR1–22	Khanuy / 2006	B-222533 AMS	Tooth collagen	2790 ± 40 BP	BP 2970–2780
Hort Azuur DS 2, F3	Erkhel / 2006	B-240691 AMS	Charcoal	2690 ± 40 BP	BP 2870–2750

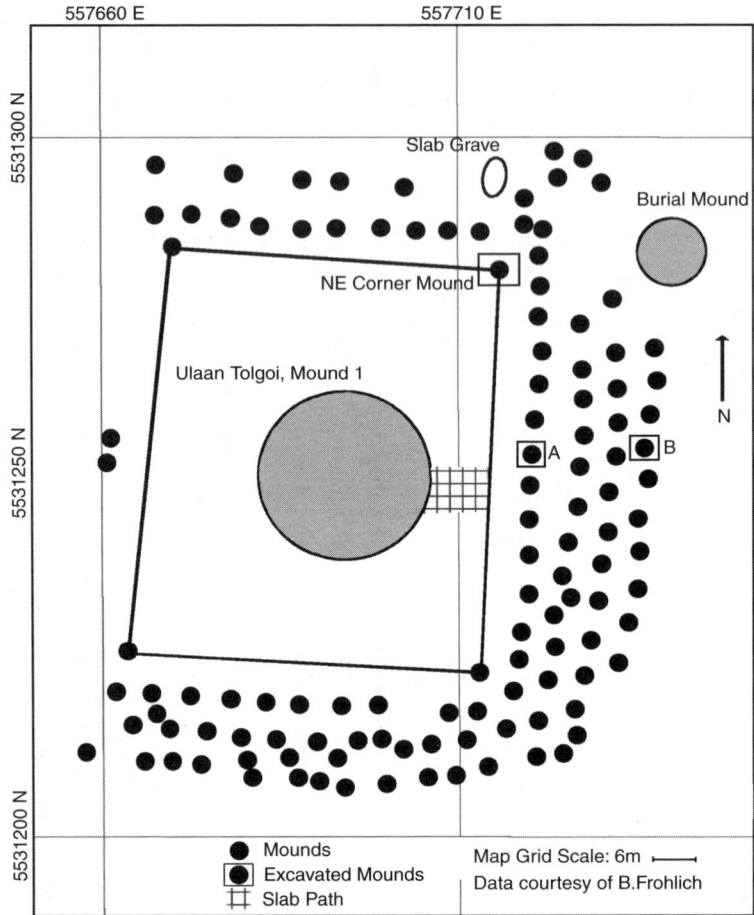

Figure 20.9. Ulaan Tolgoi Mound 1 showing corner pavement and two horse head burial feature excavations (map data courtesy Bruno Frohlich).

ranks along the south fence (Fig. 20.9). We excavated three features: a small stone mound at the northeast corner of the fence that would allow for dating the fence construction, a horse mound (A) in the middle of the first row outside the east fence, and a small mound (B) in the fourth and outer rank of horse mounds. As at Urt Bulagyn, we hoped these samples would provide a chronological key for fence construction, early horse sacrifice, and the latest phase of horse sacrifice activity at the site.

The corner mound did not contain horse remains, and a caprid bone found among the mound's upper tier rocks, which dated to cal. 1900–1720

BP, is considered intrusive. Satellite mound A, situated outside the east fence, produced no artifacts but contained parts of a weathered horse skull and mandible. This horse, which was dated to cal. 3080–2870 BP, had died long before burial and was not a freshly killed sacrifice – such as those encountered for the other burials noted earlier. Mound B produced a complete east-facing horse head, like those found at DS 4, which was dated to cal. 3320–2940 BP. Statistically, both could have been contemporary events.

These studies reveal similarly dated horse rituals associated with hearth rings, deer stones, and human burials in khirigsuur mounds. Jacobson's view that the deer stones may be a "late Bronze Age intrusion into ground ritually sanctified by Bronze Age predecessors" and that the deer stones at Ulaan Tolgoi are a post-khirigsuur "cultural afterthought" (Jacobson-Tepfer 2001: 51) is not supported by our data, although it remains to be seen exactly which dates and specific geographic distributions characterize the deer stone–khirigsuur relationship. While it is certain that these impressive landscape features continued to play a role in the spiritual lives of successive generations, even to the present as herders frequently deposit the remains of deceased animals inside khirigsuur fence lines, our data indicate single events for the ritual associated with the erection of DS 4 and 5 at Ulaan Tolgoi. These events are contemporary with the construction of khirigsuur mound burials.

Darkhad Valley Finds

SURVEYS AND excavations also have been extended north of Muren into the Darkhad Valley west of Lake Khovsgol. Only a few deer stones had been reported previously for this region, which was not surveyed by Volkov, and its deer stones do not occur in the numbers, large clusters, or large sizes known from regions further south. It is perhaps significant that the present population of the Darkhad Valley, while largely Mongolian, is ethnically and linguistically distinct from the rest of Mongolia because of geographic and physiographic barriers. This region is isolated by a range of high hills separating it from the steppe to the south and the Sayan Mountains that impede access to Russian Tuva to the west and north. Physically, the Darkhad is a large drained preglacial lake that formed a sister lake to Lake Khovsgol before it drained in the early Holocene into the Little Yenesei, leaving a flat grassland with marshes and lakes underlain by permafrost. Today the lake bed provides extensive pasture for ethnic Darkhad Mongolian herders, while its nearby mountain regions support Tuvan-speaking (Dukha) reindeer-herders. Darkhad's geographic situation may explain the lower frequency of deer stones and the smaller number of large khirigsuurs, but the issue remains: is its deer stone complex a pre-cursor, co-eval, or a derivative from the deer stone–khirigsuur complex of the central Mongolian steppe to the south?

Our work here, which is still exploratory, has produced some surprises. In 2004 we excavated a site at Tsatstain Khoshuu that had a single deer stone and no khirigsuurs. The stone was roughly worked and its only deer stone "insignia" was an engraved circular ring near the peaked top of the stone. A few meters away we recovered three horse heads buried in shallow pits. All three heads were oriented to the east, and one had vertebrae and hooves included with it. No artifacts were recovered, but the horse teeth from F1 produced a date of cal. 3160–2920 BP and F2, cal. 3330–3060 BP, making this simple deer stone the earliest known in Khovsgol Aimag.

In 2005 we found two stones bearing deer stone elements at Evdt-2 about 30 kilometers south of Tsatstain Khoshuu. Both had fallen and were partly buried. DS 1 had an engraved circle groove at its top, similar to the Tsatstain Khoshuu stone, and DS 2 had a rudimentary encircling belt groove but no other motifs. This stone was surrounded by a series of oval features, one of which contained an end scraper of dark flint made on a thick prismatic blade. This is the only chipped stone tool recovered from a deer stone or khirigsuur site in the Hovsgol region, and if it is not intrusive, it raises questions about the chronology and persistence of stone tool traditions. Occasionally, micro-blades and chipped lithics have been found at khirigsuur and slab burial sites in the Egiin Gol and the Baga Gazaryn Chuluu region in the Gobi, and in these cases the lithics are believed to be intrusive (W. Honeychurch personal communication). A charcoal sample from the stone ring with the scraper at DS 2 has been dated to cal. 3350–3090 BP.

The 2006 season produced other unusual finds from the northern Darkhad. Two rock art sites on the north bank of the Shishged River contained images of the iconic deer stone emblem, one on a huge panel containing hundreds of Bronze and Iron Age naturalistic images of other animals (Fig. 20.10). In addition, several sites were found to have deer stones of exceedingly small size, less than 75 centimeters in length overall, carrying only rudimentary carvings and associated with datable horse skulls and charcoal samples. One of these, Hort Azuur, has a north-south alignment of at least four minimally decorated deer stones.

Figure 20.10. Mongolian Deet image from Avtiin rock art panel, Shishged Gol, northern Khovsgol Aimag.

One of these is carved with an iconic deer image and a spotted leopard, which is a rare element in Khovsgol deer stone art. Charcoal associated with this deer stone has been dated to cal. 2870–2750 BP during the latter part of the deer stone chronology for the Khovsgol region. Further dating and stylistic comparison of the Darkhad deer stones with the classic Mongolian and nearby Sayan and Baikal forms may produce insight into the chronology and relationships of the northern Mongolian deer stone complex. Whether the rudimentary nature of Darkhad deer stone carving is an early stylistic form (as might be suggested from the early dates), reflects geographic marginality, or is simply aberrant needs clarification. Nevertheless, their simple designs, frequent absence of the deer image, and presence of rings, slashes, and necklace pits and belts are characters that resemble elements of both Type 2 and Type 3 deer stones.

Conclusion

AT THIS point it is too early to offer many conclusions. However, excavations confirm earlier reports that deer stones are not found with human remains and have few artifact associations. Their frequent occurrence with large khirigsuur complexes suggests a ceremonial and ritual connection with honoring dead leaders. Because two of the most characteristic features of Type 1 Mongolian deer stones are the variability of their belt assemblages and the different size, placement, and organization of deer images, deer stones may represent specific people depicted with their own weapons and body tattoos. In other words, deer stones may be cenotaphs memorializing real warriors, chiefs, or heroes whose souls needed to be sent off to the upper world with the assistance of their personal deer-bird helping spirits.

The presence of a shared ritual involving horse head burials and oval feasting rings, identifiable radiocarbon ages, and linkage between deer stones, khirigsuurs, and rock art demonstrates that all three expressions are part of a single Late Bronze Age culture and ceremonial complex. More work is needed to understand the beliefs and rituals of this time. We have found two deer stones at Ulaan Tolgoi to be ringed by east-facing horse heads packaged with the cervical vertebrae and one or more hooves. These features are placed in a circular arrangement around these deer stones, and each appears to represent single events dating to approximately cal. 2900 BP. Inspection of many other deer stone sites in Khovsgol Aimag and other areas of northern Mongolia reveals surface traces of rock features commonly clustering around deer stone settings. Fist-sized pecking stones found among the horse head features indicate that horse burials occurred at the same time as the carving of the stones. It seems likely that small open hearths represent feasting activities associated with the deer stone dedication ritual. The same concentric pattern of horse mounds and feasting hearths outside khirigsuur fences suggests that similar rituals and ceremonies occurred with the construction of both types of features.

Similar spatial patterning between the architecture of individual khirigsuurs and khirigsuur locations on the east-facing slopes of prominent geographic features like Ulaan Tolgoi suggests the existence of an underlying belief linking individuals and whole communities in a shared set of religious beliefs and cosmology. One of the striking features of this system is the hierarchical geographic replication of social and political structures represented at different social, spatial, and cosmological levels. Horses are sacrificed and buried facing east in small mounds on the east sides of khirigsuurs. Commanding leaders have ostentatious burials in prime valley locations with numerous horses sacrificed, whereas other members of society have poorly appointed graves on the eastern slopes of "sacred hills." Deer stone sites demonstrate a similar although reduced range of variation, seen in size, extent, and quality of ornamentation, presence or absence of emblematic motifs, and the surrounding arrangement of horses sacrificed and buried around them with their heads pointing east. Both khirigsuurs and deer stones have feasting hearths around the outskirts of the monuments. Widening scales of social and political power may also be indicated by the increasing number and size of mounds, deer stones, and associated horse sacrifices as one proceeds from the more isolated and economically marginal Darkhad region to the relatively more central and fertile Khanuy Valley, whose people likely had greater opportunities for trade, social interaction, and economic prosperity.

Given the variation in deer stone styles and regional variation in those styles, further dating and stylistic study should reveal where deer stones first appeared and how the concept spread and developed into the classic Type 1 Mongolian deer stone. If the Darkhad dates continue to be the earliest of this type, we may presume a northern origin and a subsequent expansion into the classic form along the northern Mongolian steppe. Other deer stone types like the Type 2 Sayan-Tuvan form with independent floating animals, and the Type 3 Altai style with few animals but the presence of circles, slashes, "necklace" lines, and cup marks need to be radiocarbon-dated before their relationship with the Type 1 stones can be determined. As many scholars have noted, roots of Scythian art are present in the form of the Mongolian deer stone image and one suspects that related art forms must have been produced in wood, textile, and felt that have not been preserved. We may also wonder about the eastern ramifications of such a cultural system and whether its influence reached beyond the Amur and Siberia to Korea, the Okhotsk, and the coasts of the North Pacific and Bering Sea, where it has long been suspected that Bronze Age art and shamanistic ceremonialism influenced the development of early Eskimo art and culture (Collins 1951, 1971; Schuster 1951; Schuster and Carpenter 1986; Chard 1958, 1974; Arutiunov and Fitzhugh 1988; Fitzhugh 1993, 2002, 2005; Fitzhugh et al. 2005).

How this cultural wave takes shape and develops cannot be determined with current data. However, its rapid movement and the considerable cultural change that occurred would have been facilitated by mobile, technologically advanced, and politically and socially flexible societies we now recognize as characteristic of steppe pastoralists of the late second and early first millennia BCE. Absence of any trace of permanent habitation sites in the form of middens, repeated occupations, or substantial deposits of debris clearly indicates the existence of a nomadic pastoral economy that was even more transient than found in rural Mongolia today. The absence of grave goods associated with the deer stones and their appearance in the immediately succeeding slab grave complex suggests that a major population movement and cultural transformation occurred immediately following the deer stone period, between cal. 2800 and 2600 BP, when the display of individual wealth began to transform the previous Spartan approach to the afterlife. Tsybiktarov (2003: 93) links this with environmental change and a culture and demographic clash between Caucasian khirigsuur and deer stone builders and an expanding biologically Mongoloid-based slab burial culture. Whatever the nature of these changes, a shift toward an even more elaborate style of burial occurred at Pazyryk about the same time. The slab grave was a radical departure from the materially poor but architecturally rich ceremonialism of the deer stone–khirigsuur complex. Likewise, looking farther back in time, the deer stone complex marshaled social and religious values into a cultural system that seems to have had no obvious prototype in Mongolia. The mechanism by which this was accomplished may have been military and economic, but its expression in the deer stone and khirigsuur complex suggests imposition of a hierarchical social and political system whose order is evident in highly organized and codified mortuary ceremony, burial architecture, and public art.

Acknowledgments

Many Smithsonian colleagues have participated in this research during the past several years, to all of whom I owe a debt of thanks, especially Paula DePriest, Bruno Frohlich, William Honeychurch, and Daniel Rogers. Francis Allard shared information and participated in our joint 2006 field program. I also thank my Mongolian partners, especially Ochirkhuyag, J. Bayarsaikhan, T. Sanjmiatav, and Adiyabold Namkhai, and the late director of the National Museum of Mongolian History, Idshinnorov, and current director, Dr. Ayudain Ochir. Our research has also been facilitated by the Santis Foundation, American Center for Mongolian Studies, Dr. D. Tseveendorj, director of the Institute of Archaeology, Drs. Enktuvshin and Galbaatar of the Mongolian Academy of Sciences, Mongolian officials in Khovsgol Aimag, and officials of the U.S. Embassy. Funding has come from Ed Nef and the Santis Foundation, Robert Bateman Arctic Fund, Trust for Mutual Understanding, U.S. Department of State Ambassador's Fund, Smithsonian's National Museum of Natural History, the National Geographic Society, and the Arctic Studies Center. William Honeychurch made helpful suggestions to improve the final draft. Helena Sharp, Christie Leece, Bruno Frohlich, Bill Honeychurch, and Marcia Bakry assisted with maps and illustrations; Bill Honeychurch provided the translations for Russian article titles; and Igor Krupnik provided transliteration for Russian bibliographic citations.

References

Allard, F., and D. Erdenebaatar. 2005. Khirigsuurs, Ritual, and Mobility in the Bronze Age of Mongolia. *Antiquity* 79: 547–563.

Amartuvshin, Ch. 2003. Egiin gol dakh' dorvoljin bulshny sudalgaany zarim asuudal (Some Questions concerning the Study of the Slab Burials of Egiin Gol). In C. Idshinnorov and D. Tseveendorj (eds.), *Mongol-Solongosyn Erdem Shinjilgeenii Ankhdugaar Simpoziumyn iltgeliin emkhtgel* (Collected Papers of the First Mongolian-Korean Academic Symposium). Ulaanbaatar: Mongolian National Museum, Academy of Sciences, Korean National Museum, pp. 83–94.

Arutiunov, S., and W. W. Fitzhugh. 1988. Prehistory of Siberia and the Bering Sea. In W. W. Fitzhugh and A. Crowell (eds.), *Crossroads of Continents: Cultures of Siberia and Alaska*. Washington, DC: Smithsonian Institution Press, pp. 117–139.

Barfield, T. 2001. The Shadow Empires. Imperial State Formation along the Chinese-Nomad Frontier. In S. E. Alcock, T. N. D'Altroy, K. D. Morrison, and C. M. Sinopoli (eds.), *Empires: Perspectives from Archaeology and History*, Cambridge: Cambridge University Press, pp. 10–41.

Bayarsaikhan, Jamsranjav. 2005. Notes on Meaning of Deer Stone Iconography. In W. Fitzhugh, J. Bayarsaikhan, and P. Marsh (eds.), *The Deer Stone Project: Anthropological Studies in Mongolia, 2002–2004*. Washington, DC, and Ulaanbaatar: Arctic Studies Center, Smithsonian Institution and National Museum of Mongolian History, pp. 35–41.

Chard, C. S. 1958. The Western Roots of Eskimo Culture. In *Actas del XXXII Congreso Internacional de Americanistas*. San Jose, Costa Rica.

1974. *Northeast Asia in Prehistory*. Madison: University of Wisconsin Press.

Chlenova, N. L. 1962. Ob olennykh kamniakh Mongolii i Sibirii. In S. V. Kiselev (ed.), *Mongol'skii arkheologicheskii sbornik*. Moscow: Nauka, pp. 27–35.

Collins, H. B. 1951. Origin and Antiquity of the Eskimo. *Annual Report of the Smithsonian Institution for 1950*. Washington, DC: Smithsonian Institution, pp. 423–467.

1971. Composite Masks: Chinese and Eskimo. *Anthropologica*, n.s., 13(1–2): 271–278.

Desroches, J.-P. (ed.). 2003. *Mongolie, le premier empire des Steppes*. Paris: Acts Sud/ Mission archeologique française en mongolie.

Di Cosmo, N. 1999. State Formation and Periodization in Inner Asian History. *Journal of World History* 10: 1–40.

2002. *Ancient China and Its Enemies: The Rise of Nomadic Power in East Asian History.* Cambridge: Cambridge University Press.

Dikov, N. N. 1958. *Bronzovyi vek Zabaikal'ya.* Ulan Ude: Nauka.

Dobzhanski, V. N. 1987. K voprosu o khronologii i kul'tlurnoi prinadlezhnosti olennykh kamnei Mongolii. In A. I. Martynov and V. I. Molodin (eds.), *Skifo-sibirskii mir.* Novosibirsk: Nauka, pp. 99–102.

Erdenebaatar, D. 2002. Burial Materials Related to the History of the Bronze Age in the Territory of Mongolia. In K. M. Linduff (ed.), *Metallurgy in Ancient Eastern Eurasia from the Urals to the Yellow River.* Lewiston: Edwin Mellen Press, pp. 190–236.

Fitzhugh, W. W. 1993. Art and Iconography in the Hunting Ritual of North Pacific Peoples. In *Proceedings of the 7th International Abashiri Symposium.* Abashiri: Abashiri Museum of Northern Peoples, pp. 1–13.

2002. Yamal to Greenland: Global Connections in Circumpolar Archaeology. In B. Cunliffe, W. Davies, and C. Renfrew (eds.), *Archaeology: The Widening Debate.* Oxford: Oxford University Press, pp. 91–144.

(ed.). 2003. *Mongolia's Arctic Connections: The Hovsgol Deer Stone Project, 2001–2002 Field Report.* National Museum of Natural History, Smithsonian Institution. Washington, DC: Arctic Studies Center.

(ed.). 2004a. Project Goals and 2003 Fieldwork. In *The Khovsgol Deer Stone Project, 2003 Field Report.* National Museum of Natural History, Smithsonian Institution. Washington, DC: Arctic Studies Center, pp. 1–24.

(ed.). 2004b. *The Hovsgol Deer Stone Project, 2003 Field Report.* National Museum of Natural History, Smithsonian Institution. Washington, DC: Arctic Studies Center.

2005. The Deer Stone Project: Exploring Northern Mongolia and Its Arctic Connections. In W. Fitzhugh, J. Bayarsaikhan, and P. Marsh (eds.), *The Deer Stone Project: Anthropological Studies in Mongolia, 2002–2004.* National Museum of Natural History, Smithsonian Institution, and National Museum of Mongolian History. Washington, DC, and Ulaanbaatar: Arctic Studies Center, pp. 3–31.

Fitzhugh, William W., J. Bayarsaikhan, and Peter Marsh (eds.). 2005. *The Deer Stone Project: Anthropological Studies in Mongolia, 2002–2004.* National Museum of Natural History, Smithsonian Institution, and National Museum of Mongolian History. Washington, DC, and Ulaanbaatar: Arctic Studies Center.

Frohlich, B., and N. Bazarsad. 2005. Burial Mounds in Khovsgol Aimag, Northern Mongolia: Preliminary Results from 2003–2004. In W. Fitzhugh, J. Baiyarsaikhan, and P. Marsh (eds.), *The Deer Stone Project: Anthropological Studies in Mongolia, 2002–2004.* National Museum of Natural History, Smithsonian Institution, and the National Museum of Mongolian History. Washington, DC, and Ulaanbaatar: Arctic Studies Center, 57–84.

Frohlich, B., N. Bazarsad, D. Hunt, and E. Altangerel. 2005. Mass Burials at Hambin Ovoo. In W. Fitzhugh (ed.), *The Hovsgol Deer Stone Project, 2003 Field Report.* National Museum of Natural History, Smithsonian Institution. Washington, DC: Arctic Studies Center, pp. 92–104.

Frohlich, B., M. Gallon, and N. Bazarsad. 2004. The Khirigsuur Tombs. In W. Fitzhugh (ed.), *The Hovsgol Deer Stone Project, 2003 Field Report.* National

Museum of Natural History, Smithsonian Institution. Washington, DC: Arctic Studies Center, pp. 42–61.

Frohlich, B., M. Gallon, and D. Hunt. 2003. Surveying and Excavating Bronze Age Burial Mounds and 20th Century Mass Burials in Mongolia. Arctic Studies Center, National Museum of Natural History. *Arctic Studies Center Newsletter* 11: 21–23.

Gryaznov, M. P. 1950. *Pervyi Pazyrykskii kurgan*. Leningrad: Iskusstvo.

    1980. *Arzhan – Tarskii kurgan ranneskifskogo vremeni*. Leningrad: Nauka.

    1984. O monumental'nom iskusstve na zare skifo-sibirskikh kul'tur v Stepnoi Azii. *Rkheologicheskii Sbornik* (Leningrad) 25: 76–82.

Hatakeyama, T. 2002. The Tumulus and Stage Stones at Shiebar-kul in Xinjiang, China. *Newsletter on Steppe Archaeology* 1: 1–8. http://kanazawa-u.ac.jp/steppe/ sougen13.hatakeyama.html.

Honeychurch, W. 2003. Inner Asian Warriors and Khans: A Regional Spatial Analysis of Nomadic Political Organization and Interaction. Ph.D. dissertation, University of Michigan.

Honeychurch, W., and Ch. Amartuvshin. 2004. An Examination of the Khunnu Period Settlement in the Egiin Gol Valley, Mongolia. *Studia Archaeologica* 21(1): 59–65.

    2006a. States on Horseback: The Rise of Inner Asian Confederations and Empires. In Miriam Stark (ed.), *Asian Archaeology*. Oxford: Blackwell, 255–278.

    2006b. Survey and Settlement in Northern Mongolia: The Structure of Intra-regional Nomadic Organization. In D. L. Petersen, L. M. Popova, and A. T. Smith (eds.), *Beyond the Steppe and the Sown: Proceedings of the University of Chicago Conference on Eurasian Archaeology*. Leiden: Brill, 183–201.

    2007. Hinterlands, Urban Centers, and Mobile Settings: The "New" Old World Archaeology from the Eurasian Steppe. *Asian Perspectives* 46(1): 36–64.

Honeychurch, W., and J. Wright. 2008. Asia North and Central: Prehistoric Cultures of the Steppes, Deserts, and Forests. In Deborah Pearsall (ed.), *Encyclopedia of Archaeology*. London: Elsevier, pp. 517–532.

Honeychurch, W., J. Wright, and Ch. Amartuvshin. 2007. A Nested Approach to Survey in the Egiin Gol Valley, Mongolia. *Journal of Field Archaeology* 32: 339–352.

Ingold, T. 1980. *Hunters, Pastoralists, and Ranchers: Reindeer Economies and Their Transformations*. Cambridge: Cambridge University Press.

Jacobson, E. 1993. *The Deer Goddess of Ancient Siberia: A Study in the Ecology of Belief*. Leiden: Brill.

    1998. The Recreation of Landscape Settings in Petroglyphs of Northern Central Asia. *International Journal of Central Asian Studies* 3: 192–214.

    2002. Petroglyphs and the Qualification of Bronze Age Mortuary Archaeology. *Archaeology, Ethnology, and Anthropology of Eurasia* 3(11): 32–47.

Jacobson, E., V. D. Kubarev, and D. Tseveendorj. 2001. *Mongolie du Nord-Ouest: Tsagaan Salaa/Baga Oigor*. 2 vols. Répertoire des Pétroglyphes d'Asie Centrale: Fascicule 6. Paris: De Boccard.

Jacobson-Tepfer, E. 2001. Cultural Riddles: Stylized Deer and Deer Stones of the Mongolian Altai. *Bulletin of the Asian Institute*, n.s., 15: 31–56.

Jettmar, K. 1994. Body-Painting and the Roots of the Scytho-Siberian Animal Style. In B. Genito (ed.), *The Archaeology of the Steppes: Methods and Strategies.* Naples: Instituto Universitario Orientale, pp. 3–15.

Khazanov, A. 1994. *Nomads and the Outside World.* Madison: Wisconsin University Press.

Khudyakov, Iu. C. 1987. Khereksury i olennye kamni. *Arkheologiia, ethnografiia i antropologiia Mongolii.* Novosibirsk: Nauka, pp. 136–162.

Kovalev, S. V. 1987. O datirovke olennykh kamnei iz kurgana Arzhan. *Skifo-Sibirskii mir.* Novosibirsk: Nauka, pp. 93–98.

Kubarev, V. D. 1979. *Drevnie izvaianiia Altaia: Olennye kamni.* Novosibirsk: Nauka.

Lattimore, O. 1938. Geographical Factor in Mongol History. *Geographical Journal* 91(1): 1–16.

1992 [1940]. *Inner Asian Frontiers of China.* New York: Oxford University Press.

Linduff, K. 1995. Zhukaigou, Steppe Culture, and the Rise of Chinese Culture. *Antiquity* 69: 133–145.

Molodin, V. I. 2000. The Pazyryk Culture: Problems of Origin, Ethnic History, and Historical Destiny. *Archaeology, Ethnology, and Anthropology of Eurasia* 4(4): 131–142.

Novgorodova, E. A. 1989. *Drevnaia Mongoliaia.* Moscow: Nauka.

Okladnikov, A. P. 1954. Olennyi kamen' reki Ivolgi. *Sovietskaia Arkheologiia* 19: 207–220.

Polosmak, N. V. 2000. Tattoos in the Pazyryk World. *Archaeology, Ethnology, and Anthropology of Eurasia* (Novosibirsk) 4(4): 95–102.

Polosmak, N. V., and V. I. Molodin. 2000. Grave Sites of the Pazyryk Culture on the Ukok Plateau. *Archaeology, Ethnology, and Anthropology of Eurasia* (Novosibirsk) 4(4): 66–87.

Potanin, G. N. 1881. Ocherki Severo-Zapadnoi Mongolii. Vol. 2. St. Petersburg: Tip. Kirschbaumana.

Radlov, V. V. 1893. *Atlas Drevnostei Mongolii* (Atlas of Mongolian Antiquities). Trudy Orkhonskoi Expeditsii, 2, St. Petersburg: Tip. Imp. Akademii Nauk.

Rogers, J. D. N.d. The Contingencies of State Formation in Eastern Inner Asia. Unpublished manuscript, Department of Anthropology, Smithsonian Institution.

Rogers, J. D., E. Ulambayar, and M. Gallon. 2005. Urban Centers and the Emergence of Empires in Eastern Inner Asia. *Antiquity* 79(306): 801–818.

Rudenko, Sergei I. 1970. *Frozen Tombs of Siberia: The Pazyryk Burials of Iron Age Horsemen.* Berkeley: University of California Press.

Sanjmyatav, T. 1995. *Mongoliin Khadni Zurag.* Ed. N. Ser-Odzhav. Ulaanbaatar: Institute of History, Mongolian Academy of Sciences.

Savinov, D. G. 1994. *Olennye kamni v kul'ture kochevnikov Evrazii.* St. Petersburg: Nauka.

Schuster, C. 1951. Survival of the Eurasiatic Animal Style in Modern Alaskan Eskimo Art. In Sol Tax (ed.), *Selected Papers of the 29th Congress of Americanists, New York, 1949.* Chicago: University of Chicago Press, pp. 35–45.

Schuster, C., and E. Carpenter. 1986. *Materials for the Study of Social Symbolism in Ancient and Tribal Art: A Record of Tradition and Continuity.* 12 vols. Edited and written by Edmund Carpenter, assisted by Lorraine Spiess. New York: Rock Foundation.

Sementsov, A. A., G. I. Zaitseva, J. Gorsdorf, A. Nagler, H. Parzinger, N. A. Bokovenko, K. V. Chugunov, and L. M. Lebedeva. 1998. Chronology of the Burial Finds from Scythian Monuments in Southern Siberia and Central Asia. In W. G. Mook and J. van der Plicht (eds.), *Proceedings of the 16th International 14C Conference. Radiocarbon* 40(2): 713–720.

Takahama, S. (ed.). 2006. *Research into the Origin of Mounted Nomads in the Eastern Eurasian Steppe. Report of Fieldwork in 2003–5.* Kanazawa: Kanazawa University.

Takahama, S., H. Toshio, K. Masanori, M. Ryuji, and D. Erdenebaatar. 2006. Preliminary Report of the Archaeological Investigations in Ulaan Uushig I (Uushigiin Ovor) in Mongolia. *Bulletin of Archaeology* 28: 61–102. Kanazawa: Department of Archaeology, University of Kanazawa.

Tseveendorj, D., N. Urtnasan, A. Ochir, and G. Gongorjav. (eds.). 1999. *Historical and Cultural Monuments of Mongolia* [in Mongolian]. Ulaanbaatar: Mongolian Academy of Humanities.

Tsybiktarov, A. D. 1997. Khereksury Buriatii, Severnoi i Tsentral'noi Mongolii. In P. B. Konovalov (ed.), *Kul'tury i pamiatniki bronzovogo i rannego zheleznogo vekov Zabaikal'ia i Mongolii.* Ulan-Ude: Nauka, pp. 38–47.

——— 1998. *Kul'tura plitochnykh mogil Mongolii i Zabaikal'ia.* Ulan-Ude: Nauka.

——— 2002. Eastern Central Asia at the Dawn of the Bronze Age: Issues in Ethnocultural History of Mongolia and the Southern Trans-Baikal Region in the Late Third–Early Second Millennium B.C. *Archaeology, Ethnology, and Anthropology of Eurasia* 3(11): 107–123.

——— 2003. Central Asia in the Bronze and Early Iron Ages. *Archaeology, Ethnology, and Anthropology of Eurasia* 1(13): 80–97.

Vainshtein, S. I. 1980. *Nomads of South Siberia: The Pastoral Economies of Tuva.* Cambridge: Cambridge University Press.

Vasil'evski, R. S. 1985. *Drevnie kul'tury Mongolii.* Novosibirsk: Nauka.

Volkov, V. V. 2002 [1981]. *Olennye kamni Mongolii.* Ulaanbaataar: Academy of Sciences; Moscow: Nauka, 2nd ed.

Volkov, V. V., and A. E. Novgorodova. 1975. Olennye kamni Uushigiin Overa (Mongolia). In *Pervobytnaia arkheologiia Sibiri.* Leningrad: Nauka, pp. 78–84.

Wallace, E., and B. Frohlich. 2005. Bronze Age Burial Mounds in Northern Mongolia: Use of GIS in Identifying Spatial and Temporal Variation. In W. Fitzhugh, J. Baiyarsaikhan, and P. Marsh (eds.), *The Deer Stone Project: Anthropological Studies in Mongolia, 2002–2004.* National Museum of Natural History, Smithsonian Institution, and National Museum of Mongolian History. Washington, DC, and Ulaanbaatar: Arctic Studies Center, pp. 88–97.

INDEX

Abashevo, 15, 29, 52, 53, 55, 80, 82, 132, 133, 140, 157, 300, 308, 309, 317
agro-pastoralists, 237, 299
Alakul', 28, 30, 31, 32, 37, 65, 137, 138, 153, 160
Altai, 35, 38, 48, 84, 110, 111, 130, 132, 133, 138–141, 251, 265, 285, 348, 364, 369, 380, 385, 386, 387, 391, 405
Altyn-Depe', 40
Anatolia, 48, 93, 94, 95, 98, 166
Andronovo, xvii, 41, 67, 68, 81, 83, 86, 110, 136, 137, 138, 160, 217, 218, 222, 249, 250
animal style, 275, 344, 346, 380
Anyang, 140, 252
Arkaim, xvi, xvii, 2, 23, 27, 29, 30, 47, 54, 63, 64, 109, 112, 149, 151, 153, 155, 156, 190, 202, 206, 207, 209
Arzhan, xvi, 281, 359, 378, 379, 398

Bactria-Margiana Archaeological Complex, 39, 53, 65
Baikal Lake region, 379, 383, 385, 403
Balanbash culture, 80, 317
Begash, 28, 33, 34, 37, 38, 42
beifang, 168, 269
Bol'shekaraganskii cemetery, 29

carnelian, 91
Carpatho-Balkan metallurgical province, 116, 117, 120, 121, 125
Caspian, 32, 48, 96, 137, 143, 195
Catacomb culture, 53, 55, 126–131
Caucasus, xviii, 11, 48, 91, 96, 98, 100, 124, 195, 198, 209
cenotaphs, 77, 134, 327, 363, 380, 404
Central Asia, 1, 14, 27, 53, 54, 64, 65, 66, 67, 85, 101, 131, 137, 146, 151, 191, 206
central places, 325, 365, 369
Chandman-Ulangom culture, 348
chariot, xv, xvi, xvii, 6, 16, 30, 47, 53, 54, 55–62, 63, 64, 66, 67, 68, 80, 148, 149, 151, 162, 236, 252, 264, 275, 390
chariot complex, 16
Cherkaskul' culture, 137
Chernorech'ye III, 54
chiefdom, xvii, 2, 5, 12, 13, 20, 23, 24, 27, 39, 75, 87, 149, 150, 205, 238, 298, 316
Chifeng region, 242, 245, 259, 260, 276
Childe, V. Gordon, xv, 91, 147
China, ix, x, xi, xii, xiii, xvii, xviii, 1, 83, 100, 108, 110, 111, 134, 140, 141, 144, 168, 170, 171, 185, 215–224,

413